C0-AUV-069

GEOLOGY IN ENGINEERING

GEOLOGY IN ENGINEERING

ROBERT BOWEN
PhD, BSc

Visiting Professor of Geology,
Whittier College, California, USA

ELSEVIER APPLIED SCIENCE PUBLISHERS
LONDON and NEW YORK

ELSEVIER APPLIED SCIENCE PUBLISHERS LTD
Ripple Road, Barking, Essex, England

Sole Distributor in the USA and Canada
ELSEVIER SCIENCE PUBLISHING CO., INC.
52 Vanderbilt Avenue, New York, NY 10017, USA

British Library Cataloguing in Publication Data

Bowen, Robert
Geology in engineering.
1. Engineering geology
I. Title
551'.024624 TA705

ISBN 0-85334-234-2

WITH 50 TABLES AND 75 ILLUSTRATIONS

Printed in Northern Ireland at The Universities Press (Belfast) Ltd.

Preface

There are numerous works on engineering geology and geotechnical investigations which discuss relevant interrelationships between civil engineering, soil mechanics and the earth sciences and this book is not intended to add yet another such to their number, but rather to concentrate upon subsurface work and demonstrate by appropriate case histories how the several disciplines have been ever more integrated over the past decade or so. Basically it is divisible into two sections.

The first section (Chapters 1–5) deals with fundamental aspects of geology which are specifically germane to the subject, i.e. rocks, minerals and water (liquid and frozen) or petrology (igneous, metamorphic and sedimentary), mineralogy and hydrology/hydrogeology. Within this section are treatments of subjects intimately connected with the topics such as denudation, erosion, transportation and deposition together with descriptive material on physical (mechanical) and chemical weathering, the geological and engineering classifications of rocks, rocks in time, the deformation, faulting and folding of rocks, crystallography and crystal systems, mineral properties and groups (mafic and felsic primary and secondary minerals) plus the classification of minerals, groundwater in construction, rocks as constructional materials, their strength characteristics, permafrost, glacial and periglacial phenomena and associated engineering problems.

The second section (Chapters 5–10) is devoted to topics believed especially significant regarding geology in engineering practice. These are foundations, earth movements and non-diastrophic structures, tunnelling, large scale hydraulic structures (dams) and remote sensing. The types and selection of foundations are examined with reference to various kinds of building (dam foundations being discussed in Chapter 9) and mention is made of spread, combined, cantilever and continuous footings, mats, rafts and piles (including root piles). Various earth movements are scrutinized, mostly of non-diastrophic character, i.e. not involving vast deformations of the planetary crust (as opposed to diastrophic events producing mountain ranges, etc.). In the chapter on tunnelling the problems, rock and soils behaviour and empirical methods are discussed and a classification of tunnels provided. As regards larger scale hydraulic structures, mainly dams, the approach is to give an historical background, to appraise the geology, to describe the geophysical investigations together with geological reconnaissance and detailed work, to examine grouting and to classify dams. Attention is directed towards adverse geological situations and some results of construction are considered. Finally remote sensing is surveyed and its implications for civil engineering in the future assessed.

ROBERT BOWEN

Contents

CHAPTER 1

Rocks: Their Classification, Alteration and Uses as Constructional Materials

1.1. ROCKS

These comprise mineral aggregates, loose or consolidated, hence soils are regarded as belonging to this category of natural materials by geologists although engineers tend to distinguish between soils and rocks in actual practice. The essential mineral contents, usually making up over 95% of the total volume, are of the greatest importance, determining the type and characteristics of the rock concerned. The spatial arrangement of the minerals in a rock constitutes the texture, a classificational feature. The usual classification of rocks by geologists is genetic, relating to their modes of origin, and includes the following groups:

(a) Igneous rocks formed from ascending hot liquid material arising deep in the Earth called magma which crystallizes into the solid state as the temperature falls.
(b) Sedimentary rocks formed as result of accumulation and compaction of:
   (i) pre-existing rock fragments disintegrated through erosive processes;
   (ii) organic debris such as shells;
   (iii) materials dissolved in surface or groundwater and later precipitated in conditions of oversaturation.
(c) Metamorphic rocks formed from any pre-existing rock subjected to increases in pressure or temperature or both.

1.2. IGNEOUS ROCKS

These include a wide variety of minerals of which only eight are considered to be essential, namely quartz, orthoclase, plagioclase,

TABLE 1.1
MINERALS IN IGNEOUS ROCKS

Composition	*Quantity of silica*	*Minerals present as essential*
Acid (late crystallization from the magma with relatively low temperature)	Over 65%	Quartz, orthoclase, sodium–plagioclase, muscovite, biotite and sometimes hornblende
Intermediate	55–60%	Quartz, orthoclase, plagioclase, biotite, hornblende and sometimes augite
Basic	45–55%	Calcium–plagioclase, augite and sometimes olivine and hornblende
Ultrabasic (early crystallization from the magma with relatively high temperatures)	Under 45%	Calcium–plagioclase, olivine and sometimes augite

muscovite, biotite, hornblende, augite and olivine. High quartz content igneous rocks are termed acid as opposed to those with low quartz content which are termed basic. As well as these two subdivisions, there are intermediate and ultrabasic ones shown in Table 1.1.

As regards the textures, these are very varied reflecting the wide spectrum of possible physicochemical conditions of origin. Most igneous rocks are composed of interlocking crystals, some (a few) of perfect crystalline form and hence are described as crystalline. The size of the crystals is a basis for classification and relates to the mode of cooling of the original magma. Thus igneous rocks vary from fine-grained to coarse-grained with crystals ranging from as small as 1 mm across in the former to 3 mm or more across in the latter.

Igneous rocks are divisible into three groups according to their emplacements, namely:

I. Plutonic, which form by slow cooling at depth of enormous masses of magma; the depth of formation may extend down many kilometres and of course after cooling and through geological time when erosion goes on, such plutonic igneous rocks may become exposed at the terrestrial surface; as heat dissipates very slowly from such vast bodies of material, the crystallization process is slow and the rocks cool into a coarse-grained texture.

II. Hypabyssal, which form as small bodies of rock called dykes and sills; the former are termed discordant because they transgress layering in the host rock; the latter are termed concordant because they are injected along such layering; even though the cooling process was more rapid than with plutonic rocks, their crystals are still reasonably large so that they assume a medium-grained texture.

III. Volcanic, which form at the planetary surface and originate as lavas emitted through volcanic activity and, as these cool relatively rapidly, attain a fine-grained texture.

I and II comprise intrusive rocks, i.e. rocks intruded into the crust of the Earth, while III comprises extrusive rocks, i.e. rocks extruded on to the Earth's surface.

As well as the textures already indicated, igneous rocks may exhibit others, such as:

(a) Glassy—due to the very rapid cooling of magma which allowed insufficient time for crystals to form; the rock constitutes a natural glass and this type occurs mostly among acidic extrusive rocks.

(b) Porphyritic—in which both large and small crystals are associated; the smaller constitute the matrix for the larger, the latter being termed phenocrysts; this texture occurs in extrusive rocks, less frequently in hypabyssal rocks; it constitutes a rock type, the porphyries.

(c) Vesicular and amygdaloidal—commonest in extrusive rocks, these result from the emergence from solution during cooling of gases previously dissolved in magma; such gases form bubbles and expand as the magma ascends; they may be trapped by rapid congealment of the molten rock on the surface and form vesicles; later, minerals may form in these to produce amygdales.

(d) Ophitic—resulting from simultaneous crystallization of two minerals to produce a felted arrangement of interlocking crystals; usually this texture is found in a basic hypabyssal rock called dolerite (see Table 1.2); it is important in engineering because the rock featuring it acquires a very high crushing strength eminently suited to its use as a roadstone.

TABLE 1.2
CLASSIFICATION OF NORMAL IGNEOUS ROCKS

Type	Quantity of silica (%)	Grain size			SG
		Extrusive	*Hypabyssal*	*Plutonic*	
Acid	>65	Rhyolite Dacite	Quartz and orthoclase porphyries	Granite Granodiorite	2·67
Intermediate	55–65	Pitchstone Andesite	Plagioclase Porphyries	Diorite	2·87
Basic	45–55	Basalt	Dolerite	Gabbro	
Ultrabasic	<45	Basic olivine basalts	Basic dolerites	Picrite Peridotite Serpentinite Dunite	3·0–3·5

(e) Pegmatitic—produced by the fact that water and other fluxes sometimes occur in the final stages of magmatic cooling and lower the temperature of crystallization of minerals in consequence of which a longer growth time is available; hence, single crystals may attain a much greater size than the usual in coarse-grained rocks; occasionally such a crystal may become large enough to weigh several tons and this type of texture is found in veins of acidic igneous rock termed pegmatite.

The classification of igneous rocks taking into account emplacement is given in Table 1.2.

The schema in the table does not of course mean that the rock types listed are completely distinct. In fact many parallel each other and may grade into each other if the conditions of formation were sufficiently similar. Also it refers to normal (abundant) igneous rocks of the calcium-alkaline type in which the dominant feldspar is calcium-rich. There are other, less common igneous rocks which crystallized from alkaline magmas in which sodium and potassium were relatively common. In consequence these contain large quantities of orthoclase rather than plagioclase and under 65% silica as a rule. The fine-grained variety of these is termed trachyte, the medium-grained felsite and the coarse-grained syenite.

Figure 1.1 shows igneous rock composition in mineralogical terms. The process whereby the original magma separates into two or more

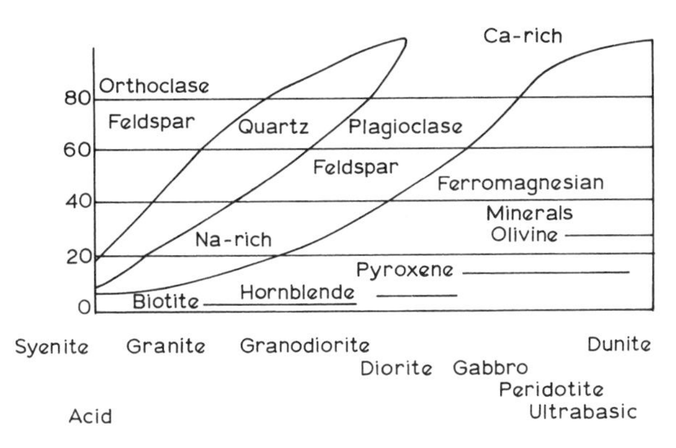

Fig. 1.1. The compositions of igneous rocks based upon the proportion of each component mineral.

fractions which later consolidate into various types of igneous rock is termed differentiation. The ferromagnesian minerals referred to in the figure contain essential iron and/or magnesium and include olivine, hypersthene, augite, hornblende and biotite. An acronymic word, femic, is derived from ferro and magnesian and applies to a particular group of arbitrarily calculated minerals in the CIPW classification. This is a system of classifying igneous rocks proposed by Cross *et al.*[19] in 1903 and based on the chemical composition. It requires the calculation of the norm as a first step. This does not necessarily coincide with the mode, the percentage by weight of individual minerals making up a rock. Since all normative minerals are anhydrous, the norm of rocks which contain noteworthy quantities of hydrated or hydroxyl-bearing minerals shows significant deviations from their modes. However free quartz in the norm usually means free quartz in the mode as well. After calculation of the norm, the relative proportions of the arbitrary minerals thus calculated are employed to classify the rock in detail.

1.3. SEDIMENTARY ROCKS

These originate from detrital material and dissolved mineral matter produced by the physical and/or chemical breakdown and decomposition of pre-existing rocks and also from the skeletal remains of plants and animals. Such products include materials constituting drift which covers many solid rock outcrops and also constitute soils with which site engineers deal continually in excavation and foundation work.

The word denudation is applied to all processes tending to lower the surface of the land and erosion is one of these. Erosion may involve the transportation of debris and this distinguishes it from weathering which is a decompositional process affecting rocks *in situ*. As weathering is so closely linked with sedimentary rocks, it is advisable to refer to it in some detail.

Weathered rock is commonly encountered as foundations are being excavated because fresh rock is rare at the surface of the Earth. Hence it is essential that the site engineer as well as the geologist should know how the process of weathering alters the properties of rocks and often makes them weaker as regards bearing capacity, durability, etc. Weathering entails the mechanical breakup and/or the chemical decomposition of rocks and may be caused by a variety of agents including Man. It is feasible to divide weathering into the two categories of mechanical and physical/chemical, the first produced mainly by rain, frost, wind, gravity and temperature change and the second produced mainly by chemicals in the atmosphere, precipitation and surface waters such as rivers. Some of these factors will predominate in certain areas and some in other areas. For instance aeolian weathering (wind action) and insolation (temperature change) are most important in arid and semi-arid desert-type environments while rain dominates in tropical forests. Ice is significant in glacial conditions. In temperate and tropical areas, humidity coupled in the latter case with heat promotes very deep chemical weathering. This is so in Sierra Leone and parts of India where laterites are widely developed to considerable depths.

1.3.1. Mechanical (Physical) Weathering

This causes rock fracture to occur and the minute particles so produced may be further disintegrated, but there is no chemical alteration of any kind. The basic reason for this type of weathering is stress change in surficial layers which results from a variety of factors such as the unloading by erosion of overlying materials, water freezing in joints, crystal growth, temperature-induced rock volume changes and plant action (tree roots are often highly deleterious to nearby buildings as well). A very important factor is unloading because it can initiate a series of weathering stages by the penetration through release joints of air, water, plant roots, etc. The unloading process reduces the confining pressure on underground rock and a slight expansion then occurs and manifests itself as tiny movements along original planes of weakness in the rock mass such as bedding planes in sedimentary rocks or joint systems in igneous rocks. Residual stress in rocks may be large so that for instance quarried blocks of granite may expand by as much as 0·1%. The fracturing of rock from unloading does not go very deep because the lower the effect penetrates, the higher is the resistance tending towards its elimination. This is a very practical matter because stress relief in tunnels may take the form of rock bursting which necessitates adequate protection for workers. As regards the freezing of water in joints, etc., the phenomenon depends upon the physical fact that water expands by about 9% on freezing and this promotes the release of latent heat with an accompanying rise in pressure. Considerable disruption may thus be induced—a pressure exceeding 200×10^3 kN m^{-2} is exerted at −22°C, a maximum pressure only reached if the system is entirely closed, air bubbles in the ice are scarce and the very low temperature is attained (at this pressure the freezing point of water is depressed to −22°C). A closed system may be created easily when water freezes initially at the surface of a rock and seals fissures (this applies to cracks in buildings as well). In addition, although the hydrostatic pressure may not reach the maximum value, it can exceed the tensile strengths of most rocks. Such a pressure may not be completely relieved by rock shattering; for instance there may be extrusion of ice from surface cracks as well. The saturation coefficient is an important parameter for this matter because if it is lower than about 0·8 (i.e. if 20% of the pore space remains unfilled) expansion on freezing can be accommodated within the rock and frost action will not occur. On the other hand, if the saturation coefficient is higher then disruption may be anticipated.

Although, in elementary reviews of weathering, disintegration is mentioned as the sole result of phase change recognized as frost action, there is another force related to the growth of ice crystals. When a saturated porous material freezes, ice crystals begin to form in the larger pores and water is withdrawn from the smaller ones. Mechanical disruption may accompany the growth of these crystals. The preferential growth of ice crystals in larger pores is explained thus: if supercooled water in a porous system begins to freeze both in a small pore and in a large one, the growth of the crystal in the former will be limited by the pore space in which it is contained. Once a difference in size has been established between crystals in the large and small pores, the larger crystal will feed on the smaller until the coarse pore space is filled. Everett[1] has developed a model to explain why, once a large pore is filled, further growth of the crystal proceeds against the constraint which is imposed by the walls of the pore, leading perhaps to disruption. The pressure which can build up in a large pore of radius R connected to a supply of water at the reference pressure by a capillary of radius r is proportional to $(1/r - 1/R)$ and failure will occur if this pressure exceeds the strength of the porous material. The pressure will ultimately depend therefore on the pore structure of the material. Honeyborne and Harris[2] showed in 1958 that pore structure is probably one of the most important rock properties governing the resistance of rock to frost action. Experiments have demonstrated that porosity and the actual percentage of material shattered are not significantly related statistically so that more refined measures of pore structure, e.g. direct measurement of the relation between micropores and macropores and indirect assessments based upon suction (dynes cm^{-2}), are utilized. Also measures of pore structure need to be combined with other variables such as saturation coefficient and the frequency of planes of weakness so as to improve the estimation of rock durability.

As regards crystal growth, precipitation that penetrates rock contains traces of dissolved substances which increase as the percolation continues. If evaporation or freezing follow, oversaturation of the solution results in the precipitation of salts. Growth of the ensuing crystals creates considerable pressure which can cause rock disintegration. Other chemical reactions also promote this, e.g. the hydration of anhydrite to gypsum which involves an increase in volume. In urban areas rainwater becomes acidic to form acid rain, one of the most devastating forms of pollution. The definition of acid rain is precipitation with a pH below 5·6 and its origins may be uncontrolled high-sulphur emissions from power plants and industrial boilers which release SO_2 and the nitrogen oxides into the atmosphere. After venting into the air, molecules of these compounds are taken up in prevailing winds and interact in the presence of sunlight with vapour to form dilute solutions of nitric and sulphuric acids, acid rain (which also can take the form of dry particles, snow and even fog). Acid rain alters the pH levels of rivers and lakes when falling directly into them. On land it is absorbed into the soil where it is capable of breaking down natural minerals such as those of calcium, potassium and aluminium, thereafter carrying them into the substrata and so leaching out a ley source of nutrients for trees and plants. The process goes on until toxic metals are contributed to nearby water bodies with concomitant damage to aquatic life. The effects are so serious that Canadian environmental officials believe that as many as 48 000 lakes will become lifeless by the end of this century if nothing is done to prevent this happening. Even now about 2000 or more lakes in Ontario are too acidified to support trout and bass. The problem is also acute in Western Europe. Acid rain, possible arising in the highly industrialized West German Ruhr, has affected about 20% of the 100 000 lakes in Sweden and on that country's west coast at the village of Lilla Edet well water has become so acidic that copper lines for plumbing have turned green with corrosion. It is estimated that the Federal Republic of Germany loses approximately US$ 800 million annually in timber losses through acid rain. Some pH levels for rain in Western Europe are shown on Fig. 1.2.

Changes in rock volume due to temperature have been alluded to and the frequency of fluctuations in temperature around the 0°C mark is a most important factor. In fact Potts[3] considers this to be more significant than the intensity of freezing in weathering.[3] Opinions differ as to whether or not the rate of freezing is important. It could be that a

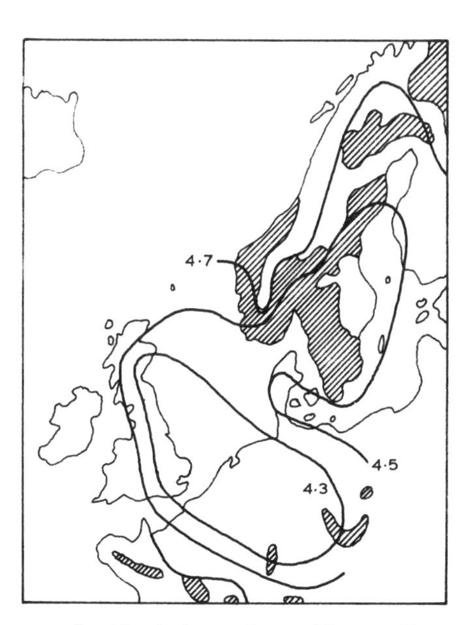

FIG. 1.2. Provenance of acid rain in northwest Europe, figures referring to pH. Shaded areas are regions of very high sensitivity.

TABLE 1.3

ROCKS AND THEIR RELATIVE RESISTANCES TO FROST ACTION (1, MOST SUSCEPTIBLE; 7, LEAST SUSCEPTIBLE)

Field studies		*Experimental studies, "Icelandic" conditions*	
Ardennes (*Potts*)	*Dartmoor* (*Potts*)	*Potts*	*Wiman*
1. Phyllite—pure schist (metamorphic)	1. Metamorphosed sediments	1. Igneous rocks	1. Slate
2. Calcareous schist (metamorphic)	2. Fine-grained granite	2. Sandstones	2. Gneiss (metamorphic)
3. Phyllite—quartz schist (metamorphic)	3. Diabase (= dolerite)	3. Mudstone	3. Porphyritic granite
4. Limestone	4. Elvan[a]	4. Shale	4. Mica schist (metamorphic)
5. Grits/sandstones	5. Tourmalinized medium-grained granite		5. Quartzite (metamorphic)
6. Quartzite (metamorphic)	6. Quartz shorl		
7. Conglomerate	7. Coarse-grained granite		

[a] Dyke of microgranite, sometimes porphyritic, hypabyssal.

rapid rate of freezing may promote the development of a closed system. Using the results of Wiman[4] and Potts,[3] comparison of various rock types in terms of relative resistance to frost action may be made with Table 1.3 which shows igneous rocks to be the most susceptible to frost action.

Salt precipitation has been referred to above and salts which may be responsible for rock disintegration fall into the following categories:

(a) salts originally present;
(b) salts formed during decomposition;
(c) salts derived from external sources.

As regards soils, salts ascend by the capillary absorption of solutions and this has been observed to attain the level of damp courses in buildings or at least to reach a high point at which the rate of evaporation and the rate of absorption are balanced. The most common salts include sulphates, carbonates and the chlorides of sodium, calcium, potassium and magnesium. As regards major disintegration processes associated with the cryptofluorescence of salt, three have been identified:

(a) thermal expansion of crystallized salts;
(b) growth of salt crystals;
(c) hydration of salts.

The hydration of salts entails forces of importance. Pressure so created can reoccur many times during a season or sometimes even within a day. The precise mechanism of anhydrous salt formation and its translation into higher hydrates varies locally. For instance when the soil temperatures during the day are high, capillary rising salt solutions may yield crystals without or low in water of crystallization. Then, during the night when temperatures fall, the anhydrous salts or lower hydrates absorb water vapour from the atmosphere when the pressure of the atmospheric aqueous vapour exceeds the dissociation pressure of the hydrate involved and higher hydrates are formed. The transformation is related to relative humidity and temperature conditions in the atmosphere and also to the dissociation vapour pressure of salts. The pressures developed by hydration can be calculated using the equation of Winkler and Wilhelm:[5]

$$P = \frac{(nRT)}{(V_h - V_a)} 2{\cdot}3 \log \frac{(P_w)}{(P'_w)}$$

where P is the hydration pressure in atmospheres, n is the number of moles of water gained during hydration to the next higher hydrate, R is the gas constant, T is the absolute temperature in K, V_h is the volume of hydrate (molecular weight of the hydrated salt divided by its density in gm ml^{-1}), V_a is the volume of the original salt (molecular weight of the original salt divided by its density in gm ml^{-1}), P_w is the vapour pressure of water in the atmosphere (mm mercury at a given temperature) and P'_w is the vapour pressure of the hydrated salt (mm mercury at a given temperature).

This equation has been used to demonstrate hydration pressures for various salts, e.g.

$$Na_2CO_3.H_2O = Na_2CO_3.7H_2O = Na_2CO_3.10H_2O$$

These show that under ideal conditions hydration pressures for some salts may approach those associated with frost weathering. Actually three conditions must apply for rock disintegration to be caused by hydration:

(a) the hydration process should be accomplished in at least 12 h or else it cannot be completed by diurnal temperature changes;
(b) the hydrating salt must not be able to escape from the pore in which it is contained;
(c) the hydration pressure must exceed the tensile strength of the rock; this is often easily accomplished because nearly all rocks have low tensile strengths.

As regards crystal growth, the pressure exerted depends upon the degree of supersaturation of the solution and may be described thus:

$$P = \frac{RT}{V} \ln \frac{(C)}{(C_s)}$$

where P is the pressure on the crystal, V is the molar volume of crystalline salt, C_s is the concentration at saturation point without pressure effect and C is the concentration of a saturated solution under the external pressure P.

Of the various salts, Na_2SO_4 is the most damaging but NaCl and $CaSO_4$ are also deleterious. Effectiveness varies with climate, the solubility of sodium sulphate increasing with rising temperature.

Turning to precipitation, this includes all liquid and frozen forms of water and thus comprises rain, sleet, snow, hail, dew, hoar-frost, fog-drip and rime. The intensity of rainfall is derived by dividing the quantity by the duration during an individual storm as well as during shorter periods. In order to assess it chart records of the rate of rainfall, i.e. hyetograms, are essential and so are rain gauges. Raindrop sizes vary with intensity and typical figures are 0·1 cm with 0·1 cm h^{-1}, 0·2 cm with 1·3 cm h^{-1} and 0·3 cm with 10·2 cm h^{-1}. Raindrops erode soils and the amount removed by erosion following the weathering process is related to raindrop energy. In fact the prevention of raindrop impact reduces soil erosion. The quantity of soil in runoff, i.e. eroded, is therefore related to raindrop energy. The median dimensions of raindrops increase with the intensity of rainfall (for low and moderate intensity falls) according to the relation $D_{50} = aI^b$, where D_{50} is the median dimension of raindrops in mm, I is the intensity in mm h^{-1} and a and b are constants. In the case of high rainfall intensities, it has been recorded that the drop dimensions diminish and the maximum sizes of drops are of the order of 5–6 mm in diameter. The actual drop dimensions influence the velocity of impact. A drop of rain which is in free fall will accelerate under gravity until its force becomes equal to the frictional resistance of air. At this point, the drop of rain attains its terminal velocity. As a consequence of the fact that fall distances to maximum velocity are short, most raindrops will hit the ground at terminal velocity. Rainfall intensities can be obtained from recording rain gauge data using a tipping bucket instrument providing a contact every time a measurement of precipitation which is prearranged collects in the funnel unit. There is a relationship between rainfall and soil loosening, separation and transportation expressed by

$$G = KDV^{1 \cdot 4}$$

where G is the weight of soil splashed in g, K is a constant for the particular soil type concerned, D is the diameter of the raindrops in mm and V is the impact velocity in $\mathrm{m\,s^{-1}}$. A compound parameter known as the EI_{30} has been derived in order to explain satisfactorily soil losses in terms of rainfall. This is the product of kinetic energy E (in foot tons per square inch) and intensity I (in inches per hour) where I_{30} specifically refers to the 30 min intensity, i.e. the maximum average intensity in any $\frac{1}{2}$-h period during a storm. Soil splashing varies greatly from site to site and becomes more significant where the vegetative cover is sparse thus providing minimum shield value. Slope is very important because where soil is horizontal such splash erosion can break up and move particles, but there will be no loss. Soil splashes can do a lot of damage, e.g. the impact of raindrops can disperse surficial clay and silt particles as well as promote the formation of surface runoff accompanied of course by runoff erosion. Runoff is an erosive rather than a weathering process, but the two are so intimately linked that it must be mentioned here. It probably does not commence immediately rain falls on to a bare or sparsely vegetated soil surface, water initially infiltrating into the ground at a rate determined by a number of factors such as type, structure and texture of the soil, moisture content and the condition of the surface. Infiltration capacity, the maximum sustained rate at which a particular soil can transmit water, varies with conditions—during a rainstorm it can be very rapid at first, later declining to a constant value. This decline results from rain packing, the in-washing of fine material, swelling of colloids and breakdown of the surface soil structure. Here maximum infiltration capacity is displayed and if rain proceeds for a time t then

$$f_p = f_c - (f_o - f_c) \exp(-kt)$$

where f_p is the infiltration capacity, f_o this parameter at maximum, f_c the parameter after decline to a constant value and k is a constant. As f_p diminishes with time during a rainstorm, runoff increases. The ability of rainwater to penetrate rock is determined partly by the rock's porosity and partly by its permeability which can be increased if fissures and fractures exist.

In research on building stones, three related measures have been found to be very useful and these are porosity, water absorption and, based on these, the saturation coefficient.

Turning now to the wind erosion system this comprises variables related to surface winds and climate, surface materials and surface conditions. Aeolian erosion in semi-arid and arid regions commences when the air pressure on particles overcomes the force of gravity on them. Initially particles are moved through the air by saltation, their impact on others promoting further weathering and erosion with more saltation or suspension or surface creep. The effect of each depends upon the availability of suitable particles together with the attainment of wind velocities at or above which particles can be moved, such movement being initiated by saltation (fluid threshold) or by impact (impact threshold).

Surface winds responsible for weathering are turbulent and their velocity increases with elevation thus

$$V_z = 5{\cdot}75\, V_d \log \frac{z}{k}$$

where V_z is the mean velocity at any height z, V_d is the drag velocity and k is a roughness constant. As saltating material becomes incorporated into the wind the momentum and therefore the surface wind velocity are reduced. The mean wind drag per unit horizontal area of ground surface, d (expressed e.g. in dynes cm^{-2}) is related to the drag velocity V_d and fluid density ρ and the relations can be expressed thus

$$d = \rho V_d^2$$

Particles abrade rocks and flowing air may erode in two ways, by picking up loose weathered material and transporting it (deflation) and by abrasion. In defining a local wind erosion factor of climate C and using it to delimit the relative wind erosion conditions, Yaalon and Ganor[6] have given the expression

$$C = \frac{v^3}{(P-E)^2}$$

where v is the average wind velocity (miles per hour at a standard elevation of 10 m), and $(P-E)$ is Thornthwaite's measure of precipitation effectiveness. This climatic index is based upon the fact that the rate of soil movement varies directly as the cube of wind velocity and inversely as the square of effective moisture which is taken to be proportional to the $(P-E)$ index.

Vegetative weathering is particularly strong in tropical and humid regions whereas in the temperate zones it is not so important but can still have adverse effects. Interestingly it has been observed that beech and larch trees are more effective weathering agents than are oak and pine.

In the Pleistocene physical or mechanical weathering was active in Europe, the climatic regimen of periglacial regions having been affected by the proximity of the enormous ice sheet. The limits of this extended as far south as Bohemia and of course the Alpine glaciers were much more extensive than now. All this caused a climate much more severe than that at present. As a consequence, much solifluction occurred and there was intense physical weathering, especially frost-induced. Permafrost conditions existed as evidenced by wedge-formed and widened joints infilled with ice in the glacial stages as well as by the surficial loosening of rocks. In periglacial areas the surface of the bedrock was disturbed to as much as 6 m depth below the original surface of the land. Periglacial weathering is very significant for the site engineer because of its effects on the design of foundations.

1.3.2. Chemical Weathering

Here secondary minerals are produced by chemical reactions which initiate changes in the chemical composition of rock components. The process takes place either surficially or at depth and in the first case is controlled by the depth of subsurface water and the rock resistance, the intensity generally increasing with depth. Chemical alteration at depth is of course quite independent of subaerial factors so that the processes acting are not related to the relief of the landscape. Deep seated alterations of this type may include the serpentinization of ultrabasic rocks such as peridotites or the propylitization of andesites, both arising from internal conditions of hydrothermal alteration. This latter is a chemical phenomenon involving heated or superheated steam and, as well as causing alteration, these may be responsible for deposition of ores in rock fissures in a concentric arrangement around a magma-formed igneous mass. Reactions may take place with the components of the host rocks also. Where hydrothermal decomposition is encountered, it usually extends to a depth.

The major agents of chemical weathering are detailed below.

1.3.2.1. Rainwater

This often percolates into the soil, descending through it into the zone of aeration proper from which it may join the groundwater system. The completion of this process depends upon the geological conditions and the climatic regimen, especially the amount of rainfall and the depth of the water table. In the zone of aeration, the pores between mineral grains are partly air-filled, partly water-filled. In addition to the above, there is a reverse process, namely capillary rise accentuated during dry periods and evaporation.

1.3.2.2. Oxidation

This affects materials originating under higher temperature and more extreme dynamic conditions than those now operating in the planetary surficial layers. It is an exothermic process forming minerals with the accompanying liberation of heat. Such minerals have a lower density and higher volume. A typical case is the oxidation of divalent to trivalent iron which can acquire hydroxyl ions and form the hydroxide $Fe(OH)_3$. Sulphides may be oxidized as well with pyrite and marcasite (FeS_2) first forming sulphate and then limonite. Sulphuric acid is liberated during the process and this is capable of dissolving minerals such as calcium carbonate in the weathered zone. The fragmentation of rocks is accelerated by the crystallization of newly-formed sulphates, such as gypsum and epsomite ($MgSO_4.7H_2O$), along bedding planes and joints in pyrite-bearing shales. For this reason such rocks are unsuitable for use as building stones. The oxidation of sulphides releases heat and hence the decomposition of pyrite and marcasite may promote the self-ignition of surficial coal seams. The liberation of heat during sulphide weathering may trigger similar fires in coal tips.

Hydrolysis affects the decomposition of complex silicates. Water can dissociate into the free ions H^+ and OH^- and the extent of this rises with increasing temperature, acid and CO_2 content (acids increase the hydrogen ion content). The end product of the chemical decomposition of silicates is clay. If the environment is alkaline, less intense chemical weathering causes the partial alteration of feldspar into clay minerals of the illite group. Weathering of bentonite under similar conditions promotes the formation of clay minerals of the montmorillonite group.

In a slightly acidic environment, potassium feldspar will alter to form kaolinite (a process termed kaolinization) if the climatic conditions are warm and humid. In tropical areas of high humidity and temperature, lateritization takes place, most of the silicic acid being removed as gel and the insoluble iron and aluminium hydroxides remaining as a concentrate in the upper layers.

Hydration entails the incorporation of water into the chemical composition of anhydrous minerals with accompanying volume increase. When anhydrite is altered to gypsum there is a volume increase of 60% which can be serious from the engineering standpoint. For instance Triassic rocks containing anhydrite occurred in some Alpine tunnels and their hydration with swelling has made the driving of these very difficult. Other minerals showing the effect include haematite which expands on conversion to limonite.

1.3.2.3. Carbonation

This is a chemical weathering process characterized by the production of secondary carbonates and bicarbonates through the reaction of CO_2 with weathering products, such as the oxides of calcium, magnesium, potassium and other elements. All surface waters contain CO_2 derived from the atmosphere. Such waters percolating through weathered rocks can precipitate limestone, mostly on bedding and joint planes. Solution processes involve rocks and minerals, the effect depending upon the presence of CO_2 and oxygen in atmospheric water which on percolating into the ground obtains more CO_2 deriving from the decomposition of organic materials below ground and the life activities of some organisms which may also yield O_2. The most soluble materials are gypsum, chlorides such as halite (rock salt), limestone and dolomite. Insoluble materials include quartz and micas which in consequence are often components of clastic (fragmental) rocks. The process of importance in foundation engineering is that in which dissolved CO_2 alters calcium carbonate to the bicarbonate because this is the starting point in the process of karstification. Associated insoluble residual soils are termed terra rossa and from their thickness can be estimated the degree of lowering of the original limestone surface. This subject is very important in dam siting because karstification impairs the bearing capacity and permeability of limestones, the associated development of subterranean caverns reducing the former and increasing the latter. Limestones are lithified rocks containing over 50% $CaCO_3$ and load pressures on a calcite assemblage normally increase its strength and stability. As they increase, depositional differences diminish and with dynamic modification calcite crystals respond to the imposed stress by assuming a preferred orientation. The karstification process involves development of underground cavities and is facilitated if the limestone is pure. Karstic galleries and caves generally develop at several horizontal levels following the development of a valley. Such levels often correlate with the altitudes of Pleistocene river terraces which therefore may constitute good indicators of the position of the major cavities. In limestone areas which have subsided, karstic cavities may occur at great depths below the level of the valley floor. In cavernous limestones, the sealing of karstic cavities by cement and cement mortar is not always successful. Cavities may be several metres across so the task is virtually impossible since cement will be washed away by flowing water. Hot asphalt has been used at some dam sites, but the approach requires a complex device to heat the asphalt in the grouting holes so that it may be fluid at the cavities to be filled. Also asphalt sealing is impermanent, usually serving as a temporary barrier to facilitate cement grouting. Washing of clay infilling out of karstic cavities and replacement with a suitable grout is difficult, necessitating a long-term grouting programme. The degree of karstification is critical in regard to uplift. The full value of this factor must be taken into account in designing gravity dams in limestones traversed by open joints. Dams are very rarely erected successfully on karstic limestones, but one such case is the Cherokee Dam on the Holston River, as discussed in Chapter 9. As regards subterranean caverns, their collapse may promote the formation of sinkholes at the ground surface as is the case at Halle, East Germany. Here there have been large ground surface deformations together with subsidence, all dangerous to existing buildings and inimical to projected structures as well. Similar effects are recorded from Cheshire in England due to subterranean salt deposits.

1.3.3. Weathering Zones

It is clear that a detailed and thorough investigation of the weathered zone is always necessary in engineering construction and the primary job is to determine the depth of weathered material as well as the extent to which weathering processes have proceeded. The depth will vary according to rock resistance and fracture state and attempts have been made to quantify field data so as to provide a numerical analysis of particular situations. This follows the work of Fookes *et al.* in 1971.[7]

They differentiated six grades based on the fracture spacing index, the porosity and the results of strength tests on irregular rock samples. In the same year Zolotarev distinguished three weathering zones, an upper (dispersion) termed I and characterized by the complete alteration of both the mineral and chemical composition and character of the primary rock, a middle (II) fragmentary zone of mainly physical weathering divided into several horizons (A, B, C, D) and a lower (III) fractured zone controlled by faults and other tectonic phenomena.[8] In 1968, Ondrasik divided grades of weathering according to macroscopic criteria as follows:[9]

(a) Complete weathering zone—complete decomposition and disintegration of rock.
(b) Intensive weathering zone—small fragments and loose minerals produced, interstices being filled with loam-like material and strong chemical weathering although original macrostructures preserved.
(c) Slight weathering zone—mechanically weathered to large fragments and blocks in primary positions. Spaces formed by widening of primary joints and infilled with loam.
(d) Partial weathering zone—physical weathering shown by the widening of primary and tectonic fractures and also by the loosening of coherence along planes of weakness facilitating rock disintegration along such planes.

It is important to observe whether the products of such processes are *in situ* or transported. Distinguishing aeolian from water weathering is facilitated because wind-blown particles and grains are more perfectly rounded than water-worn ones and also better graded. This results from the fact that wind cannot for a given velocity move particles above a critical size. For instance a wind velocity of $8\ \mathrm{m\,s^{-1}}$ can move quartz particles of 1 mm diameter whereas wind with a velocity half as great can only move grains of diameter 0·35 mm. In deserts most of the wind-blown grains have diameters between 0·15 and 0·3 mm. Winds also possess a sieving action so that fines are carried further from the source. Wind sweeps such particles along until an obstacle is encountered and then deposition to form dunes occurs. Larger grains, such as pebbles, are too heavy to be shifted, but may be smoothed and faceted by sand storms to form dreikanter (three-cornered) stones. The desert surface is worn down by aeolian processes and some oasis-containing hollows are eroded down to the water table. The most adverse action of wind is to strip dry topsoil as it did in the dust bowl areas of Kansas and Nebraska in the USA. Deposits originated by wind include coastal dunes and sandhills, desert dunes such as barchans and seifs (ridge dunes) and also loess. Topsoils may have variable depths from centimetres to metres according to the rock sources and they comprise inorganic mineral particles and vegetative humus. Acids arising from the latter may act as agents of chemical weathering. Characteristically, topsoils have high porosities and grade down into subsoils, mixtures of soil and rock fragments with ever decreasing organic contents. Below these occurs a weathered rock zone with fresh rock underneath at variable depths. Materials comprising soils may be classified by the Wentworth scale, the Atterberg (MIT) scale or by the Casagrande or Unified Soil Classification (see Table 1.4). Mechanical analysis is utilized to determine the sizes of particles present in a soil, this being effected by sieving or other means.

A vertical profile of a soil is termed a soil profile. Various zones may be distinguished such as A, the topmost zone of leaching; B, the intermediate zone of accumulation; C, the lower zone of weathered parental rock material and D, the rock zone. As noted above, heavy rainfall removes soil which, if it reaches rivers, constitutes suspended load in them. Vegetative cover protects the land from rainfall and if

TABLE 1.4

VARIOUS SOIL CLASSIFICATIONS

Wentworth		*Atterberg*		*Casagrande*	
Grade	*Size*	*Grade*	*Size*	*Grade*	*Size*
Pebbles	>2 mm	Gravel	>2 mm	Gravel, G	7·6 cm (No. 4 sieve, 0·48 cm)
Sand		Sand		Sand, S	
Very coarse	2 to 1 mm	Coarse	2 to 0·6 mm	Coarse	No. 4 to No. 10 sieve
Coarse	1 to 0·5 mm				
Medium	0·5 to 0·25 mm	Medium	0·6 to 0·2 mm	Medium	No. 10 to No. 40 sieve
Fine	0·25 to 0·125 mm				
Very fine	0·125 to 0·06 mm	Fine	0·2 to 0·06 mm	Fine	No. 40 to No. 200 sieve, 0·074 mm
		Silt			
Silt	0·06 to10·002 mm	Coarse	0·06 to 0·02 mm	Silt, M	>2 mm
		Medium	0·02 to 0·006 mm		
		Fine	0·006 to 0·002 mm		
Clay	<0·002 mm	Clay	<0·002 mm	Clay, C	<2 mm

such a cover is removed then erosion is facilitated. In the case where this process attacks soils with boulders, the result may be earth or rock pillars, i.e. thin columns of soil capped with a boulder acting to prevent the underlying soil from being washed away. Mudflows are the results of build-up of water content in the soil through heavy rainfall. After deep weathering a deep rotten cover may overlie the substrate and this is the result of chemical action. For instance in granite areas groundwater acidified by passage through soil may penetrate joints and reduce this normally hard and competent rock to a crumbled and rotten material often metres deep. 90 m deep weathering of granite has been observed in Australia and on Dartmoor in the UK weathering to 10 m depth has been observed. The formation of granite tors is connected with the frequency of jointing, these comprising masses of competent rock left standing above the general landscape by erosion along wider spaced joints defining their shapes. Reverting to sedimentary rocks, there are four major groups of these:

(a) terrigenous, clastic or detrital formed from rock fragments;
(b) chemical, formed by the precipitation of dissolved salts;
(c) organic, derived from the skeletal remains of plants and animals;
(d) calcareous, limestones and dolomites containing over 50% calcium carbonate.

Minerals of every kind may be included and while some of these are weathering products, others such as quartz are fresh (see Table 1.5). Minerals of high resistance to weathering are stable and apart from quartz include garnet, tourmaline and rutile. Alteration products are various but the ultimate product is usually clay. Some minerals originate close to where they occur, for instance if precipitated from chemical solution, e.g. carbonates, sulphides, chlorides or silica. Others may be transported over great distances. Chemically precipitated sedimentary rocks are usually crystalline, for instance:

(a) calcareous—calcite, dolomite;
(b) sulphate—gypsum, anhydrite;
(c) chloride—rock salt;
(d) silica—chalcedony.

This is not invariably the case, however—compare oolitic limestones.

Grain size is important in terrigenous (material derived from the land) sedimentary areas as an indication of distance of transportation, the coarsest being deposited nearest to the area of origin while the

TABLE 1.5

MINERALS ARRANGED IN ORDER OF THEIR RESISTANCE TO WEATHERING (MOST RESISTANT AT THE TOP) AND THEIR ALTERATION PRODUCTS

Mineral	*Alteration products*
Quartz, SiO_2	—
Micas, silicates of aluminium and potassium plus iron and magnesium in the dark varieties	Variable, including clay minerals and hydrous silicates
Feldspars, aluminous silicates of iron, sodium, calcium or barium	Soluble matter and clay minerals
Augite, a pyroxene calcium, magnesium, iron, aluminium silicate	Hydrous silicates
Hornblende, an amphibole calcium, magnesium, iron, aluminium silicate with sodium	Serpentine
Olivine, an olivine family mineral with composition R_2SiO_4 where R is iron or magnesium	Iron oxide

finest particles travel furthest. Clay particles or rock flour will be the last to settle when water passes over at negligible velocity. If settling takes place in an estuary or in the sea, flocculation may take place due to the presence of salts in the transporting medium. Whether grains are round or not depends upon their hardness and the extent of impact en route and indicates the degree of weathering to which they have been subjected during transportation. Grains of sand are fully rounded only through aeolian action which is important because of the widespread use of sand in civil engineering for making concrete, etc. Sediments of terrigenous character may be well or poorly sorted, the latter being more heterogeneous than the former. The engineering concept of a well graded soil is exactly the opposite of a well sorted one. Being well graded implies that a wide spectrum of sizes occurs. Various significant sedimentary structures occur in nature as described below:

(a) Graded bedding—the deposition of a sediment with assorted sizes of grain comprising a sequence with coarse material basally grading into fine grained material at the top. Original layering in sediments is termed stratification (bedding), each layer being a stratum or bed and the interfaces between such strata or beds constituting bedding planes.

(b) Cross bedding usually occurs in sandstones deposited in shallow

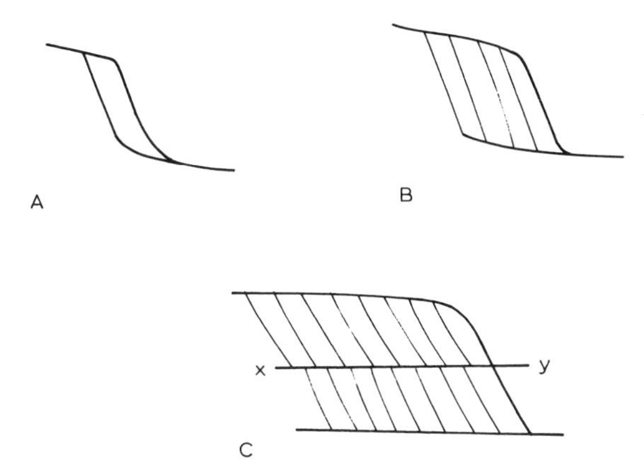

Fig. 1.3. A, Incrementation of sand by one layer; B, addition of other layers; C, initial layer truncated along xy.

water and is also found in other rocks such as volcanic ashes. In the former case successive layers form as the grains of sand sink in slow moving deeper water downstream of a sandbank or delta. Every layer slopes downstream and is at first S-shaped. Erosion of the top of the sand bank or delta by the river will leave the minor layering intact but will truncate it at the junction with the upper bedding plane as in Fig. 1.3.

(c) Reef limestones form from organic remains, e.g. corals, concentrated into reefs because the original organisms were colonial in resistance to the attacks of waves.

(d) Clastic limestones contain fragments of organic or inorganic materials or both and much of their content is of chemical origin. When well sorted, larger fragments are held in a pure calcite matrix precipitated chemically. In poorly sorted ones, these larger fragments may become cemented by calcite mud.

There are various kinds of clastic limestones such as oolitic limestones in which concentric layers of calcite have been deposited around nuclei of mineral grains or other materials so as to form small spheres. Millions of these cement together to produce the characteristic oolitic structure. Another kind is calcitic mudstone, precipitated calcite mixed with clay. A third kind is dolomite, an alteration product of limestone in which the calcite has been partly replaced by magnesite (the calcium of the calcite

TABLE 1.6
THE SEDIMENTARY ROCKS

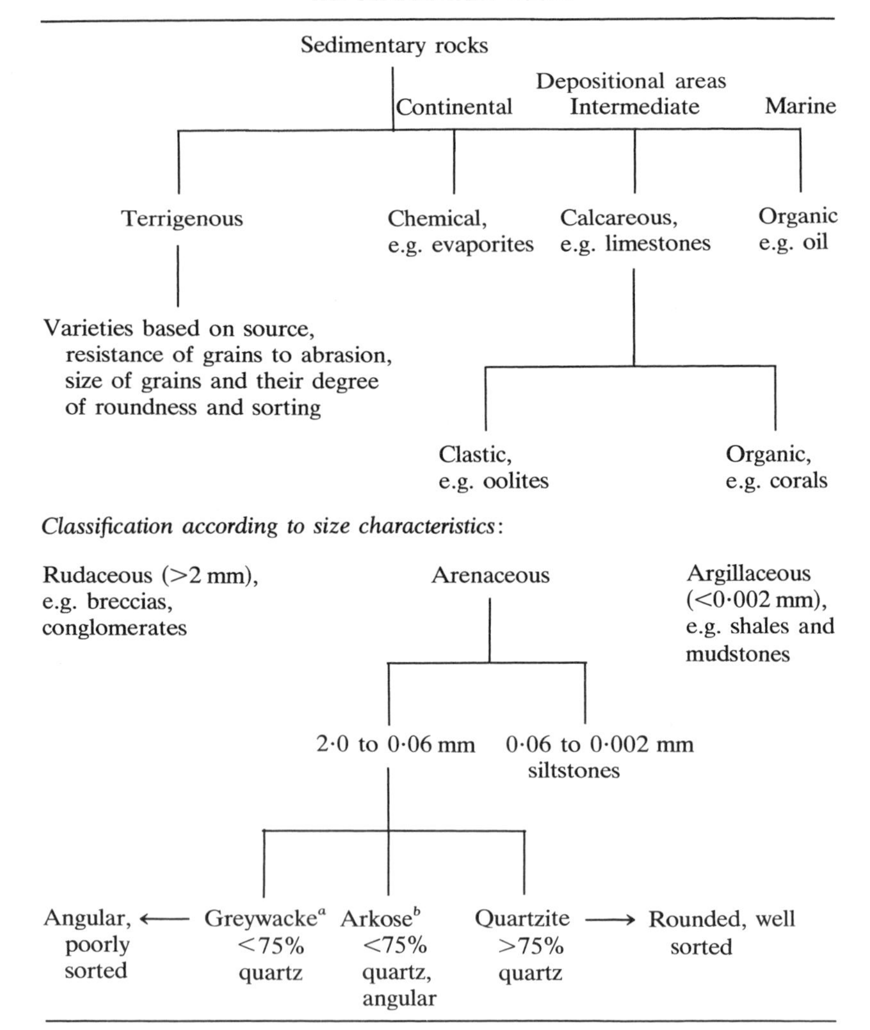

[a] Graded bedding possible; comprise lithic fragments often cemented by argillaceous materials; if subject to high grade diagenesis (low grade metamorphism), may become chloritic; common rock types in geosynclines.
[b] Cross bedding possible; contain much feldspar plus quartz plus other minerals, e.g. micas; term sometimes applied to sandstones of 25% feldspar (at <25% feldspar termed feldspathic sandstones); taken as indicative of erosion under arid conditions and rapid burial.

having been partly replaced by magnesium often with accompanying increases in the sizes of crystals). These are usually chemical sedimentary rocks. Finally shelly limestones comprise wholly or partly shelly material bonded by calcite or terrigenous materials.

A classification of sedimentary rocks is given in Table 1.6. They are sub-horizontal when deposited but later crustal tectonic movements may incline or fold them. Either way the angle made with the horizontal constitutes the dip of inclined beds and that line at 90° to this constitutes the strike.

1.4. METAMORPHIC ROCKS

These result from the subjection of pre-existing rocks to increases in pressure and/or temperature to a degree which promotes recrystallization so that a new rock is produced, one with a new texture and perhaps also a new mineral composition. The process itself is termed metamorphism and can be either:

(a) Thermal in nature, i.e. have increased temperature (with associated higher enthalpy) as the active agent of the change with the amount of recrystallization relating to this. The phenomenon occurs at the margins of a large intrusion of magma in a zone termed a thermal aureole, a zone which can extend 2 or more km from the igneous mass intruded. Near to the actual contact, the metamorphic rock possesses minerals arranged in a random manner and is called a hornfels which shows no cleavage, schistosity or parallel alignment of minerals although relict fabrics may persist. Hornfelses are described by prefixing the name of the most significant minerals or mineral groups, e.g. pyroxene–hornfels or garnet–hornfels.

(b) Dynamic in nature, i.e. have increase in stress as the active agent of change, the excess heat present being unimportant in this process. It is characteristic of narrow movement belts in which rocks on one side are under displacement relative to the other. Sometimes the rocks are completely crushed, sometimes partly recrystallized. The process of the mechanical failure of rocks associated with dynamic metamorphism is termed cataclasis and when this is intense and prolonged, the rock may assume a fine-grained, crudely foliated texture and constitute a mylonite.

TABLE 1.7
METAMORPHIC GRADE OF ROCKS

			Rocks being regionally metamorphosed			
Metamorphic grade	*Rock type*	*Grain size*	*Shales*	*Greywackes*	*Limestones (impure)*	*Basic igneous*
			INDEX MINERALS			
High	Gneiss	Coarse	Sillimanite		Augite	
Medium	Schist	Medium	Iron garnet		Calcium garnet	Magnesium garnet
				Sodium plagioclase	Hornblende	
			Kyanite		Calcium plagioclase	Calcium plagioclase
Low	Phyllite*	Fine	Chlorite, biotite	Chlorite, biotite	Chlorite and sodium plagioclase	Chlorite and sodium plagioclase

* Becomes slate when stressed.

Regional metamorphism refers to dynamothermal effects over a considerable area. The actual degree of metamorphism depends upon the conditions of temperature and pressure of formation of the metamorphic rock and can be determined by reference to certain index minerals, the result being expressed as metamorphic grade (see Table 1.7). Some further details on the rock types mentioned in the table may be relevant:

Gneiss: This possesses a texture referred to as gneissose and it is a rock type resulting from the regional metamorphism of igneous or sedimentary rocks. In it the minerals are arranged in a series of irregular bands with the commonest minerals being feldspar, quartz, mica or hornblende.

Schist: With schistose texture, this is again a product of regional metamorphism with mica, chlorite and hornblende as its commonest minerals.

Phyllite: With phyllitic texture, a rock resulting from continuous stress acting on slates with some heat. The common minerals are muscovite and chlorite.

A number of other rock types exist; one mentioned earlier was *hornfels*, a thermal metamorphic product of heat acting on sedimentary rocks. It may comprise feldspar, biotite and quartz with these minerals arranged randomly. Others which may be cited are:

Slate: With slaty texture it is formed by the metamorphism of argillaceous rocks. It turns to phyllite when more stressed.

Granulite: It has a granulose texture and no spaces between minerals. It is formed by deep seated metamorphism of quartzo-feldspathic rocks through intense regional metamorphism.

Quartzite: Granulose textured, this results from the recrystallization of sandstones through regional metamorphism or through thermal metamorphism alone. Predominantly composed of quartz.

Marble: Granulose textured and resulting from the recrystallization of limestone where contact or regional metamorphism occurs.

When regional metamorphism reaches high intensity, partial melting of the rocks may occur and result in the production of a rock in which bands of crystalline quartz and feldspar exist. These derive from melted material and leave a fine-grained mica-rich matrix called a migmatite. Such banding is termed metamorphic foliation. Metamorphic rocks can originate from any other type of rock or from pre-existing metamorphic rock hence their composition is more variable than that of any other rock type.

1.5. AN ENGINEERING CLASSIFICATION OF ROCKS

The geological classification of rocks alluded to in section 1.1 above is not so useful to engineers as one based upon their specific needs. One of the first of such systems was proposed by Lauffer[10] in 1958 in regard to tunnelling and drifting activities. It is based upon two significant civil engineering parameters, namely the unsupported stability of the rock and the effective span. Of course it is usually impossible to determine the unsupported stability directly and so the rock classification must be grounded upon the experience of the assessor, i.e. ultimately upon the informed opinions of the engineering geologist or engineer. This causes conflict occasionally, but unfortunately there is no escaping the fact that it is not feasible to directly measure the parameter itself. More recently, attempts have been made to sub-divide rocks according to quantitative parameters alone so as to reach a more objective and scientific result. Such a scheme has been proposed by Barton *et al.*,[11] the factor Q, rock quality value, being calculated from many factors taking into account jointing, water content and rock pressure. Under jointing may be included not merely joint spacing and the number of

TABLE 1.8
THE BARTON/LIEN/LUNDE PROPOSAL[11]

		Classification parameters with ranges		
		RQD	*Joint roughness number (Jr)*	*Joint water reduction factor (Jw)*
		Very poor ≤ 25 Fair = 50–75 Excellent ≥ 90	Discontinuous joints = 4 Smooth even joints < 1	Dry = 1·0 Moderate inflow = 0·7 Strong inflow ≤ 0·2
		Number of sets of joints (Jn)	*Joint wall alteration number (Ja)*	*Stress reduction factor (SRF)*
		Few = 1 3 sets = 9 Crushed rock = 20	Unaltered = 1 Slightly altered = 3 Thick intermediate clay layers > 10	Favourable < 2 Open joints, shear zones = 3 Squeezing rock pressure > 10
Rock quality	=	*Size of jointed body* ×	*Shear strength between blocks* ×	*Active stresses*
Q	=	$\frac{RQD}{Jn}$ ×	$\frac{Jr}{Ja}$ ×	$\frac{Jw}{SRF}$
Tunnelling conditions	*Q*			
Very poor	1			
Fair	4–10	$200\ \frac{RQD}{Jn}\ 0{\cdot}5$	$4\ \frac{Jr}{Ja}\ 0{\cdot}02$	$1\ \frac{Jw}{SRF}\ 0{\cdot}005$
Excellent	100			

families of joints but also their roughness and their degree of weathering. Table 1.8 embodies some of their calculations of Q and shows the weighting of the influential factors.

Beniawski[12,13] has produced rock mass and engineering classifications of jointed rock masses schemes and the RMR rock characteristic according to him may be derived from the sum of six separate parameters which take into account the rock strength, the jointing and the water content. As regards tunnels, Beniawski's classifications may be related to the influence of orientation of joints (see Table 1.9).

Both of the above-mentioned systems of classification (Tables 1.8 and 1.9) attach importance to jointing because each relates 4 of the 6 parameters utilized to joints. The innate strength of the material of the rock itself and the stresses occurring in the rock mass are considered to be less significant. Hence the parameters usually cited as relevant for describing the geological properties of a rock mass from the standpoint of engineering and based on SIA Standard 199 (1975) are clearly less valuable, the more so because although they are quantitative it is usually very difficult to quantify them precisely. For reference purposes they are appended in Table 1.10.

The inability of the observer to quantify the factors in Table 1.10 applies also to theoretically quantitatively determinable rock properties (based on SIA standard 199 1975) shown in Table 1.11.

All such attempts at engineering classifications of rocks are approaches at describing the whole rock mass by means of appropriate categories which ideally can be quantified. If a purely qualitative scheme is employed with a rock mass, it can be useful if not wholly satisfactory and it is used because site engineers and geologists are unable as yet to discover the relevant numerical quantities. Its usefulness may be summarized as follows:

(a) to help in the initial study in planning, estimation of construction costs and the devising of a schedule for the actual work of construction;
(b) to afford a basis for price calculations when tenders are solicited and also in the terminal accounting stage;
(c) to provide guidance during project work as regards the choice of correctly statically orientated and structurally adequate support measures in a subterranean opening.

An Austrian tunnel construction concept employs quality categorization for estimating requisite dimensions and deformation measurements on the underground structure and the actual rocks are taken into account. Suggested supportive measures for individual rock categories have been proposed by Detzlhofer[15] and Pacher *et al.*[16] and others. Some observations on the strength and deformation parameters of rock masses may be added and of course in the construction of such an underground structure as a tunnel, these are undertaken following a preliminary numerical procedure for its dimensions. On site and laboratory tests are used in determining the mechanical properties of the relevant rock mass and also its stresses.

Rock masses are composed of a variety of complex minerals having quite different mechanical and petrographical characteristics and usually they show fracturation. This is often present in the form of jointing. It is deleterious because it lowers the strength of the rock and

TABLE 1.9

CLASSIFICATION CHARACTERISTICS OF BENIAWSKI[12]

Parameter			Range of values			
1. Rock strength	Point-load index ISRM (1972)*		>8 MN m^{-2}	4–8 MN m^{-2}	2–4 MN m^{-2}	1–2 MN m^{-2}
	Uniaxial compressive strength		>200 MN m^{-2}	100–200 MN m^{-2}	50–100 MN m^{-2}	25–50 MN m^{-2}
	I_i	15	12	7	4	2 1 0
(* Uniaxial compression test preferred for the low range 10–25 MN m^{-2}, 3-101-3)						
	I_1			2 1 0		
2. *RQD* value (according to ref. 14)	I_2	90–100% 20	75–90% 17	50–75% 13	25–50% 8	<25% 3
3. Spacing between joints	I_3	>3 m 30	1–3 m 25	0·3–1 m 20	50–300 mm 10	<50 mm 5
4. Condition of joints		Very rough surfaces, discontinuous	Rough openings 1 mm	Rough with weak wall	Smooth or infilled 5 mm thick	Soft filling 5 mm thick
	I_4	25	20	12	6	0

Parameter							
5. Water situation		Inflow over 10 m tunnel length	Inflow absent or dry	25 litres m^{-1} or moist	25–125 litres m^{-1} or water under low pressure	125 litres m^{-1} or difficult water conditions	
	I_5		10	7	4	0	
6. Direction of strike and dip of joints	I_6		Very good	Good	Okay	Unfavourable	Bad
		Tunnel	0	−2	−5	−10	−12
		Foundation	0	−2	−7	−15	−25
		Rock slopes	0	−5	−25	−50	−60

Strike normal to tunnel axis; dip in direction of advance		Dip against direction of advance		Strike parallel to tunnel axis		Dip
$\beta = 45–90°$	$\beta = 20–45°$	$\beta = 45–90°$	$\beta = 20–45°$	$\beta = 45–90°$	$\beta = 20–45°$	$\beta = 0–20°$
Very good	Good	Moderately good		Unfavourable	Very unfavourable	Bad

Determination of the rock class:					
$\sum I_i = I_1 \ldots I_6$	81–100	61–80	41–60	21–40	<20
Class	I	II	III	IV	V
Description	Very good	Good	Moderate	Poor	Very poor
Significance of the rock classes:					
Class	I	II	III	IV	V
Average standup time, span	10 years at 5 m span	6 months at 4 m span	1 week at 3 m span	5 h at 1·5 m span	10 min at 0·5 m span
Rock mass cohesion	0·3 MN m^{-2}	0·2–0·3 MN m^{-2}	0·15–0·2 MN m^{-2}	0·1–0·15 MN m^{-2}	0·1 MN m^{-2}
Friction angle	45°	40–45°	35–40°	30–35°	30°

TABLE 1.10

RELEVANT GEOLOGICAL PROPERTIES FOLLOWING SIA STANDARD 199

Rock	*Stratigraphic discontinuity of the rock mass*	*Tectonic discontinuity of the rock mass*	*Water circulation*
Unfavourable components		Degree of fracturing	Permeability features
		Degree of foliation	
Strength	Stratification (bedding)	Type of fracture surfaces	Water circulation
Quartz content	Sequence of beds	Opening of the joint surfaces	
Engineering workability	Interlayering of shales or clay	Friction and cohesion of discontinuities	
Properties when water and humidity intrude	Friction and cohesion of the bedding planes	Generalized shape of the joint blocks	
Degree of weathering			

makes it more susceptible to weathering, the agents of which penetrate the mass through fissures and openings. The deformation properties of the rock are also affected by weathering and by the entry of water. All this correctly implies that an adequate description of the rock mass by site engineers and engineering geologists should not merely describe mechanical characteristics but also deal with all the states of the rock

TABLE 1.11

ROCK PROPERTIES

Strength characteristics	*Deformation characteristics*	*Original condition of rock*	*Other characteristics*
Compressive σ_n (MN m^{-2})	Elasticity modulus E (MN m^{-2})	Thickness of overlying load H (m)	Mineralogical composition (vol %)
Tensile σ_z (MN m^{-2})	Deformation modulus ME (MN m^{-2})	Primary stress state (MN m^{-2})	Pore volume n (dimensionless)
Shear τ_s (MN m^{-2})	Shear modulus G (MN m^{-2})	Lateral pressure ratio λ (dimensionless)	Water (pH, hardness, etc.)
Geotechnic cohesion c (MN m^{-2})	Poisson's ratio m (dimensionless)	Specific gravity γ_G (kg m^{-2})	Jointing (semi-quantitative), direction and angle of dip
Angle of friction at the joint surfaces 0 rad; 360°	Change of volume in fracturing V_b (%)	Temperature T (°C)	Spacing between joints, degree of separation

mass. This has a great bearing upon the load which can be imposed safely which will be greater in competent, homogeneous, non-fractured, impermeable, relatively unweathered rocks with low porosities.

Rock mass jointing necessitates the utilization of problem-specific discontinuity analyses as a result of the fact that the strength and deformational characteristics of a rock mass do not depend solely upon the structure or fault planes or joint planes or bedding planes but also upon the nature of the rock substance. Reik and Schneider[17] in 1980 gave a valuable idealized representation which illustrates the difference between the strength of a rock element, a completely separated joint with friction and an incompletely separated joint with friction and geotechnical cohesion (shear strength).

It is useful to add some data on the uniaxial crushing strengths of common rocks (cf. triaxial data in Chapter 9). These are listed in Table 1.12. Mineralogical composition is important in determining strength since such minerals as quartz and feldspar have much greater inherent strengths than the clay minerals, micas and gypsum. Rocks containing cleavage-possessing minerals like micas tend to split parallel to the cleavages. The strength of clastic rocks having such minerals is also

TABLE 1.12

UNCONFINED CRUSHING STRENGTHS OF COMMON ROCKS

Rock type	*kN m^{-2}*	*bars*
1. *Clastic rocks*:		
Calcareous mudstone	55 200–193 200	552–1 931
Dolostone (dolomite)	62 100–351 900	621–3 517
Limestone	4 830–200 100	48–2 000
Sandstone	10 350–234 600	103–2 345
Shale	6 900–227 700	69–2 276
Siltstone	27 600–310 500	276–3 103
2. *Igneous and Metamorphic rocks*:		
Basalt	179 400–276 000	1 793–2 759
Gneiss	151 800–248 400	1 517–2 483
Granite	41 400–289 800	414–2 897
Marble	48 300–234 600	483–2 345
Phyllite (micaceous)	6 900–17 250	69–172
Quartzite	207 000–365 700	2 069–3 655
Schist	7 590–138 000	76–1 379
Slate	96 600–324 300	966–3 241

influenced by the grain size and the geologic age—old clastic rocks have higher strengths than newer ones probably because of recrystallization, compaction and cementation.

The term crystalline rocks is applicable to igneous rocks (except glassy ones), metamorphic rocks and sedimentary rocks which have recrystallized from a previously clastic state.

1.6. ROCKS AS CONSTRUCTIONAL MATERIALS

Constructionally valuable rocks have to be located by geologists in connection with many engineering activities such as making highways or erecting large scale hydraulic structures. They must be obtainable economically, i.e. with minimum transportation costs and continuous supply throughout the lifetime of the project. Existing quarries and exposures have to be investigated and if excavation has to be undertaken, there must not be too much overburden to be removed. If this condition is satisfied, it may be practical to open a new quarry near to a construction site where a big job such as building a dam is concerned because transport will not constitute a problem. In fact quarries of this type may be abandoned later and hence serve only a temporary purpose. Clearly the rock itself must be of a type suitable for the needs of the contractor and this can mean, for example, a specific grain size or a certain degree of purity. As noted above the engineering geologist will have to find the relevant source of material and to do this a multistage exploration programme is required. An initial prospection for mineral resources is followed by reconnaissance of any particular deposits located. Then there will be detailed examination and finally perhaps a research phase to consider the working of the deposit. As for the materials utilized for engineering purposes, these may be divided into the following categories:

(a) materials necessary for certain mineral features such as clays (used in bricks and pottery), asbestos, gypsum, ore minerals, fuels such as coal, bitumen and the basic components of cement (limestone and a silica-aluminous deposit such as clay);
(b) materials necessary for structural features such as aggregates and building stones with which suitability is determined not merely by size and quality, but also by their properties in the rock mass, e.g. rift and grain in granites.

Water is a requisite in almost all constructional work and may be needed with some or all of the above at some time or other or alternatively may exert a deleterious influence on some of them occasionally.

The above-mentioned multistage exploration programme proceeds step-by-step, the first essential being a positive preliminary report. Without this there will be no need to proceed, but if it is forthcoming then the usual geological and geotechnical testing follows. This should be the time to evaluate the rock quality, its ease of excavation, its petrographic characteristics and also the general hydrogeology.

It is important to indicate whether access roads are available for transport to and from the actual construction site or not and this should not be too far away from the mineral source. Much basic information is obtainable from geological maps and reports of the relevant area, but field inspection by an engineering geologist is vital. This is because the geological map normally shows the stratigraphic distribution of rocks, drift, etc., but does not have enough data for engineering purposes. For instance the identification of a bed as limestone does not indicate its purity or the presence or absence of karstification and so on. Hence an engineering geology examination of the site of the quarry is made and this includes such details as landforms, the hydrogeology, etc. Laboratory and borehole tests will be included in associated reports. All this will provide an estimate of the size of the material reserve and facilitate the production of a plan for exploitation. Some of the natural materials of use to the civil engineer for construction purposes will be discussed.

1.6.1. Building stones

The natural building stones fall into three categories, namely

(i) dimension stone;
(ii) roofing stone;
(iii) ornamental stone.

Selection in construction depends upon availability and expense as well as inherent properties, especially strength and durability.

Dimension stones include granite, limestone and sandstone, all good quality and usually obtainable in sizes sufficient for structural utilization. Granite is highly suitable for heavy construction work, e.g. a granite block of dimensions 0·5 m^2 section can carry a load of 7·5 tons. Consequently many walls of docks have been built of this rock, for instance the harbour at Colombo in Sri Lanka for which imported Cornish granite was employed from 1884 onwards, Nowadays granite is also used for facing concrete structures, such panels being anything up to 10 cm in thickness. They are aesthetically more pleasing than concrete and also they are very resistant to the chemically corrosive atmospheres of large cities (especially the acid rain, as dealt with above). Granite has a very low absorption (0·09–0·3% by weight) so it also resists the disintegrating action of frost.

As regards limestones, their chemical and structural variability necessitates care in selection for building purposes. They are indurated rocks, have hence undergone diagenesis and thus had their fabric altered. Such lithification implies diagenesis, the different stages through which aragonite and calcite muds reach equilibrium, but not necessarily recrystallization. The original matrix in limestones is usually composed of fine-grained aragonite crystals of micron dimensions. It encloses discrete elements which are divisible into fossils, oolites and intraclasts. Diagenesis normally converts the matrix into a stable aggregate. Depending upon p_{CO_2}, nucleation and accretion speed, a microcrystalline mud may remain of micron size and form an interlocking mosaic of crystals (micrite). Recrystallization leads to much larger crystals, however, and may occur simultaneously with sedimentation ('alpha sparite' *partim*) or later ('gamma sparite'). Another situation can arise where closely packed discrete elements lack micrite matrix at the time of deposition, forming a mechanically stable assemblage which may or may not be cemented by percolating solutions (in the latter case forming a void-filled sparite). Excluding pressure and temperature changes, limestones may range from those with homogeneous fabrics to those of heterogeneous nature. In the former mosaic texture of micron size may occur and in the latter grain size differences may be of the order of 1 to 10^5. Load pressures on a calcite assemblage increase its strength and stability. The primary effects of deformation are compaction, pore reduction and the development of contact surfaces between grains. These grains become xenomorphic (anhedral) and the order of magnitude of their size differences decreases to an average of 1 to 10^3. Increasing pressure diminishes the depositional differences and if it goes on increasing, with dynamic modification, calcite crystals will respond to the stress by acquiring a preferred orientation. Limestones break irregularly so that there is little chance of shearing between rock and concrete. In thin-bedded limestones shear may take place along bedding planes and the shear strength may be impaired by interbedded clayey shales and clays. The disposition of such potential sliding surfaces must be assessed.

One of the finest limestones for construction is the Portland 'freestone' of Dorset in England which is free of bedding planes for great thicknesses and has been quarried for over three centuries. The principal joints are north–south and east–west in orientation and are very widely spaced so that large blocks can be obtained and later split and trimmed. The only difficulty is the great thickness of overburden which has to be removed—up to 18 m thick. Limestones are susceptible to chemical pollution (acid rain) and hence in cities may exfoliate.

Sandstones also vary considerably depending upon the grain sizes and the cementing material. In fact the strength of a sandstone depends largely upon the cohesive strength and amount of the cement, the optimum being silica or iron compounds. The rock must have a low porosity as well since water absorption is governed by porosity and can range between 2% and 12% by weight in sandstones. Sandstone must not be used with limestone in adjacent courses of stonework because atmospheric sulphur can react with the latter to produce calcium sulphate which then may percolate into the surface layers of the limestone and penetrate the underlying course of sandstone to promote the growth of gypsum crystals. This results in a very rapid deterioration of the sandstone. Dense mortars must be avoided also (these include Portland cement), mortar for stonework requiring a low content of lime. With any sandstone emplacement the bedding should be placed horizontally. If the bedding is placed vertically (face bedding), there is a tendency for the rock to flake with resultant patches of spalled rock which contribute to weakness and also look unaesthetic. In choosing a quarry for dimension stone excavation, it is necessary to seek one with no excessive overburden, suitable stone, appropriate joint and bedding surface spacing and adequate drainage. Also the haulage distance to the site should be reasonable. In the actual excavating, black powder (charcoal, potassium nitrate and sulphur) is utilized because the fact that it burns slower than other explosives means that the rock is not so liable to shattering. Dock wall stone is termed 'armour stone'; since it is required to take the force of waves, a suitable material is granite which may also be employed in constructing breakwaters.

Roofing stone is usually slate which, though widely used in former times, has been rivalled and supplanted by artificial products such as tiles nowadays.

1.6.2. Roadstones

The uppermost layer of a road is its surfacing and this rests upon a base and a sub-base. Depending on the base the surfacing can be composed of crushed rock of small grade size (chippings) with a binder of bitumen or it may be the top layer of a concrete base or may be cemented stone or stabilized gravel. All road surfaces have to withstand abrasion as well as resist impact so any rock used in surfacing must be fresh and strong. Fragments should be angular and without any oxidized coatings on their surfaces. The 'polishing' properties must be low because of high vehicular velocities and this can be obtained when for instance sedimentary rock with hard minerals set in a soft matrix is used or an igneous rock with minerals of varying hardnesses is employed. The physical properties of crushed stone are analysed, the most important ones being crushing strength, abrasion resistance, resistance to impact and attrition as well as aggregate crushing strength, water absorption and polishing qualities—there are appropriate British Standards (BS812) and US Standard specifications. Optimum rocks for roadstone applications come from the basalt group and include dolerite, which shows high impact and low attrition values, low water absorption and suitable crushing strength if fresh (in the range 186 000 to 225 000 kN m^{-2}). These values diminish with weathering. Rocks in the gabbro group are weaker, but those in the granite trade group are satisfactory from this point of view although they do not bind as well as dolerite with tar or bitumen. Hornfels may be useful because it can have a very high crushing strength (one case of 393 000 kN m^{-2} (57 000 psi) is recorded). Those derived from shales by metamorphism are particularly valuable as roadstone and may be employed as ballast also. Gritstone group rocks have a low polishing coefficient which makes them excellent for the surfacing of roads. Some limestones are valuable too because they bond very well with tar, the product being widely utilized in macadam surfaces. Afterwards a 'wearing course' of chippings of hard rock is spread over the macadam. Limestone may also be used for granular base waterbound macadam. An aggregate of mixed grades (fines included) is laid to form a road base. Alternatively the fines may be left out and added at a later stage after the laying of the stone by utilizing a vibrating roller so as to vibrate them into the basal voids. Incorporation of vibration into the laying process produces a good interlocking structure; the surface is then tar-sealed after laying. Such a road base is permeable and of course it is necessary to supply adequate drainage. Limestones have been tested with dyed resins by impregnation which strengthens the rocks. As subsequent sectioning has clearly revealed voids, it may be stated that engineering properties have been found in such cases to be correlated with the voids content (in a set of limestones from Jamaica, for instance). Such petrological examination is an aid to the selection of suitable roadstone materials from various limestones.

1.6.3. Aggregates

These are essential components of concrete and may be gravels, crushed stone, sand, etc. There are specifications for concrete aggregate (in the UK incorporated in BS882 and BS1201) and these are of great importance because choosing the correct aggregate is vital. For instance some materials contain minerals which react chemically with concrete, others have unfavourable swelling coefficients so that the introduction of water will affect the strength of a mix. Since aggregate makes up 75% or more of the volume of concrete, the significance of these remarks is apparent. Aggregates may be divided into coarse and fine categories, the latter passing through a 4·8 mm sieve, the former being retained. The eleven groups of aggregates of the trade are listed in Table 1.13. Some of the best materials for fine aggregate are sands with grain sizes between 0·06 and 2 mm, the former also being the

TABLE 1.13
AGGREGATE TRADE GROUPS

Group	*Rocks, etc.*
1. Artificial	Slag and crushed brick, breeze, etc.
2. Basalt	Basalt, andesite, porphyrite, dolerite, hornblende-schist
3. Flint	Flint, chert
4. Gabbro	Gabbro, diorite, norite, peridotite, picrite, serpentinite
5. Granite	Granite, gneiss, granodiorite, pegmatite, quartz-diorite, syenite
6. Gritstone	Grit, agglomerate, arkose, breccia, conglomerate, greywacke, sandstone, tuff, pumice
7. Hornfels	Contact-altered rock except marble
8. Limestone	Limestone, dolomite, marble
9. Porphyry	Porphyry, aplite, felsite, granophyre, microgranite, quartz-porphyry, rhyolite, trachyte
10. Quartzite	Quartzite, ganister, siliceous sandstone
11. Schist	Schist, phyllite, slate, several altered rocks

upper size limit of coarse silt. Such sands derive from the weathering of rocks of various types and the grains may be individual mineral particles, usually quartz, which are highly resistant to abrasion and solution as well as chemically unalterable. It is seen therefore that deposits employed in fine aggregates have a simple mineral composition although the grains occasionally possess coatings such as iron oxide which can colour them. Gravels include many of the materials alluded to in Table 1.13 such as flint, chert, quartzite, granite, gneiss, etc. Some need screening, crushing or washing before use. Screening separates larger from smaller fragments, crushing imparts angularity to previously rounded grains and washing cleans the material. Much aggregate is quarried rock which has been crushed and thus comprises such angular fragments that have rough surfaces. Actually the shapes of fragments are covered by BS812 in the UK using the following terms:

(i) rounded—usually waterworn;
(ii) irregular—partly rounded by attrition or subangular;
(iii) angular—with well-defined edges and planar faces;
(iv) flakey—material possessing small thickness relative to other dimensions;
(v) elongated—fragments with length much exceeding other dimensions;
(vi) flakey and elongated—fragments with length exceeding width and width exceeding thickness.

Surface textures are described using the following non-geological terms:

(i) glassy—usually with conchoidal fracture;
(ii) smooth—waterworn;
(iii) granular—fracture surface showing rounded grains;
(iv) rough—rough fracture of medium or fine-grained rock with no visible crystalline constituents;
(v) crystalline;
(vi) honey-combed—visible pores or cavities.

The most adverse constituents in an aggregate are flat or elongated components such as shale, slate or schist. This is because they tend to split preferentially parallel to the flat surfaces so that their strength is weaker in that direction than at right angles to it. Also shale has low strength and high water absorptive capacity. Additional undesirable constituents include clay, coal, hydrated iron oxides, sulphides, sulphates, salts and organic materials.

In the UK tests for the determination of dust, silt and clay in aggregates are covered by BS812 and this also applies to tests for water absorption, organic impurities and moisture contents together with tests for mechanical properties such as crushing strength, resistance to abrasion, etc. Clay in fine aggregate should not be greater than 3% by weight, for instance. As regards minerals which react with cements, these include opal, chalcedony and chert, mortars made with these showing a rapid expansion within half a year (a process which can cause microscopic and later large scale cracking which may extend to the surface). Opal and chalcedony occur in lava vesicles in rhyolites and andesites and sometimes also in the cracks in limestones as infillings. Where good quality aggregates are unavailable, low grade ones may be utilized in construction—e.g. lateritic gravels in Sierra Leone and Uganda, soft limestones, gravelly soils and, in the Bahamas, crushed coral limestone as recorded by Hosking and Tubey in 1969.[18]

1.6.4. Other Materials of Construction

In this general category may be grouped permeable and impermeable fill, clays for core and cutoff work, rocks for rock-fill dams, embankment materials, rip rap, raw materials for cement manufacture, etc.

Rock-fill dams utilize broken rock which compacts under its own weight. Hand packed rubble may be used to form a layer on the upstream face parallel to the slope of the structure and sometimes an extra layer is added, this being made of concrete or masonry. In the case of the Salazar Dam in Portugal, a flexible steel membrane with expansion joints was used as a facing on a thin layer of concrete itself resting upon masonry above the sloping face of the rock-fill. The arrangement was excellent because it reduced leakage through the facing to almost nothing. Embankments and earthfill dams are constructed of clay or shale placed suitably with layers overlying a core-wall acting as an impermeable barrier to the passage of water through the central line of the structure. The actual core-wall may be made of puddle clay or, if clays are unavailable, of concrete. Below it a cutoff wall extends vertically down so as to extend the flow path of impounded water under the dam. The layers comprising the embankment are rolled into place and the vibration inseparable from this process compacts the material. Rip rap may be placed on the outer slope as a protection. Sometimes a gradation from soil to rock-fill is employed, successive sand layers being succeeded by fine gravels and then coarse gravels which act as filters and prevent the soil penetrating into the rockfill. This approach has other benefits, for instance in the case of the Usk Dam supplying Swansea in South Wales. Here the structure is 30·5 m in height and pore water pressure meters in the earth bank recorded very high values later confirmed by water levels in tubes driven vertically into the then uncompleted structure. This pressure was relieved by incorporation of layers of coarser materials with higher permeability into the dam embankment.

The materials geologically obtained for Portland cement manufacture are limestone (for lime), water and a suitable silica–alumina deposit such as a clay. Alluvial muds and chalk are an excellent combination where available. The main constituents of Portland cement are:

(i) tricalcium silicate;
(ii) dicalcium sulphate;
(iii) tricalcium aluminate;
(iv) an aluminoferrate, $4CaO.Al_2O_3.Fe_2O_3$;
(v) minor constituents such as magnesium, titanium, phosphorus, Na_2O, K_2O (these latter comprising less than 1·5% of the total except in high alkali cements which can cause the deterioration of concrete by reaction with minerals such as opal).

High alkali cements can possess contents of Na_2O and K_2O up to 3·5% and they are used where a concrete must attain a strength quickly or where foundation concrete is emplaced in a chemically hostile soil. A sulphate-resistant cement is produced by adding a small amount of tricalcium aluminate to ordinary Portland cement, this additive being capable of combining with sulphate solutions which could otherwise be adsorbed by just such material in a concrete foundation.

Hydraulic cement is made from calcareous nodules in shale (septaria) and has the advantage that it produces a concrete which can set very fast under water. The nodules contain about 20% silica, 15% alumina and calcium carbonate, the carbon dioxide being driven off by calcining. Where Portland cement sets, the process is exothermic, a process which may adversely affect dams; therefore, low heat cements have been developed from artificial pozzolans (and also for that matter from pozzolans). Pozzolan is mixed with Portland cement to the UK specifications of BS1370. The product not only sets with less heat emission, but also resists chemical attack better.

REFERENCES

1. Everett, D. H., 1961. The thermodynamics of frost damage to porous solids. *Trans. Faraday Soc.*, **57,** 1541–51.
2. Honeyborne, D. B. and Harris, P. B., 1958. The structure of porous building stone and its relation to weathering behaviour. *The Colston Papers*, **10,** 343–65.
3. Potts, A. S., 1970. Frost action on rock: some experimental data. *Trans. Inst. Brit. Geog.*, **49,** 109–24.
4. Wiman, S., 1963. A preliminary study of experimental frost weathering. *Geog. Ann.*, **45,** 113–21.
5. Winkler, E. M. and Wilhelm, E. J., 1970. Salt burst by hydration pressures in architectural stone in urban atmosphere. *Geol. Soc. Am., Bull.*, **81,** 567–72.
6. Yaalon, D. H. and Ganor, E., 1966. The climatic factor of wind erodibility and dust blowing in Israel. *Israel Jour. Earth Sci.*, **15,** 27–32.
7. Fookes, P. G., Dearman, W. R. and Franklin, J. A., 1971. Some engineering aspects of rock weathering with field examples from Dartmoor and elsewhere. *Quart. Jour. Eng. Geol.*, **43,** 139–85.
8. Zolotarev, J. S., 1971. Sovremennye zadachi inzhenerngeologischeskogo izucheniya processov i kor vyvetrivaniya. Voprosy inzhenernogeologischeskogo izocheniya processov i kor vyvetrivaniya. *Izd. Mosk. Univ., Moskva*, 4–25.
9. Ondrasik, R., 1968. Inzinierosko-geologicky vyskum kory zvetravania skalnych hornin na Slovensku. *Acta geol. et geogr., Univ. Bratislava*, Comenianae, **II,** 5–24.
10. Lauffer, H., 1958. Gebirgsklassifizierung für den Stollenbau. *Geologie und Bauwesen*, **1**(24), 46–51.
11. Barton, N., Lien, R. and Lunde, J., 1974. Engineering classification of rock for the design of tunnel support. *Rock Mechs*, **6**(4), 189–236.
12. Beniawski, Z. T., 1973. Engineering classification of jointed rock masses. *Trans. S. Afr. Inst. Civ. Eng.*, **15**(12), 235–44.
13. Beniawski, Z. T., 1977. Rock mass classification in rock engineering. In: *Proc. Symp. on Exploration for Rock Engineering, Johannesburg, 1976*, Vol. 1, ed. Z. T. Beniawski, Balkema, Rotterdam, pp. 97–106.
14. Deere, D. U., 1973. Technical description of rock cores for engineering purposes. *Felsmech. und Ing.-geologie*, **1**/1, 16–22.
15. Detzlhofer, H., 1974. Erfährungen und Vorschläge für die Gebirgsklassifizierung beim Bau von Wasserstollen. *Bundesministerium für Bauen und Technik Strassenforschung, Wien*, **18,** 37–50.
16. Pacher, F., Rebcewicz, L. v. and Golser, J., 1974. Zum derzeitigen Stand der Gebirgsklassifizierung im Stollen und Tunnelbau. *Bundesministerium für Bauen und Technik Strassenforschung, Wien*, **18,** 51–8.

17. REIK, G. and SCHNEIDER, H. J., 1980. The determination of quantitative rock characteristics for tunnelling and drifting. *Natural Resources and Development*, **II,** 24–43.
18. HOSKING, J. R. and TUBEY, L. W., 1969. Research on low grade and unsound aggregates. *Road Research Laboratory Rept LR293*, Crowthorne, Berkshire, England.
19. CROSS, W., IDDINGS, J. P., PIRSSON, L. V. and WASHINGTON, H. S., 1903. *Quantitative Classification of Igneous Rocks*, University of Chicago Press.

CHAPTER 2

Rocks in Time and Space

2.1. ROCKS IN TIME

Basic to the understanding of this concept is the principle of uniformitarianism, i.e. that those processes active today are similar to those operating in the past to form existing rocks. Application of it involves extrapolation of knowledge derived by an examination of these existing rocks so as to construct the geological sequence of events and from this to reconstruct the geological history of an area. Every layer of sedimentary rock records past events and the principle of superposition is applicable, i.e. if layer 2 overlies layer 1, it is younger unless it can be shown that earth movement has caused an inversion of their original position. Other observations can give information from which relative ages may be inferred; for instance if say pebbles of gabbro are found in a conglomerate, then they are obviously older. In some sedimentary rocks, fossils occur, i.e. traces and hard parts of formerly living organisms. Study of these constitutes palaeontology and its importance is partly due to the fact that different animal and plant groups characterize sedimentary layers of different age. Thus it is possible to correlate such layers in different areas. It is also possible to recognize biofacies, i.e. sedimentary rock types with particular faunas. The particular rock type itself constitutes a lithofacies. A lithofacies may be diachronous, i.e. not all formed at the same time, and in this case the phenomenon is detectable through fossil evidence. Lateral change of facies often occurs. As seen, fossils are useful in deriving the ages of sedimentary rocks and their utility is at a maximum when they are numerous and have existed for only a relatively short period of geological time, either becoming extinct or evolving into some new species or genus. Microfossils often meet these requirements and also may be found in cores; hence they are especially useful to the petroleum industry. The persistent change of life with time comprises evolution, the cause of which is usually accepted as being a modified version of Darwin's theory of natural selection. That branch of geology concerned with the analysis of rocks in time, particularly layered rocks, is termed stratigraphy and aspects of it significant to site engineering include the following:

(a) an understanding of the effects of geological history on rock distributions and subsequent estimation of the rock types to be expected under the site;
(b) an understanding of the method of presentation of geological data on maps and in reports in which rocks are classified chronologically.

2.2. THE GEOLOGICAL TIME SCALE

In order to understand this properly, stratigraphical nomenclature must first be considered. The applicable terms are listed in Table 2.1. Rock strata with a closely defined fossil content may be attributed to a zone which is taken as equivalent to a 5th order rank. Various types of such zone are recognized and include:

(a) Coenozones—strata characterized by a particular faunal assemblage.
(b) Acrozones—strata defined by the range in time of a particular species.
(c) Epiboles—strata in which a particular species attains it acme.

TABLE 2.1
RECOMMENDATIONS OF THE INTERNATIONAL SUBCOMMISSION ON STRATIGRAPHIC NOMENCLATURE (FROM 1952 ONWARDS)

Rank	*Geological time units*	*Chronostratigraphical units*
1st order	Era	Erathem
2nd order	Period	System
3rd order	Epoch	Series
4th order	Age	Stage
5th order	Time	Substage

As noted above, many lithologies are diachronous, hence terms applied to them do not correlate with those in the above table. Recommended terms include:

(i) group;
(ii) formation;
(iii) member;
(iv) bed.

The geological time scale is given in Table 2.2 together with the age of the commencement of each period (in millions of years before the present, i.e. B.P.). As well as the relative dating of sedimentary rocks by their fossil contents, rocks may be dated absolutely. Methods of doing this are based upon the process of radioactive decay. Unstable isotopes of elements have unstable nuclei which disintegrate. This is a random process so that the time of decay of any particular nucleus cannot be forecast. From statistical laws, only the probability of disintegration of nuclei in a particular time interval can be stated. This probability cannot be influenced by physical or chemical means. The probability of disintegration per unit time interval is called the decay constant λ and it is characteristic of the particular mode of decay of the radioactive nuclide. If a very large number of radioactive nuclei are considered then, since they decay at random, the disintegration rate is proportional to the number of active nuclei present N and

$$-\frac{dN}{dt} = \lambda N$$

which, after integration, becomes

$$N = N_0 \exp(-\lambda t)$$

where N_0 = number of active nuclei present at time $t = 0$. The number of active nuclei decreases exponentially and if a semilogarithmic plot is used, the slope of the line measures the decay constant.

For practical reasons the half-life $T_{\frac{1}{2}}$ of a radionuclide is usually referred to in place of its decay constant. This is defined as the time taken for the number of active nuclei to decrease by half. When $t = T_{\frac{1}{2}}$, then

$$\frac{N}{N_0} = \tfrac{1}{2} = \exp(-\lambda T_{\frac{1}{2}})$$

TABLE 2.2
GEOLOGICAL TIME SCALE

Era	*Period*			*Age*	*Events*
Cainozoic[a]	Holocene[b]			0–2	
	Pleistocene[b]		Glacial and interglacial stages comprise:		
			Alps	*Northern Europe*	*North America*
		Glacial	Würm	Weichesl	Wisconsin
		Pluvial	Riss/Würm	Eemian	Sangamon
		Glacial	Riss	Saale	Illinoian
		Pluvial	Mindel/Riss	Needian	Yarmouth
		Glacial	Mindel	Elster	Kansan
		Pluvial	Günz/Mindel		Aftonian
		Glacial	Günz		Nebraskan
		Pluvial	Donau/Günz		
		Glacial	Donau		Pre-Nebraskan?
			Epoch		
	Tertiary			65	
			Pliocene		Deterioration of climate leading to ice ages.
			Miocene		
			Oligocene		Climax of Alpine orogeny.
			Eocene		General temperature increase over the surface of the Earth.
			Palaeocene		
Mesozoic[a]	Cretaceous			136	Duration about 71 million years, i.e. from 136 to 65 my. Alpine orogeny. Chalk deposited in Upper Cretaceous. Dinosaurs became extinct.
	Jurassic			195	Duration about 59 million years, i.e. from 195 to 136 my. First birds appeared.
	Triassic[c]			225	Duration about 30 million years, i.e. from 225 to 195 my. Variscan orogeny.
Palaeozoic[a]	Permian[c]			280	Duration about 55 million years, i.e. from 280 to 225 my. Southern hemisphere glaciated. Evaporites widely deposited.
	Carboniferous			345	Duration about 65 million years, i.e. from 345 to 280 my. Coal measures formed.
	Devonian			395	Duration about 50 million years, i.e. from 395 to 345 my. Climax of Caledonian orogeny.
	Silurian			435	Duration about 40 million years, i.e. from 435 to 395 my. Final stages in the infilling of Lower Palaeozoic basins of deposition.

TABLE 2.2—*contd.*

Era	*Period*	*Age*	*Events*
	Ordovician	500	Duration about 65 million years, i.e. from 500 to 435 my. Widespread vulcanicity.
	Cambrian	570	Duration about 70 million years, i.e. from 570 to 500 my. First unequivocal shelled fossil remains. Usually unconformable at base.
Precambrian (Proterozoic and Archaean)		4 600	Duration about 4 030 million years (=about 4 aeons, 1 aeon = 10^9 years). Many orogenies. Almost 90% of the total history of the Earth, mostly crystalline rock.

[a] Palaeozoic, Mesozoic and Cainozoic are often grouped together as the Phanerozoic.
[b] Holocene and Pleistocene may be grouped together as the Quaternary.
[c] Triassic and Permian together are often termed Permotriassic.

and therefore

$$\lambda = \frac{\ln 2}{T_{\frac{1}{2}}} = \frac{0{\cdot}693}{T_{\frac{1}{2}}}$$

Thus

$$N = N_0 \exp\left(-(0{\cdot}693^t/T_{\frac{1}{2}})\right)$$

or

$$\frac{N}{N_0} = 2^{-t/T_{\frac{1}{2}}}$$

Each isotope has a characteristic half-life when it is unstable and this permits identification. Measurable half-lives vary from fractions of a second up to about 10^{17} years. In disintegration series, the original element is termed the parent and the products about which the isotope dating geologist is concerned are termed the daughters. The accuracy of measurement of the ratio between parent and daughter is maximum when they occur in comparable quantities, so the radioisotope should have a half-life similar to the suspected age. In stratigraphy therefore, isotopes appropriate for radiometric age dating possess half-lives of the order of hundreds or thousands of millions of years. Also, such radioisotopes should be common in rock-forming minerals and the daughter should not escape in gaseous form from the mineral. Table 2.3 lists isotopes which are unstable and commonly used in dating. Allusion to the methods listed in the table is by no means a complete survey of the subject and many more details will be found in *Surface Water.*[1] However, in order to point out possible pitfalls it may be mentioned that accumulation of argon-40 (a rare gas) in rocks is possible only if the minerals therein are not heated above 300°C during Earth history. Leaking away of such a gas is always a potential hazard in applying this dating tool which in principle is very valuable and can be used for the range of about 100 000 years ago to almost as far back as the beginning of the Universe.

TABLE 2.3

RADIOISOTOPES

Uranium: ^{238}U yielding ^{206}Pb and $8\,^{4}He$	$T_{\frac{1}{2}} = 4\,498 \times 10^6$ years
$^{235}U \leftarrow {}^{207}Pb$ and $7\,^{4}He$	$T_{\frac{1}{2}} = 713 \times 10^6$ years
Thorium: ^{232}Th yielding ^{208}Pb and $6\,^{4}He$	$T_{\frac{1}{2}} = 13\,900 \times 10^6$ years
Potassium: ^{40}K, 11% yielding ^{40}A	$T_{\frac{1}{2}} = 11\,850 \times 10^6$ years
^{40}K, 89% yielding ^{40}Ca	$T_{\frac{1}{2}} = 1\,470 \times 10^6$ years
Rubidium: ^{87}Rb yielding ^{87}Sr	$T_{\frac{1}{2}} = 50\,000 \times 10^6$ years
For dating recent materials:	
Carbon: ^{14}C yielding ^{14}N	$T_{\frac{1}{2}} = 5\,730$ years
Tritium, ^{3}H, $T_{\frac{1}{2}} = 12{\cdot}26$ years; silicon, ^{32}Si, $T_{\frac{1}{2}}$ not precisely known but between 60 and 710 years[a]	

[a] Discussed by the author in detail in *Surface Water.*[1]

Brief reference must be made also to the matter of plate tectonics in view of the importance of the deformation of rocks to site engineering. Structural changes in the crust of the Earth cause marine transgressions and regressions, conversion of oceanic deeps into mountain ranges. The term applied to the resultant major structural features of regions is tectonics—its study is related to the investigation of orogenies (mountain-building movements). As noted earlier, many orogenies have occurred during the history of this planet.

A better understanding of tectonic activity was achieved in the 1960s through the acquisition of a great deal of new evidence regarding the ocean floors and also from analysing the magnetic properties of appropriate rocks. Small magnetite crystals in igneous rocks assume a rather strong permanent magnetization as these cool and so are transformed into magnets indicative of the position of the magnetic poles at the time when they were formed. Magnetic poles are known to be linked with the positions of the true poles on the Earth's axis of rotation. Hence, effectively, the the original magnetization of rocks indicates the original position of their locality when formed and it has been found that specimens from successive layers often show a slow drift of the locality with time. Also, the polarity is not constant, but changes with time—the north and south magnetic poles appear to exchange positions at roughly half million year intervals. This periodic reversal of polarity is the basis of magnetic reversal dating. Magnetic field measurements are made at sea using magnetometers with gyro-stabilized platforms towed behind ships or aircraft. Practically all of the magnetism so tested comes from magnetite-bearing oceanic floor basalts. Data acquired constitute a record of magnetic anomalies recorded graphically on continuous traces and they represent minute departures (in milligauss) from the mean value of the total magnetic intensity measured in the direction of the geomagnetic mean of the Earth. In the late 1950s, a persistent north–south alignment of such anomalies was discovered in the eastern Pacific Ocean and on east–west traverses these demonstrated a very characteristic and oscillatory pattern of positive and negative values about the mean. They were thought to delineate elongated bodies of magnetite-bearing rock (inferred to be basalt) aligned north–south, having magnetization patterns sharply contrasting with those of their east–west neighbours. Similar patterns were also found to occur in the Atlantic Ocean.

In 1963 Vine and Matthews, in a classic paper,[2] analysed the problem and proposed a solution which marked a giant step forward for the earth sciences. Computed magnetic profiles assuming normal magnetization of the ocean floor did not resemble the observed profiles and so an alternative model was used with 50% of the crust in alternating bands being reversely magnetized. If the suggestion of Hess,[3] that the sea floor spreads, was correct then basaltic magma ought to well up in the axes of oceanic ridges. It would become magnetized in the direction of the prevailing geomagnetic field as it cooled below the Curie temperature, thereafter forming a massive dyke subsequently spreading laterally away from these axes. Repetition of the process with the dyke splitting axially would cause the creation of a series of blocks with alternately normal and reversed magnetization and become progressively older as the distance away from an axis increased. This was actually shown to occur and so the

Vine–Matthews hypothesis became generally accepted. It constituted a potential magnetic recorder for the determination of the velocity of the oceanic conveyor belt involved in the spreading of the sea floor and provided an accurate time scale for reversals as Heirtzler pointed out in 1968.[4]

The oceanic ridges referred to above relate to a continuous system of mountain ranges arising from the deep ocean floors. The mid-oceanic ridges are the sites of shallow earthquakes and periodic vulcanicity. They separate, i.e. sea floor spreading takes place, at rates of around 10–20 mm $year^{-1}$. Ocean floor spreading has been incorporated into an attempt to explain the overall tectonics of the Earth, the theory of plate tectonics. This states that the tectonic activity of the past 200 million years and probably earlier can be ascribed to movements of a few huge and rigid plates, into which the outer 70 km thick shell of the Earth is broken, relative to each other. The movement of such plates is possible because there is a zone of weaker and almost molten rock, the asthenosphere, of about 100 km thickness below them. Each plate contains oceanic and continental crusts and at the mid-oceanic ridges, plates are separating with the addition of new material at their margins. A compensatory plate destruction process is taking place where plates collide and one plate subducts the other, i.e. is forced beneath it. Destructive zones are present around the periphery of the Pacific Ocean, in the Caribbean Sea and elsewhere and they are associated with deep oceanic trenches, arc-shaped mountainous island chains, great mountain ranges such as the Andes, major earthquake belts and volcanic areas (in the case of the Pacific constituting the Pacific Girdle of Fire). Data accumulated by the Worldwide Seismograph Network (established in the early 1960s) have been invaluable and led to the obtaining of evidence of contemporary tectonic activity. Various categories of margin for the various plates have been identified and these are:

(a) destructive plate margins described above and involving the descent of cold lithosphere into the asthenosphere and down to depths as great as 700 km to create what are termed subduction or Benioff zones;
(b) constructive plate margins at which new crust is created, essentially the mid-oceanic ridges alluded to earlier;
(c) conservative plate margins at which crust is neither created nor destroyed, the plates sliding past each other laterally and charac-

terized by transform faults frequently offsetting the oceanic ridge axes; since the transform faults may have inactive extensions beyond the offset ridge axes, these may define circles of rotation for the previous motions and hence provide a possible key to the direction, if not the rate, of past plate movements.

The place of meeting of two plate boundaries is a junction; for three plates it is termed a triple junction, the sole manner in which the boundary between two rigid plates can end. Two types of mountain building (orogeny) resulting from plate movements are known:

(a) the island arc–Cordilleran type which depends upon plate edges for its development and the process occurs above subduction zones and is characterized by metamorphic belts and divergent thrusting;
(b) collision type forming after the impact of continent upon continent or continent upon island arc and showing dominant metamorphism of the blue schist type, thrusting being mainly towards and on to the consumed plate.

Orogenic belts provide important clues to the creation and disappearance of oceans throughout long intervals of time, but much remains to be learned about them. For instance some geologists think that Precambrian orogenic belts may be fundamentally different from younger ones arising subsequently. There may be some truth in this idea if the planetary crust was thinner then than now. The main post-Cambrian orogenies are listed above in Table 2.2 and there is a major problem with the Variscan (Hercynian) which attained its maximum about 250 million years ago. Firstly it lacks extensive ophiolitic complexes. Ophiolites are basic and ultrabasic lavas and minor intrusions associated with the infilling of a geosyncline, a basin of elongated shape which fills with sediments and later through orogenic forces is converted into a mountain chain. Various kinds of geosyncline have been identified and may be listed:

(i) eugeosyncline, possessing thick sediments and an abundance of volcanic rocks, forming some distance from the kraton or shield which is a large stable mass of ancient rock forming part of a plate;
(ii) miogeosyncline, with thinner sediments and no volcanic rocks, formed adjacent to a kraton;
(iii) taphrogeosyncline, a synonym for a rift valley, an elongated,

fault-bounded trough along which coincide a tectonic depression and a geomorphological valley; also termed a graben;
(iv) parageosyncline, a geosyncline lying within a kraton;
(v) zeugogeosyncline, a parageosyncline with marginal uplifts;
(vi) autogeosyncline, a parageosyncline without marginal uplifts.

Taking into account all difficulties that have arisen with regard to the theory, there can be no doubt that plate tectonics has been very successful in explaining lateral movements of the crust.

A simple model of plate dynamics and mantle convection has been proposed recently by Hagen and O'Connell.[5] This attempts to relate quantitatively plate tectonic motions with thermal convection in the mantle which is deemed to drive them. The mantle is that portion of the interior of the Earth lying between the Mohorovičić and Gutenberg discontinuities, i.e. from a depth of about 35 km to 2 900 km, the density being 3·3 at the Mohorovičić discontinuity and 5·7 at the Gutenberg discontinuity (see Fig. 2.1). The mantle may be composed of olivine material together with sulphides in the upper part and some nickel–iron below. Hagen and O'Connell suggested that the cooling and thickening of lithospheric plates with age and subduction cause large scale horizontal density contrasts tending to drive both plate motions and mantle flow. They quantified these driving forces to determine whether they suffice to account for observed plate motions. First two-dimensional models were computed in order to evaluate the effects of assumed rheologies and boundary conditions. They were unable to obtain plate-like behaviour in viscous models with traction-free boundary conditions. However the piecewise uniform velocities distinctive of plate motion can be imposed as boundary conditions and

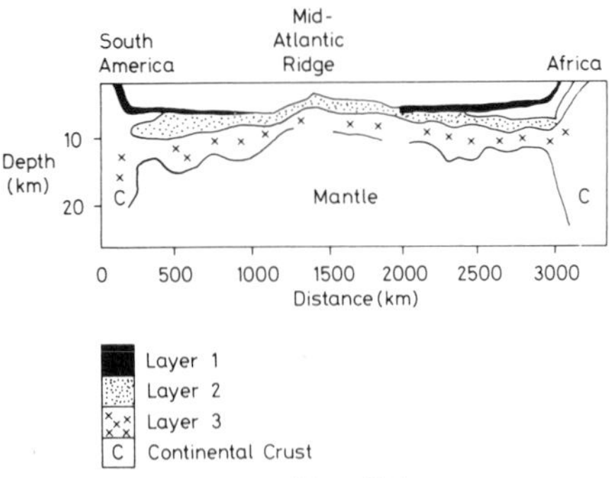

FIG. 2.1.

the dynamic consistency of the models evaluated by determining whether the net force on each vanishes. If the lithosphere, the outer rigid part of the planetary crust which has a high strength compared with the asthenosphere, has a Newtonian viscous rheology, the net force on any plate is a strong function of the effective grid spacing used, leading to ambiguities in the interpretation. Incorporating a rigid-plastic lithosphere which fails at a critical yield stress into the otherwise viscous model removes these ambiguities.

The model is extended to the actual three-dimensional (spherical) plate geometry. The observed velocities of rigid-plastic plates are matched to the solution of the viscous Stokes' equation at the lithosphere–asthenosphere boundary. Body forces from the seismically observed slabs, from the thickening of the lithosphere obtained from the actual lithospheric ages and from the differences in structure between continents and oceans are included. Interior density contrasts such as those resulting from upwellings from a hot bottom boundary layer are assumed to occur on a small scale compared to plate dimensions and are not included. The driving forces from the density contrasts within the plates are calculated and compared to resisting forces resulting from viscous drag computed from the three-dimensional global return flow and resistance to deformation at converging boundaries; the root mean square residual torque is ~30% of the driving torque. The density contrasts within the plates themselves can reasonably account for the plate motions. Body forces from convection in the interior may provide only a small net force on the plates. At converging boundaries the lithosphere has a yield stress of ~100 bars; drag at the base of the plates is ~5 bars and resists plate motion. The net driving forces from subducting slabs and collisional resistance are localized and roughly balance. Driving forces from lithospheric thickening are distributed over the areas of the plates as is viscous drag. The approximate balance of these two forces predicts plate velocities uncorrelated with plate area as observed. The model represents a specific case of boundary layer convection; the dynamical results are stated to be consistent with either upper mantle or mantle-wide convection.

2.3. ROCKS DEFORMED

This subject constitutes structural geology and it can be of great importance in site engineering. Any geological map will have cross-sections showing strata in various attitudes such as flat or tilted, folded or faulted. Sometimes igneous rocks are present, occasionally extruding on the surface as well (as lava flows). These phenomena result from processes operating during formation and producing primary structures as well as later processes (mechanical deformation and/or chemical reconstitution) which may be referred to as penecontemporaneous. Yet later structures may arise and these are stated to be secondary in nature. Consideration of these events shows that the aspects now described are important.

2.3.1. Stratification (Bedding)

This is the occurrence of recognizably different strata or beds in sedimentary succession. Many beds are homogeneous, but some show grading, i.e. pass upwards from coarser to finer particles. Separation planes part strata of sequences of uniform lithology and most beds possess an inner structure, the stratification or bedding planes. These may afford planes of ready splitting. Where the bedding planes in a bed are inclined more or less regularly to the separation planes between beds, cross-bedding (or cross-stratification) exists. Various top and bottom conditions of strata may be found, for instance mud cracks may form on the top of clayey sediments where these are exposed to a drying atmosphere for a time. Ripple marks may occur, particularly in sands in which the (non-cohesive) grains are free to move in water or air. Sole markings are referred to in Chapter 7. Tracings from fossils may also be found and include tracks, trails, castings and burrows useful in the determination of facing.

When stratified rocks are mapped, they must be identified not only at one exposure, i.e. at one place where they occur *in situ*, but possibly at many exposures. This may be done using some unique characteristic of colour, texture, mineral or fossil content or a combination of these features. Of course some formations comprise many beds which, although distinguishable at one exposure, are so similar to each other that no one bed can be identified elsewhere. An excellent instance of such strata are the flysch deposits on the northern margin of the Alps. Among such rocks, any truly distinctive bed is very valuable in mapping and is termed a marker bed or a key horizon. In mapping, especially in bore log correlation, structural utilization can be made of the arrangement of beds of different lithologies and thicknesses in a definite sequence. This is the basis of the correlation of glacial varves, sediments deposited annually in glacial meltwater lakes. It is also the basis of borehole correlation based upon the electrical resistivities of strata, the resistivity being a function of the porosity, mineralogy and water-bearing properties of the beds. The emission of gamma rays resulting from the decay of radioactive elements in rocks can be used in a similar manner. Repetition of a sequence of lithologies may cause errors in correlation. This phenomenon is perhaps best illustrated by the cyclothems of the Carboniferous coal measures of the northern hemisphere and an ideal sequence has been selected in Table 2.4 from the Yoredale series of Yorkshire in the UK (cf. for instance Allen[6]). The shapes of strata may be variable, ranging from uniform to lateral thinning (i.e. lensing or wedging out). Strata may show splitting, especially in coal seams and strongly wedge-shaped formations may occur where deposition has occurred against steep slopes such as coral reef flanks or around volcanoes or against fault scarps. A scarp constitutes part of a cuesta which in turn may be defined as a geomorphological unit composed of a gently sloping surface parallel to the dip of the strata with an escarpment or scarp face steeply inclined in the opposite direction to this dip slope and transgressing the bedding planes.

TABLE 2.4
SEDIMENTARY RHYTHMS: YOREDALE SERIES

	Eroded limestone surface
Limestone	(v) Algal phase or chert bed
	(iv) Shale with a modified limestone-type fauna
	(iii) Limestone with normal fauna such as corals and brachiopods
	(ii) Pseudo-breccias
	(i) Coral phase
	Break
Sandstone	Sandstone or coaly-shale
	Coal
	Fireclay or ganister, an arenaceous earth found below coal seams; pure silica sand
	Cross-bedded sandstone
Shale	Unfossiliferous shale
	Calcareous shale with normal shale fauna
	Limestone conglomerate
	Break
Limestone of the preceding rhythm	

Stratigraphical breaks are mentioned in Table 2.4 and these may be of various types as noted below:

(a) Diastems are unconformities produced by periods of non-deposition which are localized and short in duration. They may be distinguished by sharp surfaces separating beds and result from exposure of beds as part of the land surface through epeirogenic movements (i.e. vertical sense uplift or sometimes depressive movements which may also produce tilting and even faulting).
(b) In disconformities, younger beds overlie older ones in an approximately parallel geometry, but an erosional interval is represented by the junction. Major disconformities are rare because study usually demonstrates that, instead of parallelism, a slight angular unconformity occurs.
(c) Unconformities, i.e. angular unconformities, involve younger rocks overlying older ones which have been tilted or folded and subsequently eroded before deposition of the younger strata took place. Of course there are usually differences in both dip and strike between the two sets of beds. Unconformities may represent many millions of years of time so that the differences between the two sets of strata may be considerable.

The term non-conformity has been applied to a buried erosion surface separating an intrusive complex from overlying sediments and the basic concept here is that prolonged erosion must have taken place to expose the intrusive materials prior to burial (but of course dip and strike relationships are not involved at all). However the usual term unconformity may be applied to such cases as well.

2.3.2. Non-diastrophic structures

These are discussed in Chapter 7, and involve compaction and diagenesis of sediments and the effects of gravitational forces upon them. The word diagenesis applies to processes affecting sediments at or near to the surface of the Earth, i.e. at low temperature and pressure, and these are to be distinguished from dynamothermal alterations occurring to form metamorphic rocks although some grading of the former into the latter may take place. Subaerial weathering must also be excluded from the scope of the word.

2.3.3. Diastrophic structures

Produced by large scale deformation, they include the results of this acting upon the planetary crust to cause orogenies, etc. Some important concepts must be defined:

(a) *Dip*: This is the angle of inclination of a bedding plane from the horizontal measured in a vertical plane at right angles to the strike. Measured in any other position, an apparent dip is read and this is less than the true dip.

(b) *Strike*: This is the direction of the intersection of an inclined geological plane with an imaginary horizontal plane and it is determined on appropriate exposures by laying a spirit-level on the outcrop of the plane and measuring the bearing of the line along which the level rests.

(c) *Plunge*: This is the inclination of a linear structure-element measured from the horizontal in a vertical plane containing the line. The trend of a plunge (the strike of the vertical plane containing the line) and the sense in which the downward inclination takes place are also recorded.

These features are illustrated in Fig. 2.2.

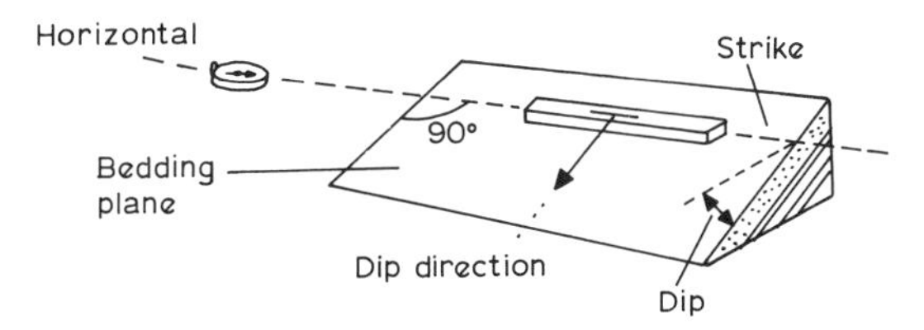

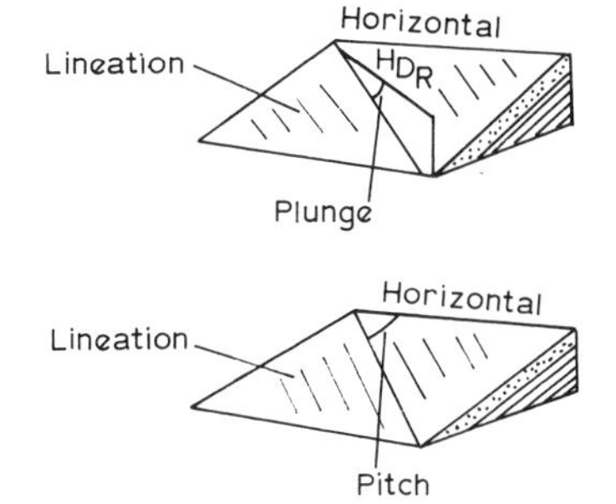

FIG. 2.2. Dip, strike, plunge and pitch.

2.3.4. Joints

These may be considered because they comprise the commonest of secondary structures in rocks and can be very variable in type. There is no visible displacement in a direction parallel to the plane of a joint and this distinguishes it from a fault. The regional mapping of joints usually shows a geometrical relationship to faults, also to folds, warps, intrusions and other tectonic elements and this suggests a genetic connection with the stress distribution and strain pattern. Joints occurring in a parallel or almost parallel array comprise a set or family and usually several sets are found in an area; they constitute the joint system of that area. Joints conforming with an overall pattern in such an area are referred to as systematic (those which do not comprise local, irregular joints). Two sets thought to represent complementary shear sets comprise conjugate sets. Master joints sometimes occur and these penetrate several strata and extend over great distances (up to many kilometres). They are very important in regional tectonic analysis. Minor joints relate to localized structures and in particular strata can be interesting in terms of detailed structural interpretation. Near the surface, joints may be weathered into open channels or cemented by secondary mineral deposits, say of opaline silica or calcite. Major jointing is frequently reflected in the drainage pattern as prominent joints are followed by streams and gullies. In fact the jointing of particular strata may be identificational features for the beds in aerial photographs. The intensity of jointing is usually greater near to the structures determining it and can be useful in the determination of buried salt domes and other subterranean structures. Most joints are tectonic in origin, representing either tension cracks or incipient shear fractures, but jointing may occur also in practically undisturbed flat lying sediments as well as in folded and faulted beds. Such jointing may represent fatigue cracks formed by small alternating strains set up by earth tides.

Joints may be classed as shear or tension joints. Shear joints are incipient shear fractures or even incipient faults and, following the work of Lensen,[7] Lu and Scheidegger[8] applied the concept of isallo-stress lines to general shear fracture analysis. The angle in plan between two conjugate joint systems is termed the compressional angle γ and from lineament study it is feasible to determine this quantity and plot it on a map. Lines of equal γ are termed isallo-stress lines and their values and pattern are indicative of regions of strike-slip, normal or reverse faulting and the direction of principal stress.

The joint pattern in a region is probably cumulative, recording all the events which have produced stress differences great enough to cause fracturation. Older joints may be inclined or displaced or dragged out of alignment through faulting. Younger joints may develop through stress relaxation.

Jointing is significant with regard to the strength of rocks in engineering works, in quarrying and mining as well as in connection with the flow of fluids such as water leaking from a reservoir or oil flowing from a fractured coral reef. Of course unexpected changes in the joint pattern may be met within engineering work, some lateral and some at depth, a few beneficial but most detrimental.

2.3.5. Faults

Where rocks have been observably displaced along a macroscopic shear or fracture plane in the earth, they are said to be faulted. The word fault refers both to the plane of faulting and the displacement. Fault planes may be shear fractures along which the internal cohesion of the rocks has been destroyed or slip surfaces resulting from plastic yielding across which cohesion remained during movement. Under certain conditions, tension cracks may constitute fault planes, e.g. in ring fractures associated with igneous activity. However most fault planes are either shear fractures or slip planes.

Faults are usually normal, reversed or strike-slip, but there are other types to which allusion will be made including transform faults. A normal fault is one in which the displacement of the block lying above an inclined fault plane is downward relative to the lower block. A reverse fault is one in which the displacement is upward. A strike-slip fault is one in which the movement is horizontal along the strike of the fault. In all faults, the movement may be somewhat oblique on the fault plane but it is the dominant direction of movement which determines the designation of the fault. So a strike-slip fault may show hundreds of metres of downthrow, but kilometres of lateral shift, thus establishing its category. In the case where the vertical and lateral displacements are of the same order of magnitude, the resultant fault may be called an oblique strike-slip fault. In bedded rocks, a fault is termed a strike fault if it strikes in the general direction of the strata, a dip fault if it strikes approximately parallel to the dip and an oblique fault if it is markedly oblique. The effects of faulting on stratified rock outcrops or exposures depend upon the relative movements of the blocks involved, also upon whether the faults are strike, dip or oblique in nature. The fault plane, which may be curved, is often called simply the fault. In place of one shear plane, there may be many parallel shears and the fault movement is then distributed in a shear zone.

The term fault zone implies that zone in which movements occur and these may comprise shearing, jointing, fracturation (causing brecciation), bending of strata by drag, etc. If the fault is inclined, the face of rock above it is the hanging wall and that below is the footwall. The fault is said to hade in the direction of dip of the fault plane, the angle of hade being the complement of the dip, i.e. the angle between the fault and the vertical.

The relative displacement between two blocks is termed the slip and comprises the resultant of two components, namely that along the strike of the fault and that parallel to its dip. As regards apparent displacement on faults, these as seen in cross-section at a particular place may be totally different from the actual movement on the fault. It is necessary to recognize the relative position of the parts of a faulted bed and in general the distance separating comparable portions of a bed, e.g. top to top, as measured in any particular plane selected at any locality constitutes the separation in that plane. The vertical separation is the parameter measured in a vertical direction. The horizontal separation may be measured in a horizontal plane in any direction desired, say north–south or east–west. The stratigraphic separation is the true thickness of strata normally separating a particular bed from that horizon with which it is brought into contact by the fault. The stratigraphic separation is related to the vertical separation by the formula $S = V \cos \delta$, where S is the stratigraphic separation, V the vertical separation and δ the true dip of the beds. The stratigraphic separation is sometimes termed the stratigraphic throw.

Throw is a word which has been used in two senses, one referring to the displacement of fault blocks relative to each other, the other to the apparent displacement of a bed in cross-section at right angles to the fault (which may be utterly different from the actual movement on a fault, as noted above).

Heave is a word used to refer to the horizontal distance, measured normal to the fault, separating the parts of the faulted bed.

Minor structures associated with faults include drag. Beds within a fault zone often show flexures which give the appearance as if the strata have been dragged back by frictional resistance on the fault plane. This constitutes drag and the ensuing flexures do not constitute drag folds in the normally accepted sense of that term.

Other minor structures are associated with mylonitization in which strongly sheared, crushed and cemented mylonite develops on fault planes with great low angle reverse faults. Mylonite is of cataclastic origin and forms in cold rocks under pressure as is evident from the development in it of structures resembling flow planes due to the shearing. Obviously a lot of frictional heat must be generated during this process. Mylonite also forms in rocks in a condition suitable for metamorphic recrystallization if rapid displacement is involved.

In softer rocks, a tough, leathery ground rock layer of dark colour may line the fault plane and this is termed fluccan, gouge or pug. It is associated especially with carbonaceous slates and is often polished and striated by the movement of the fault.

Crush breccias and crush conglomerates may develop. These must be distinguished from fault breccias in which there exists a mass of fragments jumbled together with a large number of voids. Crush breccias possess angular, lozenge-shaped fragments bounded by intersecting shear surfaces while crush conglomerates involve fragments rounded off by prolonged movement, the long axes of these lying parallel. Fault breccias form under conditions involving low normal stresses across the fault while crush breccias and conglomerates show the operation of strong normal stresses across the fault, two complementary shear directions occurring and slicing the rock into lenses between which there is a fine-grained matrix (cf. for instance Thurmer[9]). Crush conglomerates and mylonites can be associated with any kind of fault, but most often they are found on reverse faults.

Other minor structures include bulging and reverse drag. In the Wonthaggi coalfield in Victoria, Australia some of the coal seams are bulged up for a metre or so as a fault (downthrown in the opposite sense) is neared. Salt dome experiments show similar structures and these are ascribed to flowage of mobile strata under differential pressure adjacent to the fault. Rise of beds in the upthrown block as a normal fault is approached is in the reverse direction to the normal drag and similar reversed drag is to be observed in both upthrown and downthrown blocks of the Gulf Coast oilfields in the USA. Experiments were performed by Cloos[10] in 1968 simulating Gulf Coast structures and he was able to reproduce reverse drag (caused by antithetic faulting).

Feather joints and tension gashes are developed adjacent to major faults and these have also been reproduced in model experiments. Shear and tension joints developed in the deformational zone between crustal blocks which move relatively to each other along a fault constitute what are termed feather joints or federklüfte. This name has been used because such joints together with the associated fault resemble the barbs and shafts of a feather in cross-section. Feather joints may be divided into pinnate shear joints and pinnate tension joints.

Slickensides are striations on a fault plane or on the mineral deposits lining a fault and they are formed in the direction of relative movement of the fault blocks. Two or more sets usually intersecting at an oblique angle may occur, especially in layers of laminated quartz indicating successive movements in slightly different directions or perhaps a sudden deviation in the movement during one displacement.[11] Slickensides can be formed under superficial conditions, for instance on clay coatings of rock blocks involved in soil slides. In hard and brittle rocks, slickensides are frequently polished and smooth with their grooves deeper at one end and becoming shallow towards the other in the direction of displacement of the adjacent block. They are developed on conical shear surfaces in a structure called cone-in-cone which is best observed in calcareous rocks and coal.

Cone-in-cone comprises two arrays of cones, one set with the bases on a bedding plane and the apices upwards, the second on a parallel bedding plane with the apices directed downwards. The flanks of such cones are slickensided. In calcareous rocks they also have concentric rings around their circumferences. The origin of cone-in-cone is ascribed to compressive forces acting between two parallel bedding planes, the solid angles of the cones probably representing the dihedral angle between shear planes in the rock (although the angle varies considerably in fact, from 15 to 100°).

Although stress analysis of faulting must be associated with the regional tectonic setting of the relevant area in order that the mechanism generating the stresses can be understood, it may be pointed out that of the three principal stresses involved, two are horizontal and the third is vertical and gravitational in a model wherein there are three sets of conditions with all stresses compressional. In each case the faults are complementary slip surfaces or shear fractures with the acute dihedral angle between them enclosing the greatest principal stress axis. The mean stress axis is the direction of intersection of the planes. The following observations are germane:

(a) Maximum stress horizontal: minimum stress horizontal: mean stress vertical. Producing complementary strike-slip faults, this situation has one set with a dextral sense of movement and the other with a sinistral sense.

(b) Maximum stress vertical: minimum stress horizontal: mean stress horizontal. Produces normal faults dipping at more than 45°.

(c) Maximum stress horizontal: minimum stress vertical: mean stress horizontal. Produces reverse faults. The section of faulted rocks is shortened in the direction of maximum compression and such faults dip at less than 45°.

Some general observations on the various fault types with appropriate examples will now be cited.

2.3.5.1. Normal Faults

It is assumed that with these gravity is the maximum compressive force with dips exceeding 45° and so such faults may be termed gravity faults. Of course the dip may become vertical or even drop down to the near horizontal. In cylindrical faults, the movement of the upthrown block is actually upwards with the fault passing into a high angle reverse fault below. Subdivision of the terrestrial crust into blocks by normal faults constitutes block faulting and where the faults are complementary shears, the blocks are long and narrow, some remaining high as fault ridges and others depressing to form fault troughs (approximately horsts and grabens). A set of parallel faults all thrown in the same direction produce step faulting and if the relevant blocks are tilted, these are termed tilt blocks.

2.3.5.2. Fault Troughs

These constitute graben or rift valleys. Rift (bruch) implies tensional separation which indeed is often the case, but in certain instances major strike-slip faulting takes place, e.g. in the Dead Sea rift. The Rhein graben is more typical and forms part of a wider development of faults with which Cainozoic vulcanism is associated in North Africa and Europe. Many grabens occur in plateaux or in elongated mountain belts like the Andes in associations termed keystone faults with the main graben lying along an axis with bifurcated ends.

The famous East African rift valleys comprise a part of a worldwide fracturation system best displayed in the oceanic ridges, referred to earlier in this chapter. Their essential origin is tensional and resulted in the splitting apart and separation of crustal plates, but since these took place on a continent they were influenced by pre-existing factors such as foliation and folding with movements commencing as long ago as the Mesozoic. The Red Sea is part of this system of rift valleys and constitutes a typical graben with Precambrian rocks faulted in some localities against thick Cainozoic sediments. There is a marked positive gravity anomaly under a central trough and the whole feature relates to separation of an Arabian block from a North African one. The Dead Sea rift is an extension northwards of the Gulf of Aqaba, a branch of the Red Sea with negative gravity anomalies. An allusion may be made also to another but very interesting part of this rift valley system and that is to the Great Dyke of Zimbabwe which may represent a smaller, older and infilled tensional crack. At the northern end the Red Sea bifurcates into the Gulf of Aden and the Gulf of Aqaba, the latter comprising part of the Dead Sea trough and originally believed to be a ramp valley, a fault trough bounded by reverse faults, now known not to be.

Ramp valleys, however, have been found elsewhere, e.g. in Australia where in the east there are many grabens and uplifted horsts. Ancient grabens of considerable lengths up to 1000 km became infilled with younger rocks and afterwards deformation occurred and such features were termed aulacogenes by various workers and Bogdanoff[12] has cited the Donbas trough as an instance.

2.3.5.3. Reverse Faults

These may be expected to dip at somewhat less than 45° and many of the very greatest approach horizontality. The Moine thrust in the northwest Highlands of Scotland is a classic instance of thrusting associated with regional metamorphic rocks, in fact it is a thrust-nappe. The translation with involved faulting exceeds 16 km towards the north-west.

2.3.5.4. Strike-slip Faults

These can be determined most easily where they cross the trends of folds and they have been called transcurrent faults, transcurrent 'thrusts', tear faults or wrench faults. An excellent instance is afforded by the faulting, with a lateral shift of 2·4 m, of the upper Crystal Springs earth dam which took place during the San Francisco earthquake of 1906. The San Andreas Fault is traceable for at least 1000 km and constitutes a dextral strike-slip fault which is active and has had a long past history. It is now regarded as a transform fault originating in the mid-Oligocene when the American plate began to encounter the mid-Pacific plate. However there are difficulties in reconciling the age relationships with ocean floor tectonics. Physiographic evidence for strike-slip faulting is plentiful, e.g. slicing of river terraces, closure of valleys by faulted spur-ends, offsetting of stream courses, etc. It may be added that the term transform fault was proposed by Wilson[13] in 1965. Growth or diminution of a surface area is implicit in it and where two simple transforms are joined end to end, the fault between the moving blocks is also a transform fault. Wilson referred to six types of transform fault, each capable of having dextral or sinistral displacements.

2.3.6. Folds

These are flexures in rocks, i.e. a change in the quantity or direction of dip in the strata. The inference is that an original planar object has been deformed to incorporate some type of plication. The causes of the phenomenon enable four categories of fold to be recognized, namely:

(a) those caused by diastrophism, deep seated earth forces either with a strong horizontal component or, if vertical, capable of producing major uplift against gravity; examples of primary tectogenesis.
(b) those due to sliding and flow of vast rock masses under gravity; instability arose because of elevation and tilting associated with primary tectogenesis; examples of secondary tectogenesis and gravitational tectonics.
(c) those due to strictly localized effects such as igneous intrusion, salt dome injection and so on.
(d) those due entirely to superficial sliding or flowage of inadequately supported masses of newly deposited sediments; slumps and slides result and these are discussed in Chapter 7.

Diastrophic folds are found to take place in groups and hence the concept of fold systems may be introduced. These require considerable periods of time in order to form and during these, they are subject to the effects of deposition or erosion. The horizontal plane is fundamental to the classification of folds and the two basic terms anticline and syncline relate the attitude of strata to this plane.

An anticline is an upwardly convex fold in which any given bed intersects the same horizontal plane in both (so-called) limbs, dipping away from the crest (summit) of the fold on either side. The reverse of this structure is termed a syncline of which the lowest point, i.e. the opposite of the anticlinal crest, is termed the trough. It may be noted that an anticline encloses older strata in its core and a syncline encloses younger ones.

As regards symmetry, in symmetrical folds the dip, as measured where the limbs intersect a given horizontal plane, is at the same angle but in opposite directions in either limb. In asymmetrical folds, the corresponding dips are unequal. Thus a line is drawn to divide the cross-section of a fold into roughly equal parts; this will be vertical in symmetrical folds (hence said to be upright) and inclined in asymmetrical folds. Obviously the two halves will be mirror images in a symmetrical fold.

Implicit in this discussion is the fact that it is absolutely necessary to look at a fold in its entirity in order for its symmetry properly to be revealed. Antiform and synform may be used for anticline and syncline and anticlinoria and synclinoria comprise complex large folds of general anticlinal or synclinal form with many minor folds on their limbs.

In practice, nearly all folds are asymmetrical and the inclination of the median line or plane is significant. Folds in which one limb is much longer than the other are termed inequant. If one limb of an asymmetrical fold is overturned, the fold is termed an overfold. In isoclinal folds, the limbs are folded closely together and are parallel. In cuspate folds, beds are smoothly curved in sectors with adjacent sectors meeting in sharp cusps. The fan fold and the box fold (kofferfalt) can be recognized in Europe and Mont Blanc is an excellent instance of a fan fold, box folds being common in the Jura. Usually both types involve upright folds. In recumbent folds, both limbs are almost horizontal and these structures are represented on a large scale in the Western Alps. Where sub-horizontal beds suddenly dip steeply for a short distance so as to produce a steplike effect physiographically and structurally analogous to a normal fault, a monoclinal flexure, monoclinal fold or monocline results. In fact a monocline links two blocks of strata which have become displaced relative to each other and may be closely related to faults.

It may be added that dip must be regarded as applicable both locally as in the monocline case alluded to and also regionally. The regional dip is the average or mean dip over a large area in which there may be many small scale changes of dip or even broad folds.

In folded sequences, the geometry of the strata is highly variable and determined by the physical properties of the strata as well as by the mechanics of folding. This leads to the distinction between competent and incompetent beds. The former bend stiffly, the latter flow into any shape required by the operational forces. Of course, competency and incompetency are relative properties and imply greater or lesser rigidity or mobility in a folded region. If the relevant strata are traced to another region, a competent bed may become more deformable. Competency is usually judged by the amplitude of folding, folds of larger amplitude characterizing more competent beds. Theoretical analyses of fold generation through the buckling of layered materials under compression in the plane of layering are effected utilizing either elastic or viscous theory and relative competency and incompetency of strata may be treated using differences of viscosity, this being appropriate to slow deformation processes (especially of deeply buried rocks). Studies of this type have illuminated the relationships between the wavelength of rapidly growing folds, layer thickness and viscosity contrasts.

In the case of an unstratified slab of rock of internal structure which may be regarded as uniform, on bending there arises a regular distribution of tension and compression into (inner) and (outer) zones with an intermediate surface suffering no change in length and termed the neutral surface (see Fig. 2.3). The folding in question may be termed

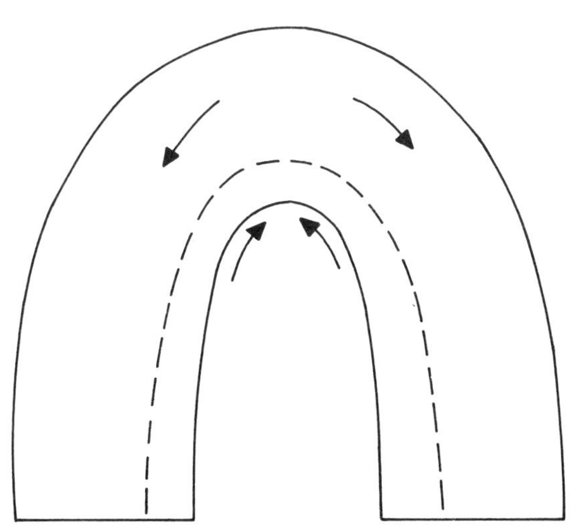

Fig. 2.3. Bending of an unstratified slab of material with uniform internal structure. (– – –) Neutral surface.

neutral surface folding and the actual position of the neutral surface is determined by the stress distribution in the slab and is of course variable. This is a consequence of the variation in possible physical properties of the material and also of the mechanical means utilized in order to bend the slab.

The material will yield to compressive and tensional forces in several ways. One way is by either positive or negative dilitation comprising partly an elastic expansion or contraction and partly a permanent dilatation (this latter is easily produced in loosely cemented and granular rocks with interstitial pore spaces); in rocks of loose nature, readjustments of the grain packing may be able to accommodate requisite strains. Another way is through distortion, either elastic or permanent. In the absence of notable dilatation, ellipses formed from original circles are roughly strain ellipses for elastic strain and for plastic deformation they are ellipses which are directly related to the strain ellipses.

Of course, not many rocks are elastic enough to allow any considerable elastic deformation over a short period. However, it is possible that small elastic strains may be dissipated or relaxed over a long period by means of readjustments of defects in crystal lattices or by actual recrystallization. Then, folding might go on very slowly without transgressing the elastic limit at any stage.

It may be interesting to consider the amount of deformation needed in neutral surface folding. Assuming this to be approximately midway through the relevant slab and assuming also that the folding results in concentric circular arcs, the extension at the outer surface will be $\frac{1}{2}\phi d$, where ϕ is the angle of bending measured in radians and d is the thickness of the slab. This is not a large deformation for many folds and can be expressed as a percentage of the length involved. For a 3 m bed folded into a semicircular arc of 1·6 km width, i.e. with radius 0·8 km, the extension and corresponding shortening is approximately 0·2% of the original length of the folded bed. This percentage will be much larger for folds of small radius and hence considerable distortion as well as dilatation will be involved in beds which are thick and closely folded. Where such distortion occurs, the position of the neutral surface is of course affected as the folding proceeds and the shape of the slab may be affected as well, specifically in regard to the parallelism of the upper and lower surfaces.

Various strain effects in bending may be mentioned and these include dilatation and bulk flow. They also include the crumpling of intrados (the concave side of a fold), such crumples (due to compressive forces acting on the concave side of the bent slab) forming between small faults or kink-planes. Tension joints may develop at the extrados (the convex side of the fold) where brittle materials are concerned. Such cracks form at right angles to the extension and thus are radial in the fold in the zone overlying the neutral surface. As they form, the slab loses coherence and hence cannot transmit tensional stress. In consequence the effective upper surface at any time is that surface joining the tips of the growing cracks. The effective thickness is therefore reduced, the neutral surface shifts downwards and may even pass to the lower surface with an accompanying cracking across of the slab.

Turning now to the folding of a stratified mass, this may be likened to a book with many thin sheets of relatively strong material which could slide over each other (as when a book is opened or closed). Although each sheet will show neutral surface folding, the distortion of a small segment in each is negligible because of its extreme thinness. Folding may be assumed to produce firstly a simple arch which is centrally fixed while the limbs rotate freely and make an arcuate fold. In an ideal situation, each sheet retains the original length and thickness unaltered and the geometry then requires that the sheets slip over each other, the slippage relative to the upper sheet being proportional to the distance from this sheet. If a succesion of such folds is considered, between any two points where the dip is zero, the lengths of the sheets are the same and the slip in one sense on an anticline is compensated in each bed by equivalent slip on a syncline. Thereafter, if the anticlinal and synclinal axes remain fixed throughout the folding, all the slip is concentrated in the fold limbs.

From the foregoing discussion it is apparent that there are two ways in which folds comprising a family of parallel curves may arise, namely through neutral surface folding and through the slipping mechanism. Each necessitates a certain kind of localized strain within the folded beds. Folding of the type mentioned for sheets is dependent upon an original lack of isotropy with low adherence between sheets of rather rigid material in which stress is transmissible without excessive plastic yielding. From a geological standpoint, this means that the adherence of grains between two adjacent strata must be low enough and that some adherence within the actual strata is high enough to ensure that the strata act as gliding units giving rise to flexural-slip folds. A well defined sandstone/shale or limestone/shale junction is able to act as a gliding surface, but in a graded bed (even with well defined stratification) the argillaceous constituents sufficiently adhere across the planes of stratification to inhibit gliding along these. van Hise,[14] as long ago as 1896, took such conditions into account in distinguishing parallel folds from other kinds. He designated other folds as similar, the geological criterion for these being that successive beds in a folded sequence assume similar curves so that the form of the fold remains the same in a vertical direction.

Shear (Gleitbrett) folding is very small scale and is demonstrated by sandy laminae in slates.

The sharp bending of foliated materials around certain planes is termed knickung and the planes knickungsebene, but in English the word kinking (and kink planes) is applicable. Kinking may occur in crystals also and in rocks it results from the rotation of foliae about planes oblique to them in such a manner that the kink planes bisect the angle between the undisturbed material and that within a kink band (this lying between two parallel kink planes).

Mitre folds are angular folds with inclined kink planes.

Folds in competent beds which have been deformed plastically and translated show exaggerated thinning on the limbs accompanied by thickening at the crests and troughs. These reflect the high mobility of the beds and are termed rheomorphic folds.

Nappe structures are important and, as well as thrust nappes mentioned earlier, there are fold nappes. The actual word nappe (German) refers to any thin and widespread sheet of rocks acting as a cover; hence it is applicable to lava flows as well as tectonic features. However, in the UK and USA, the word is confined to tectonic structures. In this sense the word may be applied to sheets or slabs of rock of large size which have been translated from their original positions for large distances, so moving over underlying rocks and finally covering them. The classic region for fold nappes is the Western Alps where large scale recumbent folding and significant low-angle thrusting are frequent.

REFERENCES

1. Bowen, R. 1982., *Surface Water*. Applied Science Publishers Ltd, London; Wiley Interscience, New York.
2. Vine, F. J. and Matthews, D. H., 1963. Magnetic anomalies over oceanic ridges. *Nature*, **199,** 947–9.

3. HESS, H. H., 1962. History of ocean basins. In: *Petrologic Studies* (a Volume in Honour of A. F. Buddington), eds A. E. J. Engel *et al.*, Geol. Soc. Am., Boulder, Colorado.
4. HEIRTZLER, J. R., 1968. Marine magnetic anomalies, geomagnetic field reversals and motions of the ocean floor and continents. *J. Geophys. Res.*, **73,** 2119–36.
5. HAGEN, B. H. and O'CONNELL, R. J., 1981. A simple model of plate dynamics and mantle convection. *J. Geophys. Res.*, **86,** 4843–67.
6. ALLEN, J R. L., 1964. Studies in fluviatile sedimentation and six cyclothems from the Lower Old Red Sandstone, Anglo-Welsh Basin. *Sedimentology*, **3,** 163–98.
7. LENSEN, G. J., 1958. Measurement of compression and tension: some applications. *NZ Jour. Geol. Geophys.*, **1,** 565–70.
8. LU, P. H. and SCHEIDEGGER, A. E., 1965. An intensive local application of Lensen's isallo-stress theory to the Sturgeon Lake South Area of Alberta. *Bull. Canadian Pet. Geol.*, **13,** 389–96.
9. THURMER, A., 1928. Entstehung von Linsen in Gesteinen. *Centralbl. für Min.*, **Abt A,** 147–58.
10. CLOOS, E., 1968. Experimental analysis of Gulf Coast fracture patterns. *Bull. Am. Ass. Pet. Geol.*, **52,** 420–4.
11. HILLS, E. S., 1972. *Elements of Structural Geology*, 2nd edn. Chapman and Hall, London.
12. BOGDANOFF, A. A., 1962. Sur certains problèmes de structures et d'histoire de la platforme de l'Europe Orientale. *Bull. Géol. Soc. France*, **7**(IV), 898–911.
13. WILSON, J. TUZO, 1965. A new class of faults and their bearing on continental drift. *Nature*, **207,** 343–7.
14. VAN HISE, C. R. 1896. Principles of North American Pre-Cambrian geology. *US Geol. Surv. 16th Annual Report*, **1,** 581–843.

CHAPTER 3

Minerals

3.1. MINERALS

A mineral may be defined as a naturally occurring inorganic substance comprising a structurally homogeneous solid possessing a specific chemical composition, the sole exception being mercury which in its natural form is a liquid. The chemical composition may be variable within limits usually expressed in terms of end members, for instance olivines range from forsterite, Mg_2SiO_4 to fayalite, Fe_2SiO_4. The structural homogeneity alluded to refers to the fundamental atomic structure which is constant and continuous through the mineral unit, e.g. the silicon–oxygen lattice in silicates. While ice is a mineral, coal, natural oil and gas cannot be considered to be such, *sensu stricto*. However there is a *sensu lato* approach which permits organic materials to be regarded as mineral, for instance bedded phosphates and the constituents of some limestones.

There are at least 2000 different minerals, but of these only about 100 are common. Of this 100, ten are abundant. It is with these abundant minerals that engineering work is mainly involved so that a construction engineer working in various places will ordinarily need to know only these ten minerals in any detail. Usually they can be identified without any special equipment—i.e. an in hand specimen, and using a hand lens ($\times 10$) if necessary—by studying their physical properties. Where more information is required, minerals in rocks in thin sections can be examined by a geologist using the petrological microscope.

Minerals may be regarded as primary or secondary according to their importance in the structure and composition of rocks. Primary minerals fall into two categories:

(a) essential, i.e. it must occur following the definition of the rock;
(b) accessory, i.e. it may or may not occur without affecting the character of the rock.

Quartz, feldspar and mica are essential components of almost all granites while zircon or apatite are accessories. Of course a mineral may be essential to one rock and accessory in another—for instance quartz, essential in granite, is merely accessory in a gabbro.

Secondary minerals arise from the alteration or reconstruction of primary minerals, are sparsely distributed and include the most durable constituents of igneous rocks. They endure after essential minerals have been destroyed by weathering. X-ray analysis reveals the atomic structure of minerals. Their groups of atoms are held together by electrical forces, those carrying positive charges comprising cations and those having negative charges constituting anions.

In essential minerals of igneous rocks, silicates are prominent and these are combinations of silicon with oxygen associated with various cations. In order of increasing size (Å), these components are: Si, 0·39 atomic radius; Al, 0·57 atomic radius; Mg, 0·78 atomic radius; Fe^{2-}, 0·83 atomic radius; Na, 0·98 atomic radius; Ca, 1·06 atomic radius; K, 1·33 atomic radius. Anions are less varied, oxygen (1·32) dominating although in some minerals hydroxyl (OH^-, 1·32) or fluorine (F^-, 1·33) may replace it to some extent. Anions are large compared with cations and the mineral structure can be considered as comprising closely packed anions (mostly oxygen) with small cations placed in the interstices between them. Two fundamental units are recognized, namely:

(i) 4 oxygen atoms closely packed around an atom of silicon to form an SiO_4 group, the oxygens occurring at the corners of a tetrahedron;
(ii) 6 oxygens in close contact and occurring at the corners of an octahedron.

From this discussion it may be seen that minerals are constructed of atoms symmetrically located and repeated three-dimensionally, the atomic arrangement of a mineral constituting its crystal structure. A crystal is a substance which has solidified in a definite geometric form and most solids if pure can be such. Non-crystal-forming substances are termed amorphous. Crystals may be classified according to the structure of their atomic lattices, those regular networks of fixed points about which molecules, atoms or ions vibrate, and these are reflected often in an external regularity of form or habit. The study of crystals, crystallography, is very important in identifying naturally occurring crystalline substances, i.e. minerals. Crystals are usually far from perfectly formed in nature, but their elements of symmetry and other characteristics facilitate identification.

3.2. CRYSTALLOGRAPHY

Crystals are usually bounded by flat, plain surfaces called faces. The line formed by the intersection of two adjacent faces is termed an edge and the point formed by the intersection of three or more faces is called a solid angle. The interfacial angle between faces is defined as the angle between the normals to the two faces so that it is possible to refer to the interfacial angle between two non-adjacent faces. One of the most important crystallographic concepts is that of the zone, a set of faces all of which are parallel to a given direction called the zone axis. Faces in a zone need not be adjacent, but when they are the edge formed by two such faces is parallel to or gives the direction of the zone axis. It is also useful to distinguish like and unlike faces. Two faces are like if they have the same size and shape in a regularly developed crystal. Sets of like faces all have the same relationships to the internal structure, the lattice. A complete set of like faces constitutes a form. Crystals may comprise combinations of two or more forms. Figure 3.1 illustrates the SiO_4 group and Fig. 3.2 shows the atomic structure of the mineral olivine.

The most important of all properties in the identification of crystalline substances is symmetry. The symmetrical arrangement of crystal faces reflects the internal symmetry of the lattice. Regular crystals, i.e.

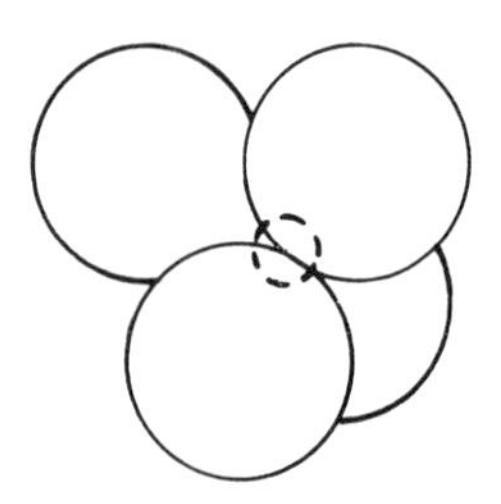

FIG. 3.1. The SiO_4 tetrahedron.

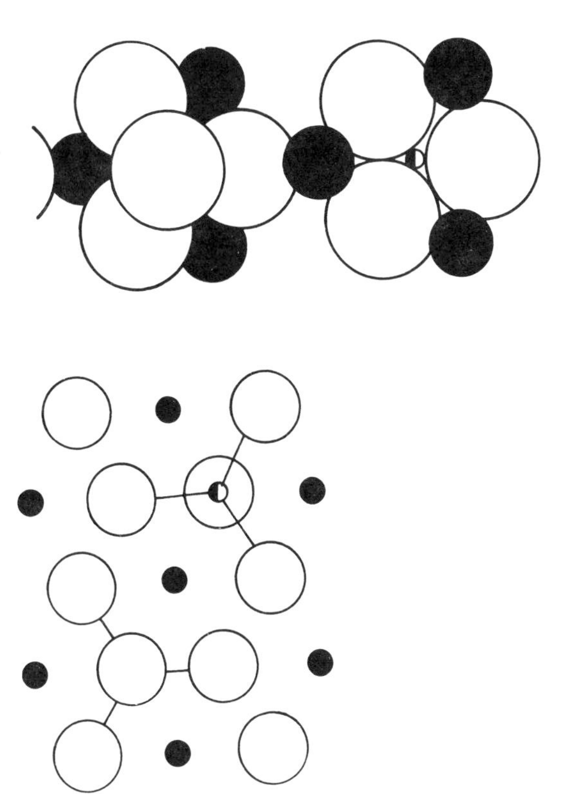

FIG. 3.2. The atomic structure of olivine. (●) Mg and Fe; (○) oxygen; (◐) silicon.

those with all like faces developed equally, are rare in nature. The common situation is one in which a crystal is irregularly developed and therefore of superficial asymmetry. However the actual symmetry will be present and may be described in all crystals with regard to elements comprising planes, axes and a centre of symmetry.

A plane of symmetry is an imaginary plane along which the crystal could be cut into two equal halves which are mirror images of each other.

An axis of symmetry is an imaginary line about which the crystal may be rotated so that congruence is attained; i.e. to an observer in a fixed position the crystal presents the same appearance after as before rotation through a given number of degrees. If a position of congruence occurs after rotation through 180°, the axis is said to be a diad or two-fold one. Other axes may be termed triad (three-fold), tetrad (four-fold) or hexad (six-fold) depending upon the rotation angle (120°, 90° or 60°, respectively).

A centre of symmetry is a point within a crystal such that straight lines could be drawn through it so that on either side of it and equidistant from it, similar portions of the crystals are encountered. Hence, a centre of symmetry is present when every face of the crystal is matched by one parallel to it on the other side of the crystal.

Considering a cube, here there are 9 symmetry planes, 13 symmetry axes and a centre of symmetry. On the basis of the symmetry axes present, crystals can be divided into seven major groups known as crystal systems. Of course there are numerous classes within these, in fact numbering 32. However, one class in each system possesses a maximum number of symmetry elements for that system and is called an holosymmetrical class.

The cube, in fact, is a member of a holosymmetrical class of the cubic system. The various crystal systems are: cubic, with four triad axes of symmetry; tetragonal, with one tetrad axis of symmetry; hexagonal, with one hexad axis of symmetry; trigonal, with one triad axis of symmetry; orthorhombic, with three diad axes of symmetry; monoclinic, with one diad axis of symmetry; and triclinic which uniquely has no axes of symmetry at all.

The morphology of crystals can be treated as a type of analytical geometry in which three dimensions are involved and it is necessary to have such axes to which planes and faces may be referred. The reference axes in question are termed crystallographic axes, not to be confused with symmetry axes (although to some extent they coincide with them). The crystallographic axes conveniently selected differ from system to system, this being a matter of convenience in the subsequent mathematical handling of the planes and faces.

(*i*) *Cubic system*: The cube may be taken as an example of this system and its faces may be referred to three axes in space, all mutually perpendicular and parallel to the edge directions of the cube. Because of the high symmetry of the cube, the three axes are indistinguishable. Therefore all may be designated by the letter a and, if it is necessary to label them specifically, the letters and subscripts a_1, a_2 and a_3 can be utilized. The cubic system is sometimes termed the isometric system. There are five classes.

Types include:

(a) galena (9 planes, 13 axes and a centre of symmetry);
(b) pyrite (3 planes, 7 axes and a centre of symmetry);
(c) tetrahedrite (6 planes, 7 axes and no centre of symmetry).

(*ii*) *Tetragonal system*: If a cube is imagined stretched or compressed along one of its crystallographic axes the resultant shape will have only two square faces left, the other four having become rectangular. This may be taken as an example of an holosymmetrical tetragonal crystal and possesses three crystallographic axes still at right angles as in the cube but no longer identical. One of the axes which may be designated c is coincident with the main symmetry axis of the crystal, the solitary tetrad. The other two axes are still like each other and they are called a_1 and a_2 as in the cubic system. They are horizontal compared with the vertical c. There are seven classes.

The zircon type has 5 planes, 5 axes and a centre of symmetry.

(*iii*) *Hexagonal and trigonal systems*: These are often grouped together and their characteristics are as follows:

Hexagonal crystals are referred to four axes, three arranged mutually at 120° in a horizontal plane with the fourth axis vertical and perpendicular to the plane. The axes are designated a_1, a_2, a_3 and c. c may be greater or smaller than the values for a. Some relevant minerals are listed in Table 3.1 (there are seven classes).

TABLE 3.1
HEXAGONAL SYSTEM, ELEMENTS OF SYMMETRY AND MINERAL EXAMPLES

Class	*Centre*	*Planes*	*Axes of symmetry rotation*	*Axes of rotary inversion*	*Examples*
1	c	7	6(ii), 1(vi)	—	Beryl
2	—	4	3(ii)	$1(\overline{\text{vi}})$	Benitoite ($BaTiSi_3O_9$)
3	—	6	1(vi)	—	Zincite (ZnO)
4	—	—	6(ii), 1(vi)	—	β-Quartz, kalsilite
5	c	1	1(vi)	—	Apatite
6	—	1	—	$1(\overline{\text{vi}})$	Not known
7	—	—	1(vi)	—	Nepheline

TABLE 3.2

TRIGONAL SYSTEM, SYMMETRY ELEMENTS WITH REPRESENTATIVE MINERALS

Class	*Centre*	*Planes*	*Axes of symmetry rotation*	*Axes of rotary inversion*	*Examples*
1	*c*	3	3(ii)	1(iii)	Calcite and other carbonates, corundum, haematite
2	—	3	1(iii)	—	Tourmaline
3	—	—	3(ii), 1(iii)	—	Quartz, cinnabar
4	*c*	—	—	1(iii)	Dolomite, dioptase ($CuSiO_3.H_2O$)
5	—	—	1(iii)	—	Only questionable instances

Trigonal crystals are referred to the same set of axes as for the hexagonal system, the Bravais axes or alternatively they may be assigned to the Miller axes; three non-orthogonals are equally inclined to the three-fold, triad, axis of symmetry and make equal angles with each other. These are parallel to three edge directions of a rhombohedron and when using them the important feature is the axial angle rather than the axial ratio. Details are given in Table 3.2 and refer to the five classes.

(*iv*) *Orthorhombic system*: An ideal orthorhombic crystal may be imagined as a shape created by stretching or compressing a cube along *two* of its crystallographic axes (cf. the tetragonal system, part (ii)). Three unequal crystallographic axes remain and these are designated *a*, *b* and *c*. There are three classes.

Typical minerals include barytes, sulphur, olivine, some pyroxenes and epsomite ($MgSO_4.7H_2O$).

(*v*) *Monoclinic system*: A monoclinic crystal may be imagined as an orthorhombic one in which the *a* axis has been tilted so that it remains normal to *b*, but not normal to *c*. *b* remains as a solitary diad axis. The amount of tilting of *a* relative to *c* varies from one crystal species to another and the obtuse angle between these two axes is referred to as beta. There are three classes.

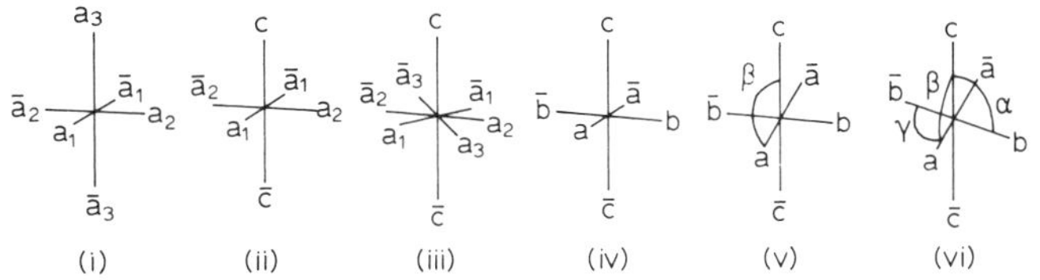

FIG. 3.3. (i) Cubic; (ii) tetragonal; (iii) hexagonal trigonal; (iv) orthorhombic; (v) monoclinic; (vi) triclinic.

Class 1 has a centre of symmetry which the others do not and it includes orthoclase, augite, hornblende, micas and chlorite.

(*vi*) *Triclinic system*: Here occurs the ultimate loss of symmetry axes and the crystallographic axes *a*, *b* and *c* are all unequal and mutually non-perpendicular. There are two classes.

Elements of the system will include the three unequal parameters and three interaxial angles. Common minerals include plagioclase, kyanite and axinite.

Some of the concepts discussed above are illustrated in Fig. 3.3. The properties of some of the terms will now be mentioned in more detail. These refer to symmetry which, as seen, depends upon the distribution of angular elements not the shape or the size of faces. The terms planes and centres are clear, but axes of rotation symmetry must be distinguished from inversion axes (axes of rotary inversion symmetry).

The axis of rotation symmetry is a line about which a crystal may be rotated so as to bring it into identical orientation (2, 3, 4 or 6 times in a rotation of 360°). The axis of rotary inversion symmetry, the inversion axis, involves a complex process. Firstly a face is rotated clockwise about an axis through a fraction of 360° and then translated through the origin of the crystal to a point equidistant on the opposite side. The origin of the crystal is its geometrical centre. The entire process is repeated until the starting point is attained. Translation refers to the carrying of a face along a straight line perpendicular to it to a similar position at the opposite end. The inversion axes are symbolized as $\bar{X}$-fold where $X = 1, 2, 3, 4$ or 6. The $\bar{1}$-fold operation entails rotation through 360°, the $\bar{2}$-fold through 180°, and so on, before translation.

The parameters of a crystal face are the ratios of the distances from

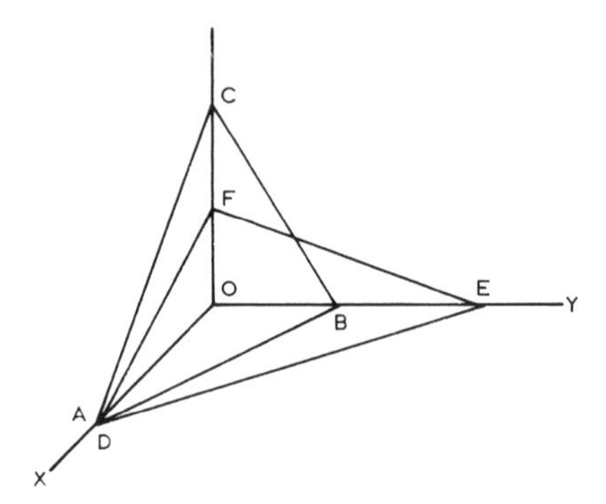

FIG. 3.4. The form (III).

the origin at which a face cuts the crystallographic axes, i.e. as shown in Fig. 3.4. Here OX, OY and OZ are crystallographic axes and two faces make intercepts on them. Considering the face ABC, the relevant parameters are the ratio of OA, OB and OC. Taking these as standard, in the case of the face DEF, OD = OA, OE is greater than OB by a factor of two and OF = 0·5OC. The parameters of DEF are therefore $\frac{1}{1}$, $\frac{2}{1}$ and $\frac{1}{2}$; this is with reference to the standard face ABC.

That form the face of which is taken as intersecting the axes at the unit lengths to be used for measuring the intercepts made by other forms on the same axes is termed the parametral form. The parameters of a parametral form are obtained by measurement and are expressible as multiples of one of them. For example gypsum has its commonest form making intercepts on the three crystallographic axes in the ratio 0·6899 : 1 : 0·4124. This is the axial ratio for this mineral. The reciprocals of the parameters are termed the indices and these are very valuable as means of writing the relation of any crystal face to the crystallographic axes, i.e. in crystallographic notation. There are in fact two systems for doing this, namely:

(a) The parameter system of Weiss in which *a*, *b*, *c* represent unequal axes, *a*, *a*, *c* two equal and one unequal and *a*, *a*, *a* three equal axes. The intercept for a crystal face on the *a* axis is written before *a* and so on, these intercepts being measured in terms of intercepts made by the parametral form. Generally therefore the expression for a crystal face in the Weiss notation is *na*, *mb*, *pc* where *n*, *m* and *p* are the lengths cut off by the face on the *a*, *b*, *c* axes. *n* or *m* is usually reduced to unity. If a crystal face is parallel to an axis the sign for infinity is placed as its parameter before the corresponding axial letter. Hence a face cutting the *a* axis and parallel to the *b* and *c* axes will have the notation $a, \infty b, \infty c$.

(b) Index system of Miller in which the indices, the reciprocals of the parameters, are employed. They are written in the axial order *a*, *b*, *c* and are stated in the simplest form, i.e. eliminating fractions. Referring to the face cutting the *a* axis and parallel to the *b* and *c* axes again, this was Weiss symbolized as $a, \infty b, \infty c$. In the Miller system this becomes 100. In the case of a face cutting the *a* axis at one unit of distance, the *b* axis at two units and parallel to the *c* axis, the Weiss notation will be $a, 2b, \infty c$. The reciprocals of the parameters are $1, \frac{1}{2}, 0$. Hence by elimination of fractions, the Miller symbols will be 210.

In the Miller system, a face parallel to an axis will contain the symbol 0, i.e. the reciprocal of infinity. Also as reciprocals are involved, the larger the number in the symbol the nearer to the origin will the face actually cut the axis in question. On the other hand the smaller the number in any given axial position in the symbol the nearer does the face approach parallelism with the particular axis. The limiting case is of course zero where the face is parallel with the axis. The general Millerian symbol may be expressed as *hkl*.

There are various conventions in the notation of crystals. For instance in indicating a crystal form the symbols are bracketed, e.g. (*hkl*). If the crystal face is indicated, the brackets are removed thus: *hkl*. Figure 3.5 indicates the form and shows the signs utilized at the ends of the crystallographic axes. A face cutting the positive end of an axis is

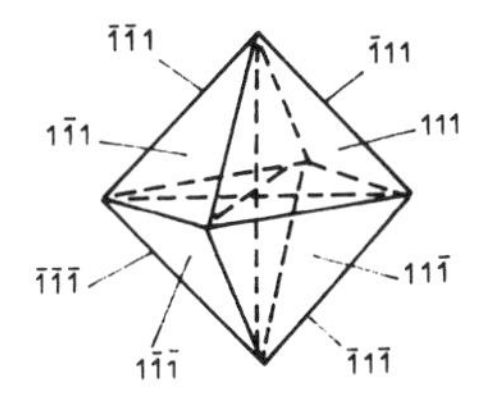

FIG. 3.5.

indicated solely by the corresponding index figure; one cutting the negative end has a negative sign placed above the index figure.

Allusion should be made to the law of rational indices which states that the intercepts which any crystal face makes on the crystallographic axes are either infinite or small rational multiples of the intercepts made by the parametral form. Hence symbols such as $2a$, $1{\cdot}75 \ldots b$, c are impermissible.

It is interesting to examine the handling of crystallographic data so as to determine the symmetry, type of faces present and axial ratios of an unknown crystal. These features may all contribute to the identification of a crystalline substance, the central preliminary objective of the mineralogist.

The stereographic projection is a commonly used system. A crystal may be imagined placed at the centre of a sphere with normals to each face projecting from the centre and intersecting the surface of the sphere. At this stage the essential angular relationships of the faces have been transferred to the sphere as a series of points. If the crystal is very small relative to the sphere, all shape irregularities disappear and only the angular relationships (constant for any given mineral) remain. Each point on the upper hemisphere of the sphere is now projected towards the south pole of the sphere and the line of projection intersects the equatorial plane at a new point. This equatorial plane is referred to as the plane of projection and it is of course a two-dimensional surface on to which the upper half of the crystal, a three-dimensional object, may be represented by a set of such points, each one a face pole. The circle bounding the plane of projection, the original equator of the sphere of projection, is termed the primitive circle. Such a diagram constitutes a stereographic projection. As described it is a stereogram of the upper faces of a crystal only and every point on it represents a direction in space, the distance between two points representing the angle between two directions. It is perfectly possible to add to the stereogram points representative of the lower faces of the crystal also. Generally, if this is done, solid points represent upper faces and open circles represent lower faces.

In more complex work and in order to avoid laborious geometrical construction, a stereographic net, the Wulff net, is utilized. This consists of a family of great circles at 2° intervals equivalent to meridians of longitude together with a family of small circles equivalent to parallels of latitude. The Wulff net can be used for many purposes, among them as a scale for measuring angles. A stereogram may be constructed either from a crystal or a crystal model.

3.3 IMPORTANT MINERAL PROPERTIES

3.3.1. Twinning in Crystals

This is a phenomenon occurring where a part of a crystal has grown or is deformed so that its atomic structure is in the reverse position to the other part and thus any crystal form it possesses is the mirror image of that of the other part. Consequently two portions are involved and they are joined in such a manner that some crystallographic direction or plane is common to both parts of the twin.

The plane dividing the twin so that one half is a reflection of the other half is termed the twin plane. The twin axis is usually perpendicular to the twin plane. The twin axis concept is just that, because in fact there has been no actual revolution in twin formation. The twin plane is always a possible face of the crystal and the twin axis is always perpendicular to some possible face or parallel to a possible edge.

A plane of symmetry cannot be a twin plane because it already divides the crystal into two halves, one of them being a mirror image of the other. A rutile twin is shown in Fig. 3.6. This is more complex than the simple twin and is termed the geniculate type.

With the mineral plagioclase repeated twinning occurs, i.e. twinning takes place repeatedly and in this particular case the twins' planes remain parallel throughout, a phenomenon termed polysynthetic twinning (see Fig. 3.7). This is diagnostic of the plagioclase feldspars.

Of course not all minerals are isotropic (i.e. with properties the same in any direction in them), some are anisotropic (properties varying internally with direction). However, because symmetry is related to the internal structure of the mineral comprising the crystal, it also indicates which of the above features occur (hence the importance of crystallography).

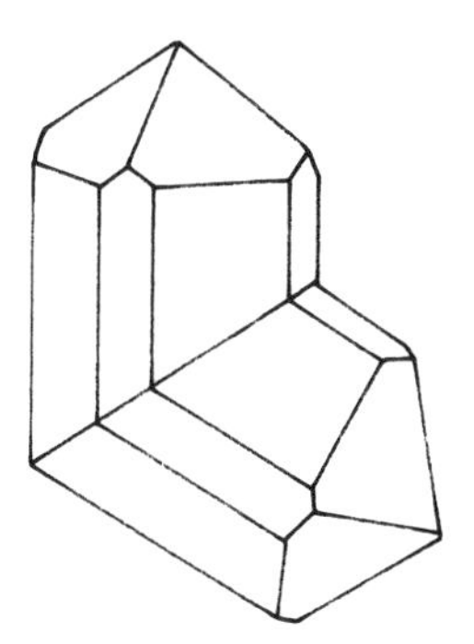

FIG. 3.6. Rutile twin.

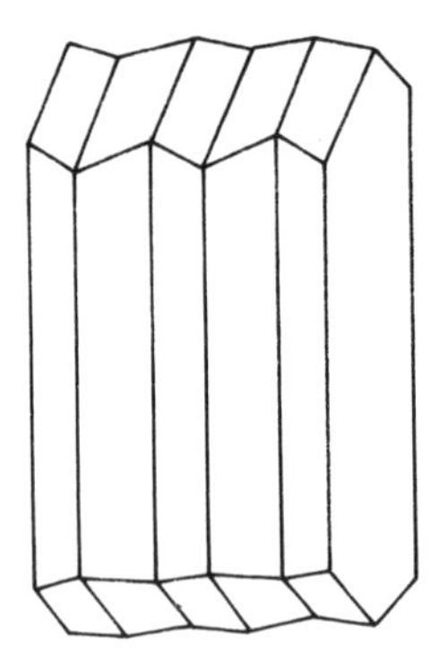

FIG. 3.7. Plagioclase repeated twin.

3.3.2. Colour and Streak

These together will facilitate identification of an in hand specimen of a mineral as indeed will all the properties alluded to below.

Colour is that observed visually and impurities influence it greatly so that it cannot be regarded as a very reliable guide except for those minerals showing iridescence, a display of prismatic colours due to the interference of rays of light in minute fissures walling in thin layers of air or liquid. Iridescence may be seen sometimes in quartz, mica and calcite. In precious opal the brilliant colours result from the presence of very thin curved or distorted layers with slightly differing optical properties. Normally quartz is colourless but it can be white, green, brown, black or pink-yellow as in milky quartz, cairngorm, morion or amethyst. Additionally it may be purple or violet (amethyst varieties), fibrous (cat's eye) or reddish (ferruginous quartz).

Streak is the colour of the powder of a mineral and may differ from that of the mineral in mass, e.g. black haematite gives a red powder produced by scratching or by rubbing the mineral on a piece of unglazed porcelain called a streak-plate.

3.3.3. Tarnish

This is due to exposure to air and may be accompanied by iridescence. It is a result of oxidation or it may be caused by the chemical action of sulphur. Tarnish is distinguishable from colour by chipping or scratching the mineral. Copper pyrites frequently tarnish to an iridescent mixture of colours.

3.3.4. Cleavage

This is a tendency to split along certain preferred planes usually parallel to one of the crystal faces. In the plane of cleavage the atoms of the mineral are more closely packed together than in directions at right angles to it. Consequently this plane is a plane of minimum cohesion. Cleavage must be distinguished from fracture, the latter being an irregular process unconnected with the crystalline structure of a mineral. Amorphous, i.e. non-crystalline, substances do not exhibit cleavage. A different phenomenon is slaty cleavage which relates to rocks such as slates which split easily into thin sheets. This is due not to planes of minimum cohesion, but rather to the effects of recrystallization due to pressure. Minerals may cleave in one, two, three or even more directions, but usually one is preferred. Cleavage is described by referring to the crystallographic direction followed by the planes of cleavage and their degree of perfection. Apropos this latter, cleavage may be termed perfect, good, distinct, poor, indistinct, difficult and so on. Minerals with perfect cleavage include calcite, galena and mica. Another is fluorspar which crystallizes in cubes. If one of these is tapped with a geological hammer, it cleaves along planes truncating the corners of the cube and if the process if done regularly an octahedron results. Hence fluorspar is said to possess perfect octahedral cleavage. On the other hand galena, also crystallizing in cubes, cleaves parallel to the faces of these and its cleavage products are cubes; consequently it is said to possess perfect cubic cleavage. Calcite always produces rhombohedral cleavage fragments. Mica has one perfect cleavage parallel to which very thin sheets of the mineral may be detached. Cleavage is very important in the recognition of minerals. Fracture is different as noted above and may be conchoidal (the mineral breaking with a curved concave or convex fracture often showing concentric diminishing undulations towards the point of percussion), even (flat), uneven (rough), hackly (studded) or earthy (as in the fracture of chalk or meerschaum).

3.3.5. Hardness

This varies greatly and is a very important property in identifying minerals. It may be tested by rubbing a specimen over a fine-cut file and noting the noise produced and the quantity of powder resulting. Soft minerals yield much powder with little noise, and vice versa for hard minerals. The scale of hardness generally used is known as Mohs'

TABLE 3.3

MOHS' SCALE OF HARDNESS

Hardness	*Standard mineral*	*Observations*
1	Talc ($Mg_3Si_4O_{10}$)	Softest
2	Rock salt or gypsum (NaCl or $CaSO_4.2H_2O$)	Can be scratched by fingernail
3	Calcite ($CaCO_3$)	Can be scratched by penny
4	Fluorspar (CaF_2)	Can be scratched by knife
5	Apatite (fluor-apatite is $Ca_5F(PO_4)_3$ or $3Ca_3P_2O_8.CaF_2$ and chlorapatite is $Ca_5Cl(PO_4)_3$ or $3Ca_3P_2O_8$)	Can be scratched by knife
6	Orthoclase feldspar ($KAlSi_3O_8$ with occasional substitution of sodium for potassium)	Can be scratched by file
7	Quartz (SiO_2)	Scratches glass
8	Topaz ($Al_2F_2SiO_4$ with occasional substitution of fluorine by hydroxyl, OH)	
9	Corundum, Al_2O_3	
10	Diamond, C	

Scale and it is given in Table 3.3. The only common mineral which exceeds 7 in hardness is garnet.

3.3.6. Lustre

This is the appearance given to a mineral by reflected light from its surface and there are various types. These include metallic, vitreous, resinous, pearly, silky and adamantine—self-explanatory words. The feature applies to metallic and mineral ores and opaque minerals, the remainder being primarily silicate minerals.

3.3.7. Transparency and Translucency

These refer to qualities of seeing through minerals clearly or indistinctly. Really a mineral capable of transmitting light but which cannot be seen through properly may be considered as translucent. This condition is common in minerals.

3.3.8. Phosphorescence and Fluorescence

These refer to interesting properties of a mineral. The former is that attribute of emitting light after subjection to such conditions as heat or rubbing or exposure to electric radiation or ultra-violet light. Some varieties of fluorspar after powdering and heating on an iron plate show a marked phosphorescence. Pieces of quartz also show it if rubbed together in a dark room. Minerals capable of emitting light while exposed to certain electrical radiations are termed fluorescent minerals.

3.3.9. Tenacity

This is a measure of how the mineral deforms if crushed or bent and encompasses several properties such as sectility (it can be cut with a knife and the resultant slices broken up with a hammer as is the case with graphite or gypsum), malleability (it can be sliced if it is flattened with a hammer as is the case with native gold, silver and copper), flexibility (it can bend, for instance talc), elasticity (it can spring back into its former shape after bending, e.g. mica) and brittleness (it cannot be cut into slices but crumbles; for instance iron pyrites, apatite and fluorspar).

3.3.10. Surface Tension Characteristics

The difference in powers of adhesion of various liquids to different minerals, have been made the basis of processes of ore separation and concentration. This depends upon the fact that the surface tension between many metallic sulphides and oil exceeds that between oil and the gangue minerals such as quartz, calcite, etc.

3.3.11. Specific Gravity (Density)

This is the ratio of the weight of a given body of a mineral to that of the same volume of water. The appropriate relation is

$$SG = W_1/(W_1 - W_2)$$

where W_1 = weight of the mineral grain in air, W_2 = weight of the mineral grain in water.

An instrument useful in the determination of the specific gravity is Walker's steelyard, a balance with a long graduated beam pivoted near one end and counterbalanced by a weight suspended to the short arm. The specimen is suspended and moved along the beam until it counterbalances the constant weight, the level position of the beam being observed by a mark on the upright. After taking the reading, the specimen is immersed in water and moved along the beam until the constant weight is again balanced and a second reading obtained. The readings are inversely proportional to the weights of the body in air and water respectively. If A is the first reading and B is the second reading, then the specific gravity is

$$SG = \frac{\frac{1}{A}}{\frac{1}{A} - \frac{1}{B}} = \frac{B}{B - A}$$

Some data regarding minerals with similar specific gravities is given:

(i) between 2·2 and 4: silicates, carbonates, sulphates, halides.
(ii) between 4·5 and 7·5: sulphides, oxides.
(iii) much greater: native metallic elements, e.g. copper (8·95), silver (10·5) and gold (19·3).

Many other methods are available for determining specific gravity such as Jolly's spring balance, the pycnometer bottle and the use of heavy liquids. The latter are effective for separating two minerals of different specific gravities if theirs is between those of the minerals in the mixture. Suitable materials are bromoform, $SG = 2{\cdot}9$, methylene iodide, $SG = 3{\cdot}33$, Klein's solution (cadmium borotungstate), $SG = 3{\cdot}28$ (diluted with water), etc.

3.3.12. Taste, Odour and Feel

Taste occurs when the mineral is soluble in water and it may be described as saline, alkaline, astringent, cooling, sweet, bitter or sour. As regards odours, some are emitted when the mineral is heated or struck or breathed on or rubbed. Odours may be alliaceous (garlicky), horse-radish, sulphurous, foetid or argillaceous (clayey). Feel may be smooth, greasy, harsh or rough and some minerals adhere to the tongue.

3.3.13. Fusibility

This is an aid to determination using the blowpipe. The following minerals are noteworthy:

(i) stibnite, Sb_2S_3, melting point 525°C;
(ii) natrolite, $Na_2(Al_2Si_3O_{10}).2H_2O$, melting point 965°C;
(iii) almandine garnet, $Fe_3Al_2(SiO_4)_3$, melting point 1200°C;
(iv) actinolite, $Ca_2(Mg, Fe)_5Si_8O_{22}(OH)_2$, melting point 1296°C;
(v) orthoclase, $KAlSi_3O_8$, melting point 1200°C;
(vi) bronzite, $(Mg, Fe)_2Si_2O_6$, melting point 1380°C.

These constitute a scale of minerals of which the temperature of fusion increases in rather equal steps according to von Kobell. Certainly stibnite can be fused easily while by contrast bronzite cannot practically be fused at all using an ordinary blowpipe.

3.3.14. Magnetism

This affects many minerals of which the most important are magnetite, Fe_3O_4, and pyrrhotite, Fe_nS_{n-1}. Minerals containing iron are generally magnetic, but not invariably. Some minerals devoid of iron may be magnetic enough to allow their separation from non-magnetic materials, for instance monazite, $(Ce, La, Yt)PO_4$, a phosphate of the rare earths. Varying the strength of an electromagnet facilitates the separation of minerals of varying magnetism such as the purification of magnetite from apatite, etc., or the separation of pyrites from blende, ZnS.

3.3.15. Electricity

This can develop in minerals either by friction or heat; in the latter case the mineral may be termed pyroelectric. An instance of a pyroelectric mineral is tourmaline. The variation in the degree of electrification among minerals is applied in the electrostatic process.

3.3.16. Radioactivity

Many minerals comprising high atomic weight elements are radioactive and emit various emanations. Many radioactive elements exist, some natural, but pitchblende or uraninite, $2UO_3.UO_2$, is probably the most famous.

3.3.17. Pseudomorphism

This is the assumption by a mineral of a form different from that to which it really belongs to constitute a pseudomorph. Pseudomorphs arise from various causes such as incrustation (deposition of a coat of one mineral on the crystals of another, e.g. quartz on fluorspar), replacement (substitution of new mineral particles for old ones removed by water or some other solvent), alteration (by chemical change altering the original chemical composition) or infiltration (when a cavity previously occupied by a certain crystal is infilled by deposition in it of different mineral matter introduced by the infiltration of a solution).

Pseudomorphs are recognisable by the blunt edges of their crystals and the dull appearance of their surfaces.

TABLE 3.4

	Rutile	*Anatase*	*Brookite*
Cleavage	Poor	Perfect	—
Streak	Brown	Colourless	Colourless
Fracture	Subconchoidal	—	Brittle
Hardness	6–6·5	5–5·6	5·5–6

3.3.18. Polymorphism

This refers to a substance which exists in two or more distinct forms which have identical chemical compositions. Many examples may be cited, for instance calcite (hexagonal) and aragonite (orthorhombic), diamond (cubic) and graphite (hexagonal, rhombohedral) and the trimorphous titanium dioxide, TiO_2. This latter has three forms, rutile and anatase (tetragonal) and brookite (orthorhombic). Their specific gravities are quite different, 4·25 for rutile, 3·9 for anatase and 4·15 for brookite. Other physical characters are also different, but the chemical composition is the same throughout, i.e. titanium dioxide. The three forms are listed in Table 3.4.

3.3.19. Reaction to Acid

This is marked in carbonate minerals. If cold dilute hydrochloric acid is applied to lime, carbon dioxide bubbles cause frothing. In some sulphide ores the gas hydrogen sulphide is produced.

3.3.20. Crystal Form

As seen above if conditions are favourable minerals take up definite geometrical forms, crystals, and the following distinctions can be made: a crystallized substance has well developed crystals whereas a crystalline substance has no definite crystals, but rather consists of an aggregate of imperfectly formed crystals; cryptocrystalline is a word which denotes the possession of traces of crystalline structure; amorphous substances lack crystalline structure, a condition arising in natural glasses, rarely in minerals.

Various forms are described by the following words:

(i) acicular—needlelike crystals, e.g. natrolite, $Na_2(Al_2Si_3O_{10}).2H_2O$;
(ii) amygdaloidal—infilling of vesicles in lavas by minerals such as zeolites;

(iii) bladed—bladelike, e.g. kyanite, Al_2SiO_5;
(iv) botryoidal—comprising spheroidal aggregations, e.g. chalcedony, SiO_2;
(v) capillary—fine hairlike form, e.g. millerite, NiS;
(vi) columnar—columned, e.g. hornblende, $(Ca, Mg, Fe, Na, Al)_{7-8}(Al, Si)_8O_{22}(OH)_2$;
(vii) concretionary and nodular—detached mineral masses, e.g. flints, SiO_2;
(viii) dendritic and arborescent—treelike, e.g. manganese oxide;
(ix) drusy or druse—crystals lining a cavity in a rock or mineral vein; called a geode.
(x) fibrous—with threadlike strands, e.g. asbestos;
(xi) foliaceous—lamellate, e.g. mica, other micaceous minerals;
(xii) granular—in grains, e.g. marble;
(xiii) lamellar—comprising separate plates, e.g. wollastonite, $CaSiO_3$;
(xiv) lenticular—with flattened balls or pellets;
(xv) mammillated—with large, mutually interfering spheroidal surfaces, e.g. malachite, $CuCO_3.Cu(OH)_2$;
(xvi) radiating or divergent—crystals arranged about a central point, e.g. stibnite, Sb_2S_3;
(xvii) reniform—kidney-shaped, e.g. haematite, Fe_2O_3;
(xviii) reticulated—cross-meshed like a net, e.g. some micas;
(xix) scaly—in small plates, e.g. tridymite, SiO_2;
(xx) stellate—starlike in form, e.g. wavellite, $4AlPO_4.2Al(OH)_3.9H_2O$;
(xxi) tabulate—with broad and flat surfaces, e.g. wollastonite;
(xxii) tuberose—with irregular rounded surfaces, e.g. flos ferri, a stalactitic coralloidal variety of aragonite, $CaCO_3$;
(xxiii) wiry or filiform—in thin wires which may be twisted like the strands of a rope, e.g. native silver and copper.

3.4 IGNEOUS ROCK-FORMING MINERALS: THE MAFIC MINERALS

3.4.1. Olivines

Under this heading may be included many important rock-forming silicates and in all members the essential atomic structure is the same, namely isolated SiO_4 tetrahedra packed together in lines parallel to the crystal axis (see Fig. 3.1). All members crystallize in the orthorhombic system and olivine itself is a common mineral although perfect crystals are rare.

The composition of olivine is $(Mg, Fe)_2SiO_4$, with more magnesium than iron. The general composition of the group may be stated as R_2SiO_4, where R is either magnesium or iron. The magnesium end-member is forsterite, Mg_2SiO_4 and the iron end-member is fayalite, Fe_2SiO_4. Olivine can be regarded as intermediate between these two. It forms at high temperature and crystallizes early from the magma.

If water occurs in any amount prior to or even after consolidation, the olivine may alter to serpentine, a hydrous magnesium silicate, $Mg_6Si_4O_{10}(OH)_8$ (olivine + H_2O + CO_2 = serpentine + $MgCO_3$). Serpentine is also green like olivine, but it has a lower specific gravity, 2·5–2·6, as compared with 3·2 for forsterite and 4·3 for fayalite. Its hardness is also lower, 3–4 as compared with 6–7 for olivine. In forming, an increase in volume occurs and in consequence in some basic igneous rocks cracks arise and radiate out from the mineral. This can weaken structurally the rock in question.

The properties of olivine-bearing rocks such as gabbros and norites change if serpentine is formed. Since these usually constitute good roadmetal aggregate, it is advisable to check the degree of alteration of the olivine before use. If this is small, it may be beneficial for bonding the aggregate with bitumen. However, aggregates with extensive alteration must be avoided.

3.4.2. Pyroxenes

These constitute one of the most important groups of rock-forming silicates and compared with the olivines, they contain a higher proportion of silica than the basics which occur and are termed metasilicates. In various members, iron, magnesium, calcium and sodium can occur in varying proportions. Also, smaller quantities of aluminium, manganese and titanium occur. The group possesses the Si_2O_6 chain structure, i.e. a linkage of SiO_4 tetrahedra in straight chains. Several have the composition $RSiO_3$ where R is magnesium, iron or calcium (occasionally manganese or zinc). In others there is a substitution of aluminium for part of the silicon giving the formula $R_2(Al, Si)_2O_6$ where R is calcium, magnesium, iron, aluminium or Fe^{+++}. The lateral linkage of the tetrahedra in augite is achieved by calcium, magnesium and iron cations. There are two subdivisions of the pyroxenes, namely:

(a) orthorhombic pyroxenes including enstatite ($MgSiO_3$) and hypersthene, $(Mg, Fe)SiO_3$;
(b) monoclinic pyroxenes including augite, perhaps the best known pyroxene.

As noted, the atomic structure of augite involves bonds between individual chains and these are weak so that cleavage planes exist, intersecting at 90° on a basal face of the crystal. This mineral is common in igneous rocks with low silica, i.e. basic rocks, and often occurs with olivine. If hydrated, it forms chlorite. This has the composition $(Mg, Fe)_5Al(AlSi_3)O_{10}(OH)_8$ and it has different properties from augite. For instance its specific gravity is lower (2·65–2·94 compared with 3·2–3·5 for augite) and so is its hardness (1·5–2·5 compared with 5–6 for augite). Crystallographically it is, however, also monoclinic.

In the principal series of pyroxenes, falling temperatures during crystallization cause a shift in composition from Mg^{++}-rich to Fe^{++}-rich types. This is a progressive iron enrichment and it may be shown in various ways. For instance crystals may be zoned with Mg^{++}-rich cores and Fe^{++}-rich outer layers; or earlier formed phenocrysts, relatively large crystals set in a finer grained groundmass and constituting porphyritic texture, are more magnesian than crystal grains in the said groundmass; or in a series of associated rocks representing successive fractions of magma, pyroxenes from early fractions are more magnesian than later ones.

Olivines crystallize at higher temperatures than orthopyroxenes, the orthorhombic pyroxenes which include enstatite, $MgSiO_3$ with up to 15% $FeSiO_3$, and hypersthene, $(Mg, Fe)SiO_3$ with more than 15% $FeSiO_3$. Early formed olivine may react with the magma by taking silica from it in amounts sufficient to convert it into the corresponding orthopyroxene. This conversion needs time to complete and if a lava is involved, the cooling may be so fast that the reaction temperature is passed too quickly for it to occur, hence some olivine may survive even if the magma has enough silica to convert it all into orthopyroxene.

3.4.3. Amphiboles

This is a group of complex metasilicates chemically related to the pyroxenes. Any species of the latter may contain the same elements as a corresponding amphibole but they are present in different proportions. A fundamental difference is the presence of hydroxyl (this is the same size and functions like an oxygen atom). The X-ray structure comprises 'bands' of linked SiO_4 tetrahedra, each constituted of two pyroxene chains united by shared oxygens and parallel to the vertical axis.

Hornblende is the best known mineral in the group and has the composition $(Ca, Na, Mg, Fe, Al)_{7-8}(Al, Si)_8O_{22}(OH)_2$ with varieties arising from the substitution of for example Fe^{+++} for aluminium. Hornblende is very common in igneous rocks containing average amounts of silica and crystallizes from magma with appreciable quantities of water (hence it is a 'wet' mineral). If excessive amounts of water are present, it may alter to chlorite (as augite does when hydrated) and it is not very stable if weathered in surface rocks. Hence it is rarely found in sediments, but does occur in metamorphic rocks. Hornblende is a monoclinic amphibole, but there are also orthorhombic ones such as anthophyllite, $Mg_7Si_8O_{22}(OH)_2$, and triclinic ones as well, e.g. cossyrite, an aluminium silicate of sodium, iron and titanium. The differences between amphiboles and pyroxenes are summarized in Table 3.5. In hand specimens hornblende may be

TABLE 3.5
AMPHIBOLES COMPARED WITH PYROXENES

Amphiboles	*Pyroxenes*
Cleavages at 124°	Cleavages at 90°
Bladed forms common	Bladed forms uncommon
Transverse sections six-sided	Transverse sections eight-sided
Twins have re-entrant angles	Twins do not have re-entrant angles
Commonly pleochroic[a]	Commonly non-pleochroic

[a] If minerals in thin section are coloured they will show a variation in colour when rotated in plane polarized light (PPL), i.e. light in which the wave vibration perpendicular to the direction of propagation only occurs in one plane, not in all directions as is the case with ordinary light. This colour variation in PPL is termed pleochroism (dichroism) and is due to the fact that a mineral showing it absorbs different wavelengths in different directions.

distinguished from augite by the marked difference between the angles of cleavage intersection. Hornblende schist is a metamorphic rock consisting essentially of orientated hornblende crystals and possessing linear schistosity.

3.4.4. Micas

This is one of the most important groups of mineral silicates and one that is chemically distinct from the groups considered above because the alkali elements are present but, unlike pyroxenes and amphiboles, micas lack calcium. The atomic structure is sheetlike, i.e. the SiO_4 tetrahedra are linked together at three corners to form indefinitely extended sheets with the atoms of various kinds arranged on an hexagonal plan. The sheets are double and staggered so that the crystal possess monoclinic symmetry although they are strongly pseudo-hexagonal. There is a perfect basal cleavage which causes splitting into thin elastic plates to take place.

In composition micas are silicates of aluminium and potassium plus Mg and Fe in dark varieties such as biotite. Some varieties contain sodium, lithium or titanium. Hydroxyl is invariably present and is commonly part replaced by fluorine.

Distinguishing the two major groups of micas involves some consideration of the optical properties of minerals in general. The indicatrix is a solid geometrical figure utilized in representing the different vibration directions of light in a mineral (see Fig. 3.8). It is an ellipsoid with three rectangular axes the lengths of which are proportional to the refractive index of any beam vibrating at 90° to it. The beam travels parallel to it in this situation. These axes are termed X, Y, Z or a, b, c or α, β, γ. If the first set is adopted, X is taken for the smallest refractive index and the 'fast' ray and Z for the largest refractive index and the 'slow' ray, Y being intermediate. If light traverses a mineral perpendicular to the plane containing X and Z then, as these show the maximum difference in refractive index, the maximum birefringence (double refraction) occurs, i.e. the maximum relative retardation. All light-transmitting crystalline substances (except cubic ones) can pass light at two different speeds corresponding to the two refractive indices. Thus a light beam passing through a thin mineral plate is split into two beams, one 'fast' and the other 'slow' as indicated above, the difference between the corresponding refractive indices varying from minimum to maximum according to the direction of cut of the plate relative to the crystallographic axes. The most characteristic feature of birefringence is the production of a double image e.g. when an ink spot is viewed through a calcite cleavage rhomb (hexagonal).

If a mineral is isotropic the ellipsoid indicatrix becomes a sphere because $X = Y = Z$. If the mineral is anisotropic, there are two possible

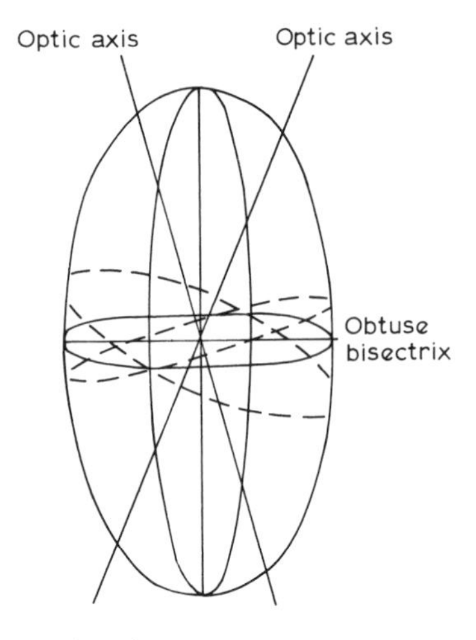

FIG. 3.8. The indicatrix. (– – –) Perimeters of the only two circular sections.

cases:

(a) $X<Y<Z$ (using refractive indices) and the mineral is biaxial;
(b) $X=Y<Z$ or $X<Y=Z$ (using refractive indices) and the mineral is uniaxial.

Now cleavage plates of micas give a biaxial interference figure in convergent polarized light and by the orientation of this figure with reference to the plane of symmetry as revealed by the percussion figure, micas may be divided into two groups (the percussion figure is that made when a blunt steel punch is placed on a cleaved mica plate and lightly struck so producing a small six-pointed star).

The two groups are the 'muscovites' in which a straight line joining the eyes of the interference figure is perpendicular to the plane of symmetry and the 'biotites' in which this line lies in the said plane of symmetry. The interference figure in the thin section of a crystal cut perpendicular to the optic axis (uniaxial case) or to the acute bisectrix (biaxial case) is observed with a microscope under crossed Nicol prisms in polarized light (convergent). In uniaxial cases it takes the form of concentric coloured rings and a black cross, the isogyre (see Fig. 3.9). In biaxial cases it is more complex with two separate sets of curves centred about two eyes and two hyperbolic isogyres, the whole pattern altering with rotation of the thin section (see Fig. 3.10).

Fig. 3.9. Uniaxial interference figure.

Coloured lines represent zones of equal retardation and to clarify this it is necessary to turn briefly to the subject of polarization colours in minerals. If a thin section of a mineral or minerals in a rock is viewed between crossed Nicol prisms (crossed polarizers) it is usually coloured. The Nicol prism operates due to the double refraction of calcite, Iceland spar. Long transparent cleavage rhombs are used and the edges ground until they make an angle of 68° to the long edge. The rhomb is cut into two longitudinally by a plane running through the two corners which have three obtuse angles. The two halves are recemented with Canada balsam. The inclination of the balsam film is such that it produces total reflection of the ordinary ray i.e., in the double image referred to above, that ray showing the ink spot in a stationary position. The extraordinary ray, i.e. that ray also showing the ink spot but this time an image rotating around the stationary one when the thin section is itself rotated, is not reflected but passes through the rhomb to produce polarized light. Either the ordinary or the extraordinary ray may be the fast ray. These rays vibrate at right angles to each other and in passing through the mineral one is of course retarded relative to the other to produce the 'fast' and 'slow'

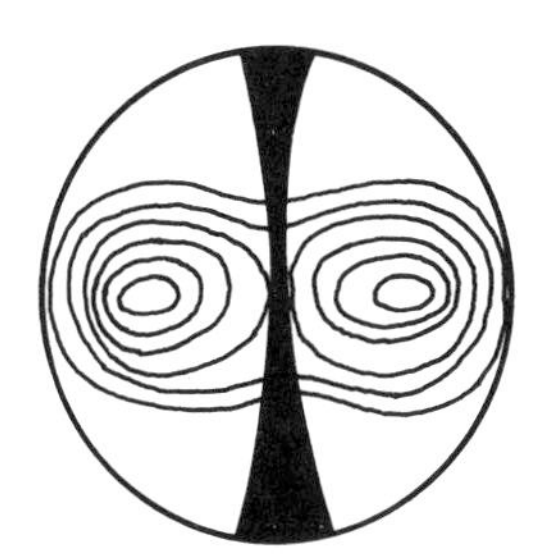

Fig. 3.10. Biaxial interference figure.

rays. This is retardation. After the two rays in question leave the mineral or rock section and enter an analyser, another Nicol prism or polarizer, they recombine into a single plane. The recombined rays interfere with each other, suppressing some frequencies and emphasizing others. This results in the production of coloured light, which constitute polarization colours. The actual colour produced in a mineral is a function of the difference in refractive index of the two vibration directions in it, the retardation

$$\Delta = (RI_{max} - RI_{min}) \times \text{thickness in microns}$$

where Δ = retardation and RI = refractive index and the usual thickness employed is 30 μ. If a wedge-shaped mineral slice is examined under crossed polarizers in a petrological microscope a range of such polarization colours appears.

This scale is divisible into various orders, each red band marking the end of a particular order. Above the third order the colours become indefinite and above the fifth constitute high-order whites. Each of the colours corresponds to a particular total retardation so that by identifying the colour observed in a mineral either alone or in a rock matrix, if the thickness of the thin section is known, a value for the birefringence can be derived.

The above remarks apply to anisotropic minerals alone because isotropic ones give no interference colours, i.e. they are black between crossed Nicols. This results from their having the same properties in all directions; hence there is no difference in velocity between any two vibration directions in the second, consequently no retardation and all frequencies are suppressed. As a matter of fact all crystalline substances other than those belonging to the cubic system are anisotropic (however, optically basal sections, i.e. sections perpendicular to the *c* crystallographic axis, of tetragonal and hexagonal minerals are isotropic).

The processes occurring in a mineral section viewed in a polarizing petrological microscope are shown in Fig. 3.11. Polarization colours occur between positions of extinction when a mineral section is rotated on the stage of a petrological microscope. Most sections of anisotropic minerals extinguish (become dark) between crossed Nicols four times during such a rotation, extinction taking place when the vibration planes of the minerals are parallel to those of the Nicol. The extinction angles, those between vibration planes and crystallographic directions in given sections of minerals, are quite useful diagnostic features, e.g.

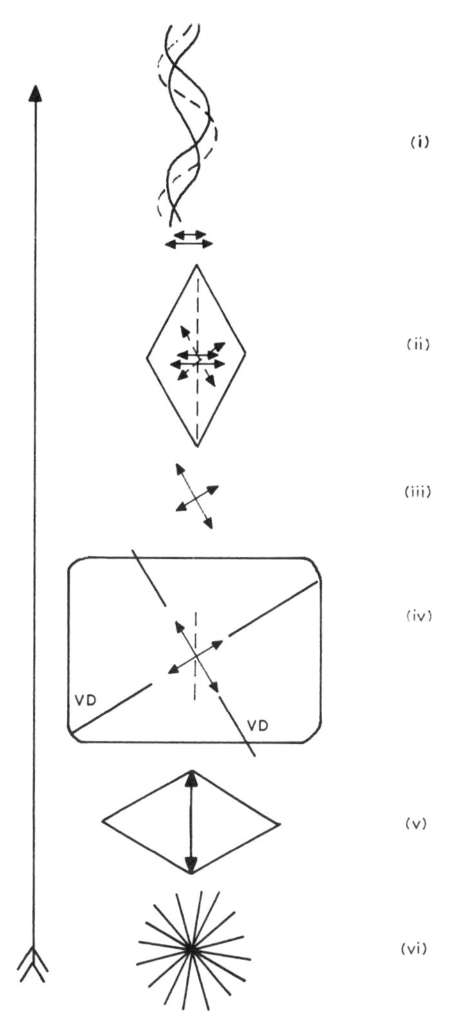

FIG. 3.11. (i) Two phase-different emergent rays; (ii) analyser; (iii) two rays emitted; (iv) mineral; (v) polarizer; (vi) source of light. VD = Vibration directions.

the extinction of hornblende occurs at an angle of 12° and of augite at 45°. Straight extinction takes place when darkness occurs as a crystallographic direction, say cleavage, parallels a cross wire in the microscope. Otherwise the extinction is said to be oblique. The extinction characteristics of the various crystal systems are given in Table 3.6.

TABLE 3.6
EXTINCTION FEATURES OF CRYSTAL SYSTEMS

System	*Extinction*
Cubic	All sections isotropic
Tetragonal, hexagonal	Basal sections isotropic, vertical give straight extinction
Orthorhombic	Pinacoidal sections give straight extinction
Monoclinic	Straight extinction except for clinopinacoidal sections which give oblique extinction
Triclinic	Oblique in all sections

Reverting to the mica family of minerals, the muscovites include muscovite or white mica, $KAl_2(AlSi_3)O_{10}(OH, F)_2$. Aluminium is substituted for silicon in the (Si_4O_{10}) group to the degree of about 1 atom in 4. In igneous rocks, muscovite is restricted to the most highly silicated types and is also widely distributed in gneisses and schists. Muscovite can originate as well from the hydrolysis of the potassic feldspars.

On the other hand if rocks containing muscovite are subjected to thermal metamorphism, the mica may be converted to orthoclase. Primary white mica formed by crystallization from a melt is distinguished from secondary white mica produced by the alteration of alkali-rich silicates. Secondary white mica is termed sericite.

The biotites or dark micas include biotite, $K(Mg, Fe)_3(AlSi_3)O_{10}(OH, F)_2$. Its properties differ slightly from those of muscovite so that, for example, its hardness is 2·5–3 (compared with 2–2·5) and its specific gravity 2·7–3·1 (compared with 2·76–3).

Usually black, biotite may also be green and muscovite may occasionally be green rather than its usual milky white (sometimes, but rarely, black muscovite is found also). Both have basal cleavage. Biotite occurs as an original component of the more basic igneous rocks such as gabbros, diorites, etc., also in biotite-gneisses and biotite-schists.

3.5. IGNEOUS ROCK-FORMING MINERALS: THE FELSIC MINERALS

3.5.1. Feldspars

These are quantitatively the most important rock-forming silicates; they are all aluminosilicates of potassium, sodium, calcium and occasionally barium, which crystallize in the monoclinic and triclinic systems. They occur in a three-dimensional network of SiO_4 tetrahedra arranged in chains interlinked by shared oxygens. These are aligned parallel to the axis in the crystal and every link comprises four SiO_2 tetrahedra. However as all oxygens are shared, the link may be stated as having Si_4O_8 in its composition. In each of such units one Si is displaced by an Al thus making the formula $(AlSi_3)O_8$ and leaving a surplus negative charge. In one group of the feldspars this is neutralized by the addition of a K^+ ion which gives orthoclase the formula $KAlSi_3O_8$.

In another group sodium ions occur in place of potassium and give albite, $NaAlSi_3O_8$.

There is a third group characterized by the substitution of calcium in place of potassium and sodium, but as this element is divalent there is an adjustment. This is effected by introduction of a second Al^{+++} in place of another Si^{++++} so that the formula becomes $CaAl_2Si_2O_8$. Rarely Ba^{++} substitutes for Ca^{++} to give celsian, $BaAl_2Si_2O_8$.

Feldspars are divisible into orthoclase feldspars and plagioclase feldspars, the latter comprising the albite-anorthite series as well as celsian. Both are white, grey or pale red in colour with a hardness of around 6 and a specific gravity ranging from 2·5 to more than 3. As regards crystallization, orthoclase and celsian are monoclinic while the albite-anorthite series is triclinic. Feldspars have two main cleavages and while simple twinning occurs in orthoclase, multiple twinning is found in plagioclases. The other main feldspar group constitutes the alkali-feldspars such as microcline and members commonly alter into sericitic mica or kaolin. They characteristically occur in more acidic igneous rocks.

Plagioclase feldspars consist of mixtures of albite and anorthite in all proportions. Thus the replacement of NaSi by CaAl results in a gradation between the two end members of the series. These soda-lime feldspars change to saussurite, a mixture of epidotes, albite and so on. Taking Ab as albite and An as anorthite, the limits of the various members of the plagioclase series are as follows:

Albite, $Ab_{100}An_0$ to $Ab_{90}An_{10}$, i.e. $<10\%$ An;
Oligoclase, $Ab_{90}An_{10}$ to $Ab_{70}An_{30}$, i.e. 10–30% An;
Andesine, $Ab_{70}An_{30}$ to $Ab_{50}An_{50}$, i.e. 30–50% An;
Labradorite, $Ab_{50}An_{50}$ to $Ab_{30}An_{70}$, i.e. 50–70% An;
Bytownite, $Ab_{30}An_{70}$ to $Ab_{10}An_{90}$, i.e. 70–90% An;
Anorthite, $Ab_{10}An_{90}$ to Ab_0An_{100}, i.e. $>90\%$ An.

The specific gravity of the plagioclases increases with the anorthite content, e.g. with artificial albite 2·6, labradorite 2·7, anorthite 2·765 (natural plagioclases are too impure for this property to be used in determination). Some percentage compositions have been selected: albite, 68·7 SiO_2, 19·5 Al_2O_3, 11·8 Na_2O, no CaO; labradorite, 55·6 SiO_2, 28·3 Al_2O_3, 5·7 Na_2O, 10·4 CaO; anorthite, 43·2 SiO_2, 36·7 Al_2O_3, 20·1 CaO, no Na_2O. Earlier it was thought that the plagioclases represent an isomorphous series, but only part of the range is isomorphous.

Plagioclases occur in almost all rocks and occasionally constitute whole rocks as in anorthosite (labradoritite). It is true to say that they are significant and often dominant constituents of many intermediate and basic rock types. Some, like orthoclase, are found in coarse-grained, slower-cooled igneous rocks while others such as sanidine, a variety of orthoclase, is a higher temperature equivalent. Another semi-precious opaline variety is called moonstone.

All feldspar crystals show zoning. As growth proceeds with the crystallization of a magma, the composition of the residual liquid changes so that layers of new material of different composition from their predecessors accumulate in concentric zones extending from calcium-rich cores to sodium-rich peripheries.

3.5.2. Feldspathoids

These are closely related to feldspars; they contain the same elements but in different proportions. Also they are noticeably poorer in silica. Nepheline is the best known and normally has the composition $K_2O.3Na_2O.4Al_2O_3.9SiO_2$, but this is variable and the composition is often $NaAlSiO_4$. The mineral is hexagonal, small glassy crystals occurring as an original constituent of volcanic rocks. It is of hardness 5–5·6 and has a specific gravity of 2·5–2·6.

It may be noted that feldspars and feldspathoids sometimes change to clay minerals if water is present. Orthoclase may change to kaolin, $Al_2Si_2O_5(OH)_4$ if all of its potassium hydroxide is removed by reaction with water. If the potassium hydroxide is not all removed then it is changed to illite, $KAl_2(AlSi_3)O_{10}(OH)_2$, this occurring where water is insufficient. Plagioclase feldspars alter to montmorillonite, $2H^+2Al_2(AlSi_3)O_{10}(OH)_2$, and this may later become kaolin if more water arrives. Thus kaolin is a product which is produced where water enters a rock any time after it consolidates. Clay minerals possess an atomic structure resembling that of the micas, but the crystals are too small to be seen except under a very high power microscope. The specific gravity of clays is 2·65, their hardness is 2–2·5 and they show one perfect cleavage.

3.5.3. Analcite

This is $NaAlSi_2O_6.H_2O$ and although chemically allied with the zeolites, being found in geodes and vesicles in lavas, it is also to be found in a wide variety of igneous rocks. Sometimes a substitution of potassium for sodium occurs. The mineral is cubic and has a specific gravity of 2·25 with a hardness of 5–5·5.

3.5.4. Melilite

This is a name utilized by a series of uncommon, complex silicates of calcium, aluminium and magnesium. Any one can be looked at as a combination of the two end members of the series, namely akermanite, $Ca_2MgSi_2O_7$, and gehlenite, $Ca_2Al_2SiO_7$.

3.5.5. The Silica Group

There is a number of mineral species comprising pure silica, SiO_2. Some occur as euhedral (fully developed) crystals while others are microcrystalline or cryptocrystalline. There is also a natural amorphous silica.

The three mineral species quartz, tridymite and cristobalite are found in both high and low temperature modifications distinguished as β- (high-) or α- (low-) quartz, etc. In all these minerals, SiO_4 tetrahedra are interlinked by all their corners spirally in both forms of quartz. However the arrangement is more symmetrical in the high- than in the low-temperature form. Hence while α-quartz crystallizes in the trapezohedral class of the trigonal system, β-quartz is a member of the corresponding class of the hexagonal system. α-quartz is the quartz of mineral veins; it is stable up to 573°C and is also common in rocks. Above this temperature β-quartz replaces it; it is said to be present in acid volcanic rocks where it is believed metastable. The mineral has no cleavage, but possesses a conchoidal fracture and a hardness of 7. Its specific gravity is 2·65.

In α-quartz, twinning is apparently ubiquitous and this is demonstrable using special optical tests.[1] Twinning does occur also in the β-variety as Drugman[2] pointed out as early as 1927.

Tridymite is the form of SiO_2 stable above 870°C. It is believed to be hexagonal holohedral and occurs only in acid volcanic rocks where it is metastable.

Cristobalite is formed from tridymite by inversion at 1 470°C. It is cubic, rarer than tridymite and is found in similar rocks. It is the highest temperature polymorph of silica.

Cryptocrystalline silica is found in a number of differently coloured and named forms, of which perhaps the best known is chalcedony which comprises fibrous or ultrafine quartz together with some opal. It is a possibility that some of the quartz has had oxygen ions replaced by hydroxyl ions.

Opal is an amorphous and hydrated variety of silica possibly originating from a silica gel. It is much softer than quartz and contains more water than chalcedony. It is usually a secondary deposit arising from the action of percolating groundwater and the silica which is deposited by hot springs (siliceous sinter and geyserite) is of an opaline nature. A siliceous sediment composed of skeletal remains of microscopic plants called diatoms is called kieselguhr or diatomaceous earth or diatomite.

As may be seen from the above paragraphs, silicates are the most important mineral group present in the crust of the Earth, possibly comprising 95% of it if the silica group be included. All silicates have atomic structures based upon the tetrahedral SiO_4 and Table 3.7 summarizes data regarding them.

It may be mentioned that talc is a hydrous magnesium silicate related to and monoclinic like serpentine with a hardness of 1, i.e. it is extremely soft with a specific gravity of 2·7–2·8. One of its varieties is soapstone, a massive form with a soapy feel to it. It occurs as a secondary mineral in gabbros, dolomites, etc. and results from the action of magmatic water or through stress in regional metamorphism or by the contact action of granitic materials. Other important secondary (accessory) minerals are discussed below.

3.6. SECONDARY MINERALS

Sometimes these may be primary, but usually they are accessory in igneous rocks.

Apart from talc, serpentine (another hydrous magnesium silicate) may be mentioned again. Its hardness is much greater than that of talc, 3–4 in fact, and its cleavage is indistinct whereas that of talc is perfect

TABLE 3.7
THE SILICATES

Designation	*Atomic structure*	*Unit*	*Examples*	*General formula*
Tekto-silicates	Three-dimensional of tetrahedra with all 4 oxygen atoms shared	SiO_2	Quartz, feldspars	SiO_2 $KAlSi_3O_8$
Sorosilicates	2 tetrahedra sharing 1 oxygen	Si_2O_7	Melilite	$Ca_2MgSi_2O_7$
Neosilicates	Independent tetra-hedra	SiO_4	Olivines	Mg_2SiO_4
Phyllo-silicates	Continuous sheets of tetrahedra sharing 3 oxygens	Si_4O_{10}	Micas, talc	$KAl_2(Si_3Al)O_{10}(OH, F)_2$ $Mg_3Si_4O_{10}(OH)_2$
Cyclo-silicates	Closed rings of tetrahedra each sharing 2 oxygens	$(SiO_3)_n$ $n = 3, 4, 6$	Beryl (six-fold), axinite (four-fold), benitoite (three-fold)	$Be_3Al_2(SiO_3)_6$ $Ca_2(Mn, Fe^{++})Al_2BO_3(SiO_3)_4(OH)$ $BaTi(SiO_3)_3$
Inosilicates	(a) Continuous single chains of tetrahedra, each sharing 2 oxygens	$(SiO_3)_\infty$	Pyroxenes, pyroxenoids	$MgSiO_3$ $CaSiO_3$
	(b) Continuous double chains of tetrahedra alter-nately sharing 2 and 3 oxygens	Si_4O_{11}	Amphiboles	$Mg_7Si_8O_{22}(OH)_2$

basal to give thin, flexible but non-elastic plates. Serpentines were grouped with chlorites by Hatch *et al.*[3] in view of their chemical affinity and frequent occurrence together. Both groups are sheet silicates having atomic structures composed of layers of linked SiO_4 tetrahedra resembling those of the micas. Serpentines are devoid of aluminium however, an essential component of micas. In addition, both serpentines and chlorites are lacking in potassium. There are two varieties of serpentine, the fibrous chrysotile and the lamellar one antigorite. Serpentine is an alteration product of olivines and pyroxenes; chlorites often also occur as alteration products of ferromagnesian minerals. Nickel-bearing serpentine is garnierite, an ore mineral. Chrysotile is one variety of asbestos. Serpentine may occur as a rock, serpentinite, formed through a process termed serpentinization and one of auto-metamorphism by a late stage hydrothermal action on ultrabasic rocks. It may be represented thus:

$$4Mg_2SiO_4+4H_2O+2CO_2=Mg_6Si_4O_{10}(OH)+2MgCO_3$$

Veins of asbestos often occur in association with serpentine in such rocks.

Other secondary minerals are discussed below.

3.6.1. Spinels

This is a very important group including magnetite and chromite. All members crystallize in the holosymmetric class of the cubic system, the octahedron being by far the commonest form. Some, including the above-named, are almost ubiquitous in igneous rocks of quite varying composition. They are also among the simplest in composition; magnetite is $Fe^{++}Fe^{+++}{}_2O_4$ and $FeCr_2O_4$.

Magnetite is the most widely distributed accessory mineral in igneous rocks and it is always opaque. It may be distinguished from other opaque ores by observing it in oblique reflected light when it shows a steely metallic sheen. It is the most strongly magnetic of all the ores of iron and indeed may be separated from them using a magnet. Its hardness is 5·5–6·5 and its specific gravity is 5·18. A slightly denser iron ore is haematite, Fe_2O_3, an hexagonal mineral not in this group.

A trigonal iron mineral is ilmenite, $FeTiO_3$, which alters to leucoxene, a variety of sphene, $CaTiSiO_5$. Among cubic minerals may be mentioned pyrite, FeS_2, which is 46·6% iron hence an ore. It is very common all over the world, e.g. in the huge Rio Tinto deposits in Spain, has a hardness of 6–6·5 and a specific gravity of 4·8–5·1. Pyrrhotite has the same chemical composition, but it is an hexagonal mineral and, unlike pyrite, it is magnetic.

Other titanium minerals include rutile, a tetragonal form of TiO_2. The most stable of the three crystalline forms of TiO_2, rutile fractures subconchoidally, has a hardness of 6–6·5 and a specific gravity of 4·2. It may be an important constituent of beach sands produced by the denudation of rutile-bearing rocks as is the case in Florida and Sri Lanka.

3.6.2. Apatite and Beryl

Apatite is a phosphate of calcium with small quantities of fluorine, chlorine and/or hydroxyl, $Ca_5(F, Cl)(PO_4)_3$ found in igneous rocks, especially pegmatites. It is an hexagonal mineral with a hardness of 5 and a specific gravity of 3·17–3·23.

Beryl is one of the beryllium minerals, actually the silicate $Be_3Al_2Si_6O_{18}$, the aluminate being chrysoberyl, $BeAl_2O_4$. Beryl is the only common beryllium-bearing mineral and structurally it is a six-membered ring of the cyclosilicates. Its hardness is 7·5–8 and its specific gravity 2·7. It is used as a gemstone under the names aquamarine (blue) and emerald (green).

3.6.3. Garnet

Garnets are silicates of aluminium, iron, manganese, chromium, calcium and magnesium with a crystal structure comprising separated SiO_4 tetrahedra with three in the unit associated with AlO_6 octahedra linked by cations in an eight-coordination. Garnets are cubic minerals having the general formula $R^{++}{}_3R^{+++}{}_2Si_3O_{12}$ where R^{++} can be Fe^{++}, Mg, Mn^{++} or Ca and R^{+++} can be Fe^{+++}, Al or Cr. They may be divided into two main groups called pyralspite and ugrandite (acronyms of the names of the three main varieties in each group listed in Table 3.8).

Garnets are commonly found in metamorphic rocks although they also occur in some granites and pegmatites as well as in lavas and pyroclastic materials (fragmental volcanic matter which has been blown into the atmosphere as a result of explosive activity and including pumice, bombs, scoriae, tuffs and welded tuffs, ignimbrites). Crystals up to 30 cm diameter have been found and the mineral is so resistant that detrital grains of it have been found in sediments. It can, however, alter to chlorite.

TABLE 3.8
GARNET TYPES

	Colour	*Hardness*	*Specific gravity*
(*i*) *Pyralspite*:			
Pyrope, $Mg_3Al_2Si_3O_{12}$	blood red	7·5	3·7
Almandine, $Fe_3Al_2Si_3O_{12}$	dark red	6·5–7·5	3·9–4·2
Spessartine, $Mn_3Al_2Si_3O_{12}$	red	7–7·5	4·15–4·27
(*ii*) *Ugrandite*:			
Uvarovite, $Ca_3Cr_2Si_3O_{12}$	green	7·5	3·42
Grossularite, $Ca_3Al_2Si_3O_{12}$	green or orange-red	—	3·5
Andradite (Melanite), $Ca_3Fe_2Si_3O_{12}$	black	7	3·75–3·78

3.6.4. Pneumatolytics

Pneumatolysis is a volatile-producing hot process associated with a late stage in the cooling of an igneous mass and capable of affecting country rocks and also the mass itself. Characteristic minerals associated with it include tourmaline and topaz.

Tourmaline, a hexagonal, trigonal mineral, is a six-membered ring cyclosilicate, a complex hydrated borosilicate of aluminium, magnesium and sodium with small amounts of potassium, lithium and fluorine. All tourmalines contain about 10% boric acid and 3·5–4% water. Tourmaline has a hardness of 7–7·5 and a specific gravity of 3–3·2. Its gemstone varieties include the red rubellite, the indigo indicolite and the blue Brazilian sapphire while the Sri Lankan peridot is honey-yellow.

Topaz is fluosilicate of aluminium, $Al_2(F, OH)_2SiO_4$, and crystallizes in the holosymmetric class of the orthorhombic system. Its hardness is 8 and its specific gravity 3·5. It is found in acid igneous rocks, also in druses and is valuable as a gemstone.

3.6.5. Zeolites

These comprise hydrated silicates of aluminium with sodium, potassium, calcium and occasionally barium so that in chemical composition they resemble the feldspars and feldspathoids. They arise from the latter by hydrothermal alteration and in fact if reheating under the same conditions occurs, they can alter into feldspars again. They appear to be pseudosymmetrical, i.e. look as if they belong to one crystal but on inspection prove to belong to another, of lower symmetry. They occur mainly as infillings of geodes, vesicles and cavities in lavas, especially basalts. They are tektosilicates with true water of crystallization and their most interesting property is their strong ability to base exchange, the basis of the original use of sodium zeolites as water softeners. The zeolites are classed as tektosilicates, but the degree of bonding between the SiO_4 tetrahedra varies considerably so that three subdivisions may be designated, namely:

(a) a fibrous group including natrolite, $Na_2(Al_2Si_3O_{10}).2H_2O$;

(b) a platy group including the stilbite–heulandite association; stilbite is $(Na_2Ca)(Al_2Si_6)O_{16}.6H_2O$ and heulandite $Ca_2(Al_4Si_{14})O_{36}.12H_2O$;

(c) an equant group including the chabazite–thomsonite association; chabazite is $(Ca, Na)(Al_2Si_4)O_{12}.6H_2O$ and thomsonite $NaCa_2(Al_5Si_5O_{20}).6H_2O$.

The felsic mineral analcite is sometimes included in the sub-division (c).

Zeolites have an atomic structure possessing open channels which can be used as a molecular sieve. They have a complex chemistry and often show twinning. They occur mostly in cavities in basic volcanic rocks and are probably hydrothermal in origin. It is important to emphasize this because some believe that zeolites result from the leaching of feldspars in primary rocks. Sometimes zeolites are arranged zonally in thick piles of lava flows, e.g. in Iceland.

3.6.6. Carbonates

These are characteristic alteration products of igneous rocks, especially basic ones. In very extreme instances there is little left except the original texture and all the component minerals have been replaced by calcite. Usually calcite arises by weathering or hydrothermal alteration of calcium-rich silicates, particularly the more basic plagioclases. Calcite is also a common associate of zeolites, chlorite and chalcedony in the vesicles and amygdales of lavas. There is a rock type, the carbonatite, which must be mentioned as a magmatic, intrusive type normally associated with nepheline syenite ring complexes. Carbonates may be divided into three categories:

(a) Hexagonal carbonates characterized by a dominant rhombohedral cleavage and including calcite, dolomite, magnesite ($MgCO_3$), siderite ($FeCO_3$) and smithsonite ($ZnCO_3$). Extensive solid solutions exist between these so that there are also members with intermediate compositions. The hardness of calcite is 3 and its specific gravity is 2·71.

(b) Orthorhombic carbonates characterized by moderate prismatic cleavage, belonging to the orthorhombic holohedral class and commonly forming pseudohexagonal triplets (due to repeated twinning). The commonest type is aragonite, an unstable mineral harder than calcite (3·5–4) and also denser (2·94) which can be distinguished from calcite crystallographically and also by means of the Meigen test. In this aragonite is stained with a solution of cobalt nitrate whereas calcite is not.

Intimate intergrowths of the two can be distinguished by means of Leitmeier and Feigl's test. In this a solution of manganese sulphate (11·8 g $MnSO_4.7H_2O$ in 100 ml water) has solid silver sulphate introduced and is heated, cooled and filtered. Then, a few drops of dilute sodium hydroxide solution are introduced and after a couple of hours the resulting precipitate is filtered off. The solution is retained in an opaque bottle. The minerals specimen is then covered by the solution after which aragonite turns grey and finally black whereas calcite becomes greyish after over an hour and remains so.

It may be added that aragonite forms at higher temperatures than calcite.

(c) Basic carbonates include malachite, $CuCO_3Cu(OH)_2$ and azurite, $2CuCO_3.Cu(OH)_2$. The former is green, the latter blue and both constitute ore minerals of copper found in the oxidized zones of copper deposits.

Calcite occurs in a wide variety of forms including nail-head spar, dog-tooth spar, Iceland spar, satin spar, stalactites and stalagmites, alabaster, tufa, chalk, limestone, marble, oolite, pisolite, etc.

3.6.7. Epidotes

This is a group of rock-forming silicate minerals which are composed of SiO_4 and Si_2O_7 silicon–oxygen units linking chains of AlO_6 (occasionally AlO_4OH) octahedra. The general formula for the group may be written $R_2^{++}\ R_3^{+++}\ O.Si_2O_7.SiO_4(OH)$ where R^{++} may be calcium, manganese, cerium or iron and R^{+++} may be aluminium, trivalent iron or trivalent manganese.

The most important mineral is zoisite, $(OH)Ca_2Al_3Si_3O_{12}$, an orthorhombic one with rosered (thulite), green or white colour, perfect cleavage, of low temperature origin like all members of the group and occurring in low to medium grade regionally metamorphosed rocks.

Epidote, *sensu stricto*, is $Ca_2Fe^{+++}Al_2O.Si_2O_7.SiO_4(OH)$.

3.6.8. Zirconium Minerals

Zircon, $ZrSiO_4$, is a well-known tetragonal mineral with cleavage parallel to the faces of the prism, hardness of 7·5 and specific gravity 4·7. It is a primary constituent of igneous rocks, especially the more acid ones such as granite, and corresponding pegmatitic forms and detrital deposits containing zircons (used as gems) occur in Sri Lanka, Burma, etc. By virtue of the fact that it is radioactive, zircon was used to estimate the ages of igneous rocks such as biotite granites containing grains of the mineral surrounded by pleochroic haloes when embedded in the dark mica.

3.6.9. Contamination Secondary Minerals

These normally occur in metamorphic rocks and also rarely in igneous ones which have assimilated xenolithic (pre-existing) sedimentary materials. They include corundum with its three varieties of ordinary, ruby and sapphire. All are crystallized alumina, Al_2O_3. Corundum is very hard, 9 on Moh's scale, fractures conchoidally or unevenly, has a specific gravity of 3·9–4·1 and is often colourless. Ruby and sapphire are red and blue in colour respectively. In thin section sapphire is light blue in colour and slightly pleochroic belonging to the rhombohedral class of the trigonal system.

Others include andalusite, Al_2SiO_5, and cordierite, $Al_3(Mg, Fe^{++})_2(Si_5Al)O_{18}$.

3.7. THE CLASSIFICATION OF MINERALS

As noted at the beginning of this chapter, minerals may be classified as essential or secondary. Niggli has used a classification based upon crystal symmetry. There are other approaches such as that of Dana which referred the various groups to their chemical compositions as follows:

(i) native elements such as sulphur;
(ii) sulphides, selenides, tellurides, arsenides, antimonides;
(iii) sulpho-salts such as sulpharsenites, sulphantimonites and sulphobismuthites;
(iv) haloids such as chlorides, bromides, iodides, fluorides;
(v) oxides;
(vi) oxygen salts such as carbonates, silicates, titanates; niobates, tantalates; phosphates, arsenates, vanadates; antimonates; nitrates; borates, uranates; sulphates, chromates, tellurates; tungstates, molybdates;
(vii) salts of organic acids such as oxalates and mellates;
(viii) hydrocarbon compounds.

Another schema is that of Rutley and Read and follows the periodic classification of the elements:

(i) (a) Li, Na, K.
(b) Cu, Ag, Au.

(ii) (a) Ca, Sr, Ba.
(b) Be, Mg, Zn, Cd, Hg.
(iii) (a) B, Al.
(iv) (a) Ti, Zr, Ce, Th.
(b) C, Si, Sn, Pb.
(v) (a) Va, Ta.
(b) N_2, P, As, Sb, Bi.
(vi) (a) Cr, Mo, W, U.
(b) S, Se, Te.
(vii) (a) Mn.
(b) F, Cl, Br, I.
(viii) (a) Fe, Co, Ni.
(b) Ru, Rh, Pd, Os, Ir, Pt.

This has the advantage that it groups together reasonably and, as far as possible, economically. However the classification adopted by the author is eminently practical from an engineering standpoint and adequately serves the needs of the engineering geologist and site engineer.

3.8. MINERAL GEOLOGY

Originally, elements were mined in a near pure state but now working is effected from deposits with a low percentage of metals in chemical combination with other elements such as oxides, carbonates, sulphides, chlorides, etc. Every such ore body is surrounded by a non-mineralized zone or one in which the phenomenon has occurred only slightly so that the material concerned does not constitute an ore from an economic point of view. Such waste (gangue) is extracted with the ore and must be removed prior to smelting.

Economic minerals originate in igneous rocks or in sedimentary rocks derived from them by weathering and erosion. They are found in metamorphic rocks as well. Yet others occur in suspension or in solution in sea water. Some workable mineral concentrations were formed within magma, perhaps by segregation from the main mass through possessing higher specific gravity. From cooling magma, gases and fluids penetrated surrounding 'country rock', passing rapidly along joints, bedding planes, etc., also metamorphosing the rock involved which itself constitutes a magmatic intrusion. The most immediate and affected area comprises the metamorphic aureole in which minerals separate out from the original gases and fluids according to their respective temperatures of solidification or condensation. Hence, around granitic circular masses (bosses)—surface expressions exposed by erosion and representing ancient igneous intrusions—a graded set of minerals occurs with tin and wolfram (tungsten) on the inner periphery and arsenic and copper further out. Details of the geology of copper are given in *Copper: Its Geology and Economics.*[4]

When surface exposure takes place, weathering and groundwater movement produce an upper layer where the mineral concentration is greater than in the interior of the ore body, i.e. a zone of secondary enrichment. Sometimes this is due to leaching which leaves a residue of economic significance, for example, with those low grade iron ores termed laterites often produced by tropical weathering on silicate rocks. More important are residues of hydrated aluminium oxide formed under these conditions, i.e. bauxite, the major source of the world output of aluminium.

It may be seen from this discussion that minerals may derive from a great variety of geological sources and it is possible to discern a general pattern of world mineral distribution. Many metallic minerals are associated with igneous activity so that they are located on shields, large structural units of the planetary crust which have been stable for at least the last 400 million years and mostly composed of Precambrian metamorphic and igneous rocks. Major instances are the vast Canadian shield, the Fenno-Scandian shield and the Siberian shield. Both structural characteristics and mineral resources are well exposed in some parts of the world, particularly where such shields have undergone glaciation. Elsewhere the solid geology may be buried by superficial (drift) deposits as in the huge alluvial troughs of Mesopotamia, the Amazonia area and the Indo-Gangetic plain.

As regards oceanic resources, these include salt (NaCl), seas providing about 30% of world requirements, magnesium (two-thirds of world production derives from the seas which contain the element in a proportion of 1 part per 800) and bromine (70% of the world's requirement of which is obtained from the oceans). The existence of manganese nodules on the seabed was recognized by the Challenger expedition in the last century, but their workability was established only in the 1960s. The brownish-black nodules occur at depths ranging from 1 500 m to 5 400 m, although they may be found occasionally in shallower waters. The chemical compositions vary considerably, but the following limits are applicable: Mn, 5–30%; Fe, 5–26%; Ni, 0·2–1·8%; Cu, 0·1–1·6%; Co, 0·1–1%. From a marine engineering standpoint there is no great difficulty involved in exploiting the nodules. Specially designed deep sea dredges or hydraulic pumps suspended from a ship could make deliveries to a land-based processing plant with a design adequately complex for handling the complex metal mix involved. A single ship dredging 10 000 tons of such nodules per day could supply more than the annual US consumption of manganese and nickel.

REFERENCES

1. Symposium, 1945. Quartz oscillator plates. *Amer. Min.*, 30.
2. Drugman, J., 1927. On β-quartz twins from some Cornish localities. *Min. Mag.*, **21,** 336.
3. Hatch, F. H., Wells, A. K. and Wells, M. K., 1975. *Petrology of the Igneous Rocks.* Thomas Murby & Co., London.
4. Bowen, R. and Gunatilaka, A., 1977. *Copper: Its Geology and Economics.* Applied Science Publishers Ltd, London; Halsted Press, New York.

CHAPTER 4

Water

4.1. WATER IN ROCKS AND SOILS

Voids, i.e. interstitial openings, are the important properties of these materials as regards water because they act, potentially or actually, as conduits for its movement and/or as places for its storage. The porosity of a rock expresses the ratio of such voids to the total rock volume as a percentage so that

$$n=\frac{V_v}{V_t}=\frac{e}{1+e}$$

where n is the porosity, e the void ratio, V_t the total volume and V_v the volume of the voids. If V_s is taken to represent the volume of solids, then

$$e=\frac{V_v}{V_s}=\frac{n}{1-n}$$

Generally speaking crystalline rocks of igneous and metamorphic character have low porosities unless secondary voids such as joints or faults occur in them. Of course such fracturation will increase the porosity of almost any type of rock. It must be noted also that primary porosity is usually greatest in young, recently deposited and poorly compacted sediments of shallow location which have not undergone compression to any marked degree. Another important property related to ease of flow in rocks is permeability. A rock is said to be permeable if water or any other liquid contacting its upper surface can pass through to its lower surface. In nature this movement is effected primarily in a semi-horizontal direction (except for the rock loess in which the movement is predominantly sub-vertical, loess being a well graded, usually non-stratified, vertically jointed aeolian-formed dust accumulation). Perviousness is a special case of permeability whereby the water or other liquid travels through mechanical discontinuities such as joints or other fractures or even bedding planes. The coefficient of permeability, K, is expressible in terms of Darcy's law:

$$K = \frac{QL}{Ah} = \frac{v}{i}$$

where Q is the amount of water per unit time, A the gross cross-sectional area through which flow occurs, h the pressure head loss, L the distance through which head is lost, v the discharge velocity, i the hydraulic gradient, i.e. the ratio between head loss and the distance in which it is lost.

K has the dimensions of velocity (distance per unit time) and some typical values are: (a) soils: clays, 0–1 m day^{-1}; sand, 10–260 m day^{-1}; gravel, up to 300 m day^{-1}.

The main factor controlling permeability is the size of the voids because the smaller they are the greater is the surface area of contact of water with mineral materials and the greater are those capillary forces which restrain flow. In the case of loose soils, the permeability increases with the square of the diameter of the grains. Figure 4.1 classifies soils with respect to permeability.

Subterranean water represents that part of the hydrologic cycle involving lithospheric circulation. The hydrologic cycle has been discussed in considerable detail by the author in *Surface Water.*[1] However it may be stated here that this water cycle is powered by solar energy and gravity. Water evaporates from oceans and continents (mostly from the former) and is transported by the atmosphere, sometimes for hundreds of kilometres, before returning to the Earth's surface as precipitation (rain, snow, sleet or hail). After its fall, precipitation may:

(i) on land:
  (a) be intercepted by vegetation;
  (b) collect on the planetary surface if this is impermeable;
  (c) seep into permeable soils or rocks;
  (d) contribute to streams which flow to lower ground and eventually reach the oceans;

(ii) at sea:
  (a) recombine with the surface water.

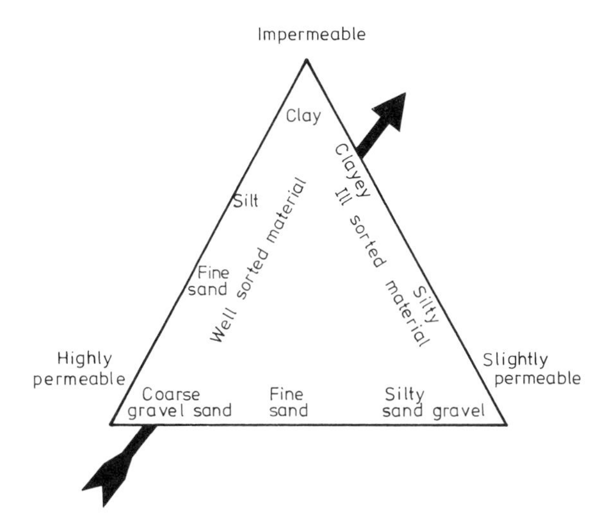

FIG. 4.1. Classification of soils with respect to permeability. (Arrow indicates direction of decrease in K.)

The main processes involved are evaporation, precipitation, interception, infiltration, seepage, storage in various water bodies, runoff and transpiration.

Where water falls on impermeable surfaces it may flow away as runoff. In fact Man's principal impact on the hydrologic cycle occurs through his interference with natural surface runoff.

Transpiration results from the drawing up of soil water through capillary action contributing, together with moisture from above, by means of roots and vegetable matter to leaves which return it as water vapour to the atmosphere.

That quantity of water going through the hydrologic cycle during any given period for a particular area can be evaluated using the hydrologic or continuity equation:

$$I - O = \pm \Delta S$$

where I refers to the total inflow of surface runoff, groundwater and total precipitation, O refers to the total outflow including evaporation and transpiration (collectively termed evapotranspiration), subsurface and surface runoff from the area and ΔS refers to the change in storage in the various forms of retention and interception. Figure 4.2 illustrates the hydrologic cycle.

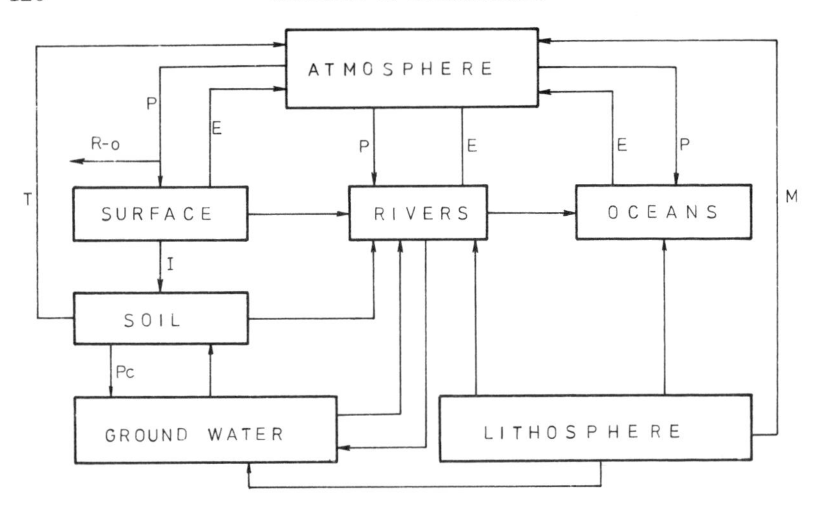

FIG. 4.2. The hydrologic cycle. R-o, Run-off; T, transpiration; P, precipitation; E, evaporation; I, infiltration; Pc, percolation; M, volcanic or magmatic water.

Lack of precise data makes quantification difficult, but it is believed that about 4×10^5 km^3 (4×10^{20} g) water are evaporated from the Earth's surface annually (about 84·4% of this from the oceans), the mean annual precipitation being 86·4 cm which must be balanced by an equal amount of evaporation. 97% of the planetary water (over $1{\cdot}234 \times 10^{15}$ m^3) is in the oceans and, according to Chow,[2] the distribution of fresh water on the Earth is:

75% in polar ice, glaciers;
14% in groundwater between 762 and 3 810 m depth;
11% in groundwater between the surface and 762 m depth;
0·3% in lakes (there are approximately three million lakes on Earth);
0·08% in soil moisture;
0·035% in the atmosphere;
0·03% in streams.

These are of course stationary estimates and it must be remembered that, although the water content of the atmosphere is relatively small at any given instant, vast amounts of water actually pass through it.

4.2. SUBSURFACE WATER CIRCULATION

Water in rocks arises from precipitation and other, lesser sources. Water originating as precipitation is termed meteoric water and may be contrasted with that coming from the other sources. These are as follows:

(a) juvenile water, a product of recent vulcanicity separating from magma;
(b) connate water, that sealed in by impermeability of surrounding rocks. If the original sediment (which was porous) was marine, then the connate water will be saline and can pollute water wells.

Precipitation shows tremendous variability around the world, ranging from almost nothing in areas such as the Sahara or Death Valley to more than 12 m annually in the Himalayas. In Great Britain the average annual rainfall is slightly in excess of 1 m.

That water infiltrating below the ground surface wets soil grains and adheres to them in the form of pellicular water and the removal of this can be effected only by evaporation or the roots of plants. Percolation of water to greater depths proceeds if precipitation continues and the soil reaches its field capacity, that point at which it cannot retain more water against the force of gravity. However the continually infiltrating water does not completely fill all available voids so that air is still present. The pellicular water and water moving under gravity constitute vadose water which occurs in a zone of aeration termed the unsaturated zone (see Fig. 4.3).

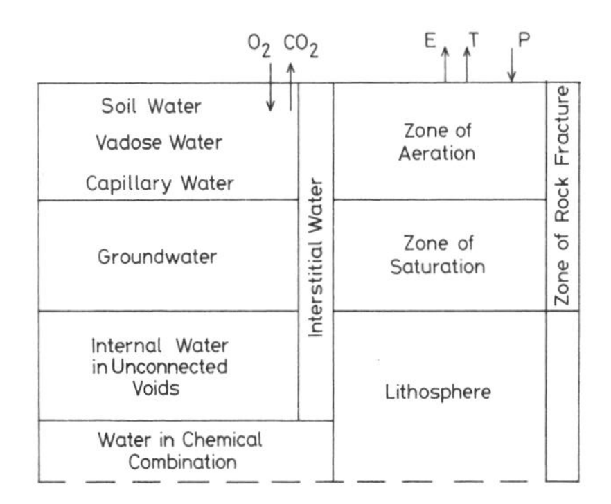

FIG. 4.3.

Eventually water penetrates down to a zone of saturation in which all voids are filled with water constituting groundwater. The upper surface of this is termed the water table and the zone continues downwards until the superincumbent load of the overburden eliminates porosity. Obviously the depth at which this takes place will vary with the geological conditions and can be as much as 10 km down in thick sedimentary deposits. In fact the water table replicates the topography overlying it and may be regarded as dynamic in nature so that this replica is modified. Water highs flatten out and gradients are reduced at a rate which depends upon the permeability of the rocks involved. Gradients vary from 1 in 100 to 1 in 10 in more hilly terrains. If a well is drilled into the groundwater body, i.e. into the aquifer, water enters it and with time fills it to a level in the water table below the top of the zone of saturation in the environmental rocks and separates from this by the capillary fringe or capillary zone. This ranges from a few centimetres to a few metres in thickness and can be as much as 10 m in thickness in fine-grained rocks with high capillary pressures. In such an open well the surface of the water is at atmospheric pressure and derives from an aquifer classified as unconfined. In unconfined aquifers the upper surface of the zone of saturation rises and falls in accordance with storage changes. If the groundwater is confined under pressure in excess of atmospheric through an overlying impermeable (confining) bed, it comprises a confined, pressure or artesian aquifer. In the case of an unconfined aquifer, water enters by the normal processes of infiltration, but in the case of an artesian aquifer, recharge of water to it takes place only in areas of outcrop of the confining bed or where this terminates underground. If subterranean termination occurs, the artesian aquifer then becomes unconfined. Rises and falls of water levels in wells penetrating confined aquifers depend on pressure changes and not upon changes in storage as in unconfined aquifers. Thus confined aquifers are agents of throughput, transmitting water from recharge areas to areas of discharge.

The concept of the potentiometric or piezometric surface is applicable. Meinzer[3] defined this as an imaginary surface everywhere coinciding with the hydrostatic pressure level of water in the aquifer. The actual water level in a well which penetrates a confined aquifer indicates the elevation of the piezometric surface at that place. If the piezometric surface lies above the ground surface, then a flowing well will occur. If the piezometric surface lies below the level of the

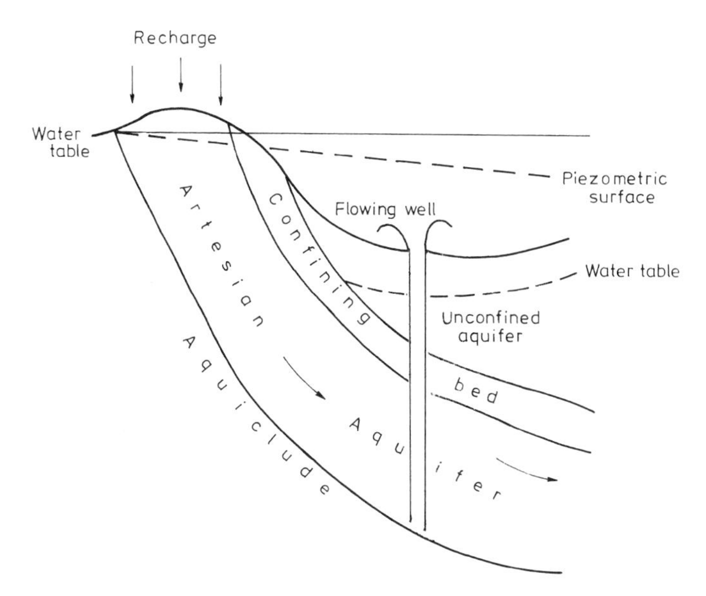

FIG. 4.4. Cross-section of a confined ground water basin showing an artesian well.

confining layer, then the confined aquifer will become unconfined. Both categories of aquifer are illustrated in Fig. 4.4.

A rather unusual instance of an unconfined aquifer is the perched aquifer shown in Fig. 4.5. This occurs wherever a groundwater body is separated from the principal groundwater system by a relatively impermeable bed of small areal extent. The latter may be diverse in

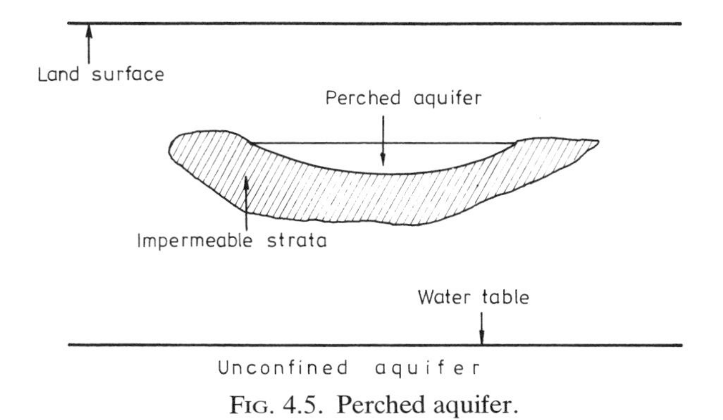

FIG. 4.5. Perched aquifer.

nature, but clay is a common occurrence, lenses of this in sedimentary deposits often impounding small perched water bodies. Of course some aquifers are partly confined and partly unconfined as e.g. the Lincolnshire limestone of eastern England. A numerical analysis procedure for such cases has been outlined by Rushton and Tomlinson.[4]

Groundwater is normally impure and contains such chemical compounds as carbonates and sulphates which have been dissolved from the rocks with which it came into contact. Hence the potability, etc., of such water relates to the aquifer in which it occurs. High sulphate groundwater usually occurs in some clays and shales and the sulphate may derive from gypsum, from sulphides or from H_2S in the groundwater reacting with various minerals.

If connate groundwater is involved, the sulphate may have been concentrated in the original sea water. Of course sulphates are important potential hazards in engineering work. For instance they are able to accelerate the corrosion of iron and react adversely with Portland and some other cements. BS4027 and other standards recommend that sulphate-resistant cement be utilized if the concentration exceeds a specified figure. Table 4.1 lists the major chemical compounds which may be found in groundwaters.

The most important hydraulic properties of rocks relate to void ratios, the amount of water which can be drained from them and the ease of flow of water through them. This latter is expressed as K, the coefficient of permeability and also by T, transmissibility, which is the rate of flow of groundwater in $m^3\ day^{-1}\ m^{-1}$ through a vertical strip of aquifer 1 m wide extending the entire length of the saturated aquifer under an hydraulic gradient of 1 in 1.

During discharge from a well, water is being removed from the environmental rocks and this produces a cone of exhaustion or depression in the water table or a cone of pressure relief in a confined aquifer in the piezometric surface. The outer limit of this cone of depression marks the area of influence of the well. Hydraulic gradients occur radially inwards around the well and induce flow into it. The relevant dimensions interrelate according to the steady state conditions of the Thiem equation thus:

$$Q = 2\pi(Kb)\frac{h - h_w}{\ln(R/r_w)}$$

where the symbols have the significance shown in Fig. 4.7. Steady state conditions are rarely if ever encountered in nature so that, dealing with

TABLE 4.1

SOME CHEMICAL CONSTITUENTS[a] OF GROUNDWATER

Common cations	*Equivalent weight*	*Common anions*	*Equivalent weight*
Calcium (relative atomic mass 40·08, valency 2)	20·04	Carbonate	30
Magnesium (relative atomic mass 24·32, valency 2)	12·16	Bicarbonate	61·01
Sodium (relative atomic mass 23, valency 1)	23	Sulphate	48·03
Potassium (relative atomic mass 39·1, valency 1)	39·1	Chloride	35·46
Other common elements dissolved in water: Carbon, relative atomic mass 12, valency 2·4; nitrogen, relative atomic mass 14·01, valency 3·5; oxygen, relative atomic mass 16, valency 2; sulphur, relative atomic mass 32·06, valency 2, 4, 6; hydrogen, relative atomic mass 1·01, valency 1.			
Minor constituents: Iron, aluminium, selenium, silica, boron, fluoride.			

[a] All may be represented on an equivalents per million scale (epm) in a shape diagram, as shown in Fig. 4.6.

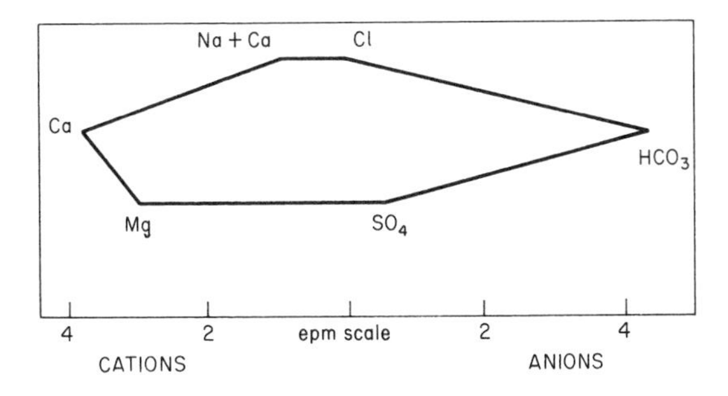

FIG. 4.6. The chemical composition of a sample of groundwater may be shown using a shape diagram as above in which cations and anions are plotted on the epm scale to produce the characteristic form.

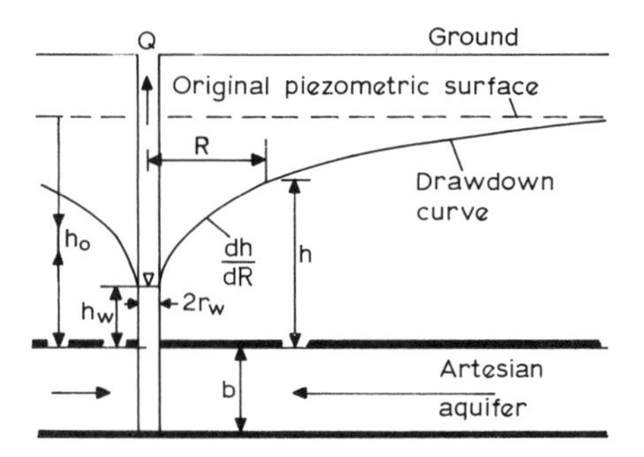

FIG. 4.7. Radial flow into a well which penetrates an extensive confined (artesian) aquifer (impermeable, sealing beds—aquicludes—shown thus: ——).

confined aquifers, the Theis equation is utilized. This may be stated as follows:

$$h_0 - h = \frac{QW(u)}{4\pi T}$$

in which Q is the discharge of the well, T the transmissibility and

$$u = \frac{r^2 S}{4Tt}$$

where S = storage coefficient, r = distance from the discharging well to the point at which the drawdown is $h - h_0$ and t = time since commencement of pumping. As to W

$$W(u) = -0{\cdot}577216 - \log_e u + u - \frac{u^2}{2\times 2!} + \frac{u^3}{3\times 3!} + \frac{u^4}{4\times 4!} \ldots$$

The Theis equation is an attempt to cope with the non-steady states normally encountered in nature and when devised in 1935 by Theis[5] was based upon the analogy between heat flow and groundwater flow. As regards the assumptions made in it, these are:

(i) isotropism of the aquifer—rarely if ever found in nature, but if large enough rock volumes are considered, appropriate characteristics are displayed;
(ii) infinite areal extent of the aquifer;

(iii) a coefficient of transmissibility which is everywhere and always constant;
(iv) water taken from storage is discharged instantaneously with decline in head;
(v) complete penetration of the aquifer is achieved and water derived throughout its thickness;
(vi) the well possesses an infinitesimal diameter—practically achieved because the well diameter is so small in comparison with the cone of depression and of course the aquifer.

Pumping tests in wells reveal many of the properties of the aquifer. The basic theory is that total penetration of the aquifer is attained because if this is not done, flow lines will be distorted and converge towards the bottom of an unscreened well. The analysis of flow conditions produces a solution for the relevant aquifer properties. Changes in these can lead to variations in overflow or alterations in head. It must be indicated that conventional analysis is often not able to provide consistent results and sometimes not all data can be utilized. This is invariably so where the drawdown in the well is a significant proportion of the saturated depth. Conventionally, analytic solutions are used as the basis of curve-fitting methods, but a discrete time–discrete space numerical approach may be utilized and for details of the method the work of Rushton and Booth[6] may be consulted. Another significant paper is that of Tomlinson and Rushton[7] on the alternating direction explicit (ADE) method for analysing groundwater flow. Both the drawdown and the recovery of water levels can be analysed and these phenomena parallel hysteresis in that confined aquifers possess a state of elasticity. However, analysing them in a recharging or discharging well is not as satisfactory as effecting the exercise in observation wells some distance away. In fact, an empirical approach is just about the only way in which the problem can be tackled. Sometimes though as Ineson has indicated,[8] computed curves can match water level data quite closely. In the case of a pumping well and two observation wells as shown in Fig. 4.8 the permeability of the relevant aquifer may be determined, assuming that after a specified period, say a day or more, steady state conditions have been approached. Using the parameters defined on the figure, the value of K is given by

$$K = \frac{Q}{\pi(A_1^2 - A_2^2)} \ln (r_2/r_1)$$

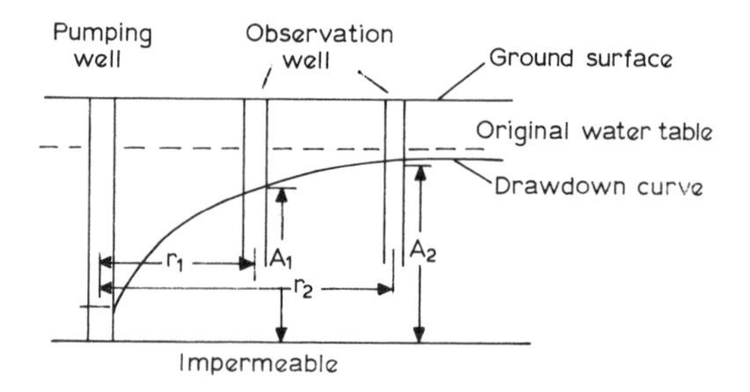

FIG. 4.8.

4.3. GROUNDWATER FLOW

Since groundwater may be regarded as a continuous contact body of fluid, if a pressure change occurs at any point there will be an effect everywhere. The actual flow of groundwater is controlled by head changes, the head at a point being a measure of the potential energy of the fluid relative to a specified state. Points possessing equal fluid potential define equipotential surfaces with the aquifer and from these it is possible to construct flow lines for the groundwater. It is possible to construct a flow net from these, i.e. a graphical representation of flow characteristics of use, when applied to the solution of hydrologic problems encountered in the vicinity of pumping wells adjacent to the boundaries of aquifers. Areal flow nets may be derived through empirical adjustment of ordinary water level contour maps and vertical ones based on data from wells along a line of section constitute flow nets showing the aquifer in cross-section.

Another approach is mathematical as Jacob[9] has indicated. This is illustrated in Fig. 4.9 which relates to a net for two adjacent wells, one discharging and the other recharging, the former a real and the latter an image well. The equipotential lines comprise circles around each well and as their radii increase the centres of the circles progressively recede from the boundary. For any potential surface

$$C = \frac{(x - x_1)^2 + y^2}{(x + x_1)^2 + y^2}$$

where C is a constant, x_1 is the distance from the boundary either to the real or to the image well and y is assumed to be zero (whereby, if

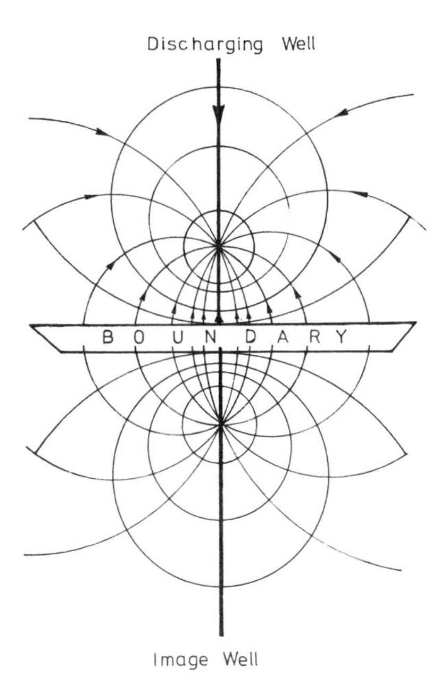

FIG. 4.9. Flow net near a pumping well adjacent to a recharge boundary.

solutions are derived for progressively greater values of x, values for C can be obtained). Flow lines *between* the two wells comprise circles centred on the boundary, the y axis.

In models, simple topography and uniform permeability are postulated with precipitation being added everywhere and steady state conditions assumed, discharge being into streams which function as sinks. Flow lines radiate at a sink and concentrate there (see Fig. 4.10). All water moves and none is stagnant. However, in nature the situation is much more complex. Permeability usually decreases with depth due to compaction and while in some cases flow is restricted to the topmost layers of soil, in others it may take place at kilometres of depth where appropriate permeability conditions exist and a high hydraulic gradient is involved. Generally the flow direction is normal to the contours of the water table; it follows topographical features and is a maximum at the near-surface zone.

Flow nets are best constructed under simple geological conditions, therefore, and in this case they are valuable in dealing with problems involving seepage of impounded water.

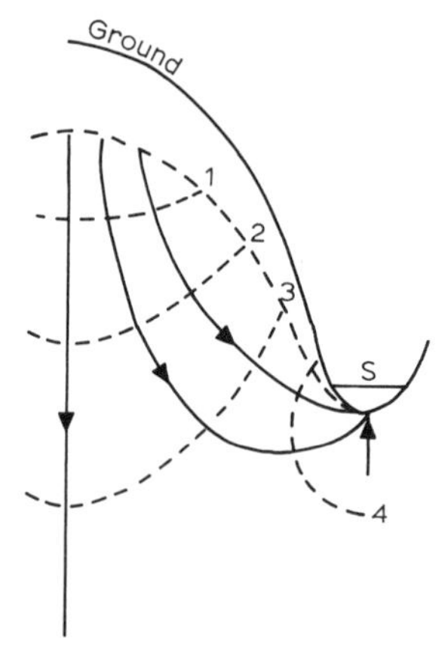

FIG. 4.10. Flow net model. Groundwater flow under a hill and valley assuming uniform permeability. The flow net is simple and regular comprising equipotential lines 1, 2, 3, 4 and flow lines orthogonal to them. Discharge occurs at S (sink).

The circulation of groundwater in the saturated zone occurs in aquifers, the movement being delimited by aquicludes (impermeable beds) bounding them. As a consequence of the presence of these and other features, groundwater basins are defined and these include areas of discharge and recharge. In them there exists an approximate balance between input and output and they vary enormously in size. It is necessary to determine the safe yield in a groundwater basin. This is the quantity of water which can be withdrawn annually without producing adverse results and the following methods are applicable to ascertaining it.

(*a*) *Hill method*: Based on the work of R. A. Hill in the USA and discussed by Conkling,[10] this takes into account annual alterations in the elevation of groundwater levels (either water table or piezometric) in a particular basin and plots them against annual drafts. The points usually fall in a straight line if the water supply to the basin is nearly constant. The draft is the safe take from an aquifer and when it is exceeded, the groundwater is being 'mined' and an overdraft situation arises (common in the aquifers of the USA).

(*b*) *Harding method*: The method of S. T. Harding[11] is based on an annual retained inflow and alteration in water table elevation. Annual values for the former are plotted against annual alterations in the elevation of the water table. Points are best fitted by a straight line and the retained inflow which corresponds to zero change in water table elevation is the safe yield.

(*c*) *Darcy's law method*: If lateral inflow occurs into a basin from a known direction, the safe yield can be derived from the long-term inflow by applying this law, as long as the average hydraulic gradient, the permeability of the aquifer and the cross-sectional area of the aquifer in a direction perpendicular to flow are known. These parameters are determinable and the method has been used by the US Geological Survey in respect of confined aquifers possessing unidirectional flow. Meinzer has outlined the technique.[12]

The above are just some of several methods of estimation of safe yield or, as it may be termed, basin yield in a groundwater basin. Nowadays basin yield may be divided into the following categories:

(i) mining yield, the situation in which groundwater is withdrawn at a rate exceeding the recharge (overdraft) and one which must be restricted in time until replenishment occurs otherwise the depletion will continue;
(ii) perennial yield, the rate at which water may be withdrawn from a groundwater basin without adverse results (this is the safe yield);
(iii) deferred perennial yield, where two pumping rates are employed; the initial rate is larger and exceeds the perennial yield so as to reduce the groundwater level. This is a planned overdraft and, after a predetermined (lower) level has been reached, a second (smaller) pumping rate is applied, one which ensures balance between water inflow and outflow.
(iv) maximum perennial yield, the maximum quantity of groundwater perennially available if all feasible methods and sources are developed for the recharge of the basin.

The natural discharge of groundwater occurs where the water table intersects the ground surface and takes many forms, such as springs, seepages, etc. Fault springs are found where previous rocks are faulted against impervious ones and may be distinguished from valley springs where the water table intersects the bottom of a valley. If the water table seasonally rises and falls, the spring will be intermittent and such a spring in England is termed a bourne. Most fresh water discharge takes place on the sea coast where it mixes with saline groundwater. Both the fresh water and the brine behave as immiscible liquids and rock may be wetted by either one.

The shape and location of a contact surface between fresh and saline water is determined by the Ghyben–Herzberg relation, which is that salt water occurs underground at a depth below sea level of about forty times the height of fresh water above sea level. This results from an hydrostatic equilibrium existing between the two fluids of different densities which may be expressed thus:

$$h_s = \frac{\rho_f}{\rho_s - \rho_f} h_f$$

where the symbols have the meanings apparent from Fig. 4.11, ρ_s is normally taken as $1{\cdot}025\ \mathrm{g\,cm^{-3}}$ and ρ_f as unity, hence

$$h_s = 40h_f$$

and this has been substantiated by field measurements in many coastal areas.

There are certain limitations on the Ghyben–Herzberg method. One is that near the shoreline there must be a seepage face for fresh water outflow and here the relationship cannot hold. If hydrostatic equilibrium fully existed, no flowage is implied but groundwater is observed to flow near coastlines.

Nevertheless under special conditions the relation is quite accurate and from it is derived the concept of a saline wedge presence at the intersection of an aquifer with the ocean, a common state of affairs.

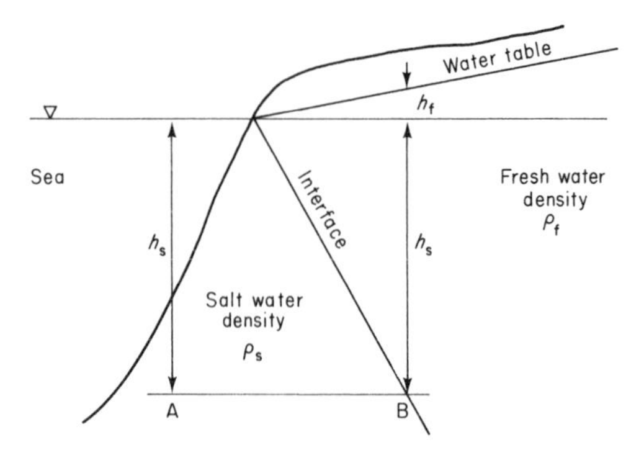

FIG. 4.11. Ghyben–Herzbèrg relation.

Examination in the field of interfaces shows that these are actually narrow zones of mixing which may be metres in width and they result from dispersion effects, fluctuations produced by tides, molecular diffusion and seasonal water table variations.

When steady the sea water–fresh water interface is parabolic, but in nature it is rarely stationary (see the work of Vappicha and Nagaraja, ref. 13). Sea water intrusions are a serious problem because they render water supplies unfit; their encroachment, therefore, must be at least arrested if not reversed. This may be done in a number of ways:

(a) by modification of pumping patterns, i.e. reducing demand and thus eliminating overdraft; for this, the cooperation of water users is necessary and very difficult to obtain;
(b) artificial recharge may be effected, i.e. natural infiltration of precipitation is augmented by pumping fresh water into wells, spreading of water on the ground, etc;
(c) a groundwater trough may be constructed, i.e. a line of wells is drilled near to and along a coastline which, when pumped, causes a depression (trough) in the level of groundwater in the aquifer; clearly such a gradient limits the intrusion of sea water;
(d) a grout curtain may be installed to contribute a subterranean, impermeable barrier in the aquifer in order to prevent the inflow of water, which can be done in shallow situations by introducing asphalt, concrete or puddled clay; a line of holes may be used for injecting grouts such as bentonite slurry or silica gel and such a barrier will be of a permanent nature.

Identifying saline waters is easily done by taste, conductivity, etc; hence the main problem in controlling them relates to their origins. For instance at Long Beach in California, there are no less than four possible saline sources, namely:

(i) the sea;
(ii) oilfield brine wastes;
(iii) irrigation wastes;
(iv) shallow connate waters.

Chemical analysis of the conventional type is hard to use to distinguish such sources because almost all of the common ions occurring in natural waters are liable to change while in groundwater transit, cations by base exchange with mineral matter, sulphate by bacterial reduction and bicarbonate by precipitation and solution. Chloride ions are not affected to any great extent, but they are far too common in nature to be employed as tracers.

Consideration may be given to isotopic analyses because isotopic contrasts may be anticipated among the various possible sources. Those isotopes of the water molecule not usually subject to what may be termed transit changes include deuterium, tritium and oxygen-18, although in the case of oxygen-18 there may be exchange with oxygen-16 in rock materials under certain circumstances. Stable isotope analysis is very useful as a result of the fact that marine waters are distinguishable from meteoric waters. Saline waters arising from evaporative concentration in lake environments are enriched in the heavier species of hydrogen and oxygen relative to precipitation and marine waters. Oilfield brines and deep fossil waters usually show oxygen-18 exchange when compared with sea water and precipitation. Isotope analyses for tritium and carbon-14 are significant because modern precipitation is labelled with these isotopes and of course these radioactive isotopes are useful dating tools for groundwaters. As regards deuterium and oxygen-18, these occur as $HD^{16}O$ and $H_2{}^{18}O$, both of which have slightly lower saturation vapour pressures than the ordinary water molecule $H_2{}^{16}O$.

It is clear that there is a slight fractionation of the deuterium and oxygen-18 in all processes of evaporation and condensation, the heavier isotopes becoming enriched in the liquid phase and depleted in the vapour phase. Dansgaard[13] has demonstrated the relationship between oxygen-18 content and mean annual temperature for precipitation at various North Atlantic stations, a relation expressed as a straight line of slope 0·69‰ per °C in agreement with the isobaric cooling model for stations in the temperature range 0–10°C when the initial temperature is 20°C. Variations in the deuterium and oxygen-18 contents of precipitation are linearly correlated (see Fig. 4.12).

In 1961 Craig[14] has given a relationship which best fits points representing the isotopic composition of samples of precipitation from all over the world and it is

$$\delta D‰ = 8\delta^{18}O‰ - 10$$

the δ‰ representing the difference in parts per thousand of the isotopic ratio in a sample with respect to that of mean ocean water (V-SMOW). This may vary slightly; e.g. in 1981, Peter Fritz *et al.*[15] gave the relationship

$$\delta D‰ = 7{\cdot}8\delta^{18}O‰ - 10{\cdot}3$$

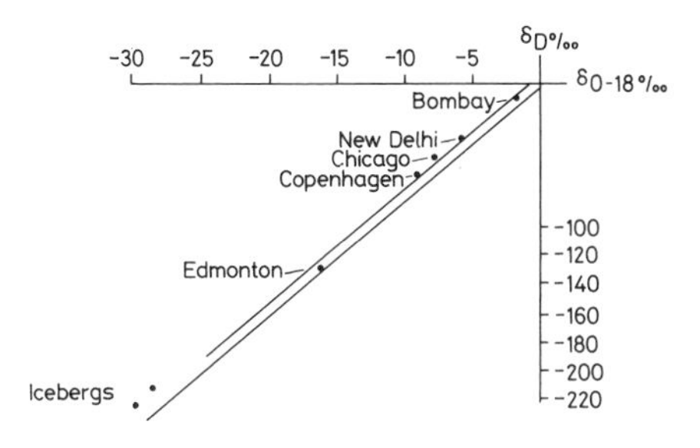

FIG. 4.12.

but it is interesting that water submitted to evaporation processes deviates systematically from this general sort of relation found in non-evaporated meteoric waters. Hence they can be distinguished.

To distinguish different sources of saline water, isotopic analyses may be effected and some instances of applications may be cited:

(a) terrestrial waters concentrated by evaporation can be distinguished from marine waters as well as from precipitation (meteoric water), being relatively enriched in deuterium and oxygen-18;
(b) modern sea water may be distinguished from connate waters because of the different origins;
(c) sea water can be distinguished from deep brines because these latter are relatively enriched in oxygen-18 with respect to deuterium due to the exchange of oxygen with rock minerals.

As regards connate waters, if these derive from Pleistocene glacial periods they are greatly depleted in deuterium and oxygen-18 relative to modern sea water. In principle, $^{13}C/^{12}C$ ratios and $^{34}S/^{32}S$ ratios may be utilized to distinguish sea water from meteoric water. Clayton and his associates[16] have used deuterium and oxygen-18 ratios in the study of 96 brines from oilfields in the Illinois and Michigan basins as well as the Gulf Coast and the Alberta Basin. Their major findings were:

(a) variation in deuterium content among groundwater basins was much larger than that noted in individual basins and showed a relationship to geographic location.
(b) the oxygen-18 contents showed a wide range in each basin and this correlated with salinity and formation temperatures.

The conclusions reached were as follows:

(a) the water is not of marine, but mainly of local meteoric origin;
(b) the deuterium contents had not undergone much exchange or fractionation;
(c) there had been quite extensive exchange between the water and the rocks of the reservoir;
(d) some samples may well have originated as Pleistocene glacial precipitation.

4.4. GROUNDWATER AND VADOSE WATER IN CONSTRUCTION AND ENGINEERING

As seen in this book, groundwater may constitute a considerable hazard in construction work. Tunnelling and large scale hydraulic structures, etc., have been examined in detail in Chapters 8 and 9 so that other aspects will be considered here.

In construction work the influx of water must be combatted and this may be done by the sinking of well points in order to lower the water table to acceptable levels. Such well points may be perforated and screened with pipes usually about 4 cm in diameter and metres long. Such a well point is fixed to a riser pipe of the same diameter and the entire unit is driven into the ground with or without jetting. Individual riser pipes are attached to a header pipe or manifold which leads to a pump. Lines of well points are arranged in conformity with the physical characteristics of the unconfined aquifer's water table surface. Here they operate as a piece of dewatering equipment. Usually such well points may be placed with their centres 0·9–1·8 m apart and the suction lifts which may be attained will range from 4·5 to 7·5 m. Obviously a lot depends upon the efficiency of the system used and the altitude at which the operation is carried out.

If very deep excavation work is being done, well points are arranged on a series of descending steps at 4·5–6 m intervals, i.e. a multiple stage set up. An instance may be cited of the influx of groundwater into the soft rock Box Tunnel excavated under the direction of Isambard Kingdom Brunel on the main line of the Great Western Railway between London and Bristol and commenced in 1836. It was planned to be 3·2 km long and its completion was delayed until June 1841, partly because of groundwater problems. The various beds encountered included clay, blue marl, the Inferior Oolite (about three-quarters of the length) and Great Oolite (the remaining quarter). Lining was carried out, about 30 million specially made bricks being utilized. It is perhaps worth mentioning, incidentally, that the sole illumination then possible for the effecting of the work was candles; 1 ton of these was utilized every week!

Sometimes lowering of the water table may have deleterious effects. An instance of this occurred in the Brooklyn area of Long Island where there is a great amount of water consumption—sewer construction and street paving work produced a drop in the level of the water table to as much as 10·7 m below sea level which promoted sea water intrusion and serious contamination of some of the wells. Continuous monitoring of the situation from its inception in the 1930s produced cessation of almost all groundwater pumping for public supply by 1947. Thereafter Brooklyn and its environs were supplied from surface water sources available through the New York Board of Water Supply. Groundwater was still pumped, but only for air-conditioning after which it was returned underground. Consequently the water table rose again. This caused some flooding of basements requiring costly remedial action.

As regards the foundations of buildings, groundwater can have a significant effect. The case of Winchester Cathedral in England may be considered. Some of the very old piles of timber under the building show decay as a result of variations in the groundwater level through history. Another case, cited by Lizzi[17] should be mentioned: the Pepoli Museum at Trapani at which works were effected in 1970 and 1973. This was constructed in calcareous tuff masonry and showed traces of stress at many points, of which the most obvious were in the northwest corner of the porch and along the west end. It was noted that the building was out-of-plumb and there was a bulging of all four sides of the porch in relation to their central sections. It was concluded after a preliminary survey that the edifice had been affected by instability movements in the subsoil arising from its nature and variations in the water level due to exploitation by nearby wells. Root piles were utilized in the remedial work. These represent a great advance in underpinning methodology and their main aspects are:

(a) they are preceded by rotary drilling through existing structures into the soil;
(b) with a concrete casing, the pile is bonded to the upper structure;

(c) they can be installed without vibrations so the stability of the existing structure is undisturbed;
(d) the construction of the root pile introduces no particular stresses in walls or the soil (excluding the small compressed air component utilized during the actual casting);
(e) concrete for the pile is composed of 600–800 kg cement m^{-3} of sieved sand and may be regarded as high strength;
(f) the reinforcement comprises a single bar for the piles having the smallest diameter (10 cm), a multibar frame or tubes for greater diameters; in underpinning work, the smaller diameters are preferred normally;
(g) the casting system gives a rough outer surface of the pile and hence marked adhesion to the soil;
(h) the settlement is of the order of a few mm even for load values up to the strength of the concrete.

Overall the most attractive characteristic of the root pile is its rapid response to any movement, however small, of the structure which it is underpinning. This may be attributed to the execution technology for which a 'palo radice' is basically a friction pile.

Various grouts may be used in underpinning; for example the lignochrome TDM employed in such work was required at Great Cumberland Place (Bilton Towers) in London for reconstruction work. Here underground garages were planned for an erection including 14 storeys and deep excavations had to be effected. Some of the nearby buildings were over two centuries old and had to be protected. The strata involved included the Taplow terraces of the River Thames, sands of various types associated with saturated gravels. These are underlaid by the London Clay (Eocene). Injection of the grout was specified and after treatment, practically uniform results were attained in sandy gravels while adequate results of lower consistency were obtained in laminated sands.

If pile rotting cannot be prevented in time, expensive remedial work may have to be carried out. This happened with Strasbourg Cathedral which was commenced in 1439 with stone footings supported on timber piles. A new drainage system was installed in the 18th century and this interfered with the local groundwater conditions and caused damage to the piles. Consequently a large scale underpinning operation for the foundation had to be effected during the early part of this century.[18] A similar problem arose with the Boston Public Library in

1929 under which about half of the piles were found to have rotted. The rotten piles were attended to and almost 40% of the building underpinned.[19]

An interesting set of factors were encountered at Milwaukee, Wisconsin, where there is a normally rather high level water table and the central part of the city is underlaid by variable soils. Close to the waterfront is the head office building of the Northwestern Mutual Life Insurance Company erected in 1912 and in 1930 it was proposed to add an extension to this which would complete a city block belonging to the company. The latter was founded on timber piles grouped under mass concrete footings. To ensure a completely satisfactory result, the architectural design provided every concrete footing with a 10 cm pile capped at the level of the sub-basement floor. Groundwater levels have been recorded in these every subsequent month. Where a drop in level occurs, water is introduced until the restoration of the usual level. The foundations have given every satisfaction.[19]

Turning to the subject of sulphates in groundwater, they can be deleterious and of course they occur in many places. An instance is now cited:[20] In 1938 the St Helier Hospital at Carshalton, Surrey, England was constructed with a capacity of 750 beds and comprising four multi-storey ward blocks with the central services in one main block, subways connecting the various parts. Concrete foundations were placed on brown clay overlying the London Clay and in the spring of 1959 some foundation concrete was exposed during maintenance work. This was observed to have deteriorated. A general study then divulged that most of the foundation was deteriorating progressively. A high sulphate content was found in the substrata (which possess quite adequate bearing capacity). Obviously the concrete had been subjected to chemical attack and this had weakened the highly aluminous cement which had also been adversely affected by a warm and damp atmosphere. A major underpinning operation had to be undertaken which necessitated the provision of new foundations for the entire building complex.

As regards excavation work of an unusual kind, the case of the New York World Trade Center may be considered. This is located in lower Manhattan and cost US$600 million. These two buildings, the tallest in the world, are supported by concrete piers lying directly on the Manhattan schist. This rock has a high bearing capacity but is steeply dipping and often shows slippage along joint planes. The actual rock level is about 21 m below street level; a six-storey basement had to reach this. By means of a survey of maps dating back to 1783, it was discovered that then the Hudson River shoreline ran two blocks inland of the currently existing shoreline. Many wharves and other buildings had been built on it and these were later covered by miscellaneous fill and lay upon mixed beds of organic silt, overlying organic silt and sand. All of these beds are water-bearing. Clearly therefore, groundwater lies quite near to the existing ground surface. Now, in the prevailing circumstances of streets and buildings as well as subterranean services, it was impossible to lower the water table around the selected site. It was decided to enclose this by means of a wall of concrete placed using a slurry trench. Of course this latter had to be very carefully excavated and was infilled subsequently with bentonite and water. Concrete was placed in the trench from the bottom upwards and a tremie pipe was employed. (Tremie injection involves insertion of large quantities of this or other grout material through open-ended pipes kept embedded in a mound of the relevant material and it is a technique utilized in cavernous rock strata in the construction of grout curtains—see Fig. 4.13.) Steel grills were also placed in the trench so that they would become encased in the concrete and reinforce the structure. As the concrete ascended, it displaced bentonite and reinforced the sides of the trench thus making it safer. After excavation

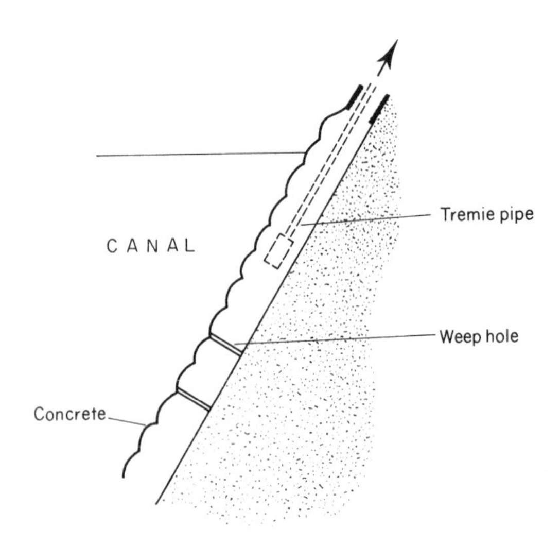

Fig. 4.13.

was effected to total depth, the walls were anchored by inclined steel tiebacks drilled into and grouted into the underlying schist. All excavated material was transported to a nearby site in the Hudson River which was enclosed by a sheet pile retaining structure and after dumping, the city had acquired a new land region stated to be worth almost US$100 million.

An extraordinarily interesting excavation project, carried out in Rotterdam, has been described by Legget.[19] As this has been discussed by the author in *Ground Water*,[24] there is no need to say more here except to mention that inner sections of subway tunnel had to be floated into place in the city centre, so as to avoid pumping, on a vast scale, of the very high water table.[2]

Another instance has been cited by Legget and is alluded to here.[19] This refers to the area including Norfolk around Hampton Roads in Chesapeake Bay, Virginia, where there are unusual traffic problems. A sunken tube tunnel, the Second Elizabeth River Tunnel, was opened in September 1962 and cost US$23 million. It is 1·275 km in length from portal to portal and comprises 12 precast concrete tubes *floated* into place. The short approach sections to the actual portals were constructed within cofferdams and the first stage of test boring occurred in 1957, drillings being done at 152 m intervals—this spacing pattern was considered satisfactory because of the rather uniform geological conditions prevailing locally. Borings were spaced at 61 m intervals on the shore however because of the greater geological complexity involved and also because meanders occurred in the relevant area. Actually this spacing turned out to be inadequate because the geological circumstances proved to be so difficult that a second stage of exploration had to be performed together with a rigorous testing of undisturbed soil samples derived from the site. Additionally extra borings were collected during construction. A soil profile along the route of the tunnel demonstrated a basal marl overlaid by clay and silt, the tunnel foundation being primarily in the first. More details of the work have been given by Legget[19] who mentions that a zone of organic silt ran for some 53 m along the centre line of the tunnel. The interesting thing is that this zone was so limited that it had not been detected by test borings which had missed it. The organic silt was very incompetent and if subjected to load would have produced settlement of about 15 cm over the long term. To avoid such a deleterious result, the zone was loaded with a 1·83 m high fill surcharge left *in situ* for 15 months. Calculations showed that during this time, excess load would promote settlement approximately the same as that anticipated under normal load. In fact the settlement achieved slightly exceeded 15 cm so that construction thereafter proceeded satisfactorily. No subsequent settlement has been seen to take place.

On the subject of groundwater and settlement, another instance is cited. Long Island is approximately 192 km long and at maximum 35 km wide; it constitutes a coastal plain bordering New York State and Connecticut. The geology is of interest because it comprises of bedrock which approaches the surface at the north but is about 600 m below the surface to the south. Overlying are a set of water-bearing beds with occasional impervious strata in them, e.g. the Raritan and Gardiner clays. Surficial glacial deposits were deposited in the Pleistocene and are quite pervious so that water entering them seeps down to underlying sands and gravels. Originally when settlement began, groundwater conditions were in equilibrium. The infiltration of precipitation maintained the water table at a high level and the balance was preserved as a result of the movement of water through springs and also in small water courses on the island. There was most likely a seepage effect into the sea as well. Under the sea the groundwater is saline so that there must somewhere be an interface, at a depth of 300 m or so. Initially the inhabitants obtained an excellent water supply from springs and rather shallow wells and used cesspools for the disposal of wastes, these draining back into water-bearing beds. Ultimately there were problems of contamination which necessitated digging deeper wells. Water was drawn from the Jameco gravel (Pleistocene) and the Magothy formation (Cretaceous) and, although pumping increased, the equilibrium was not seriously disturbed because most of the water returned into the water-bearing beds after use. As Brooklyn developed, servicing there meant that much waste water was no longer recycled but returned directly to the sea or to Long Island Sound. This resulted in a decline in the level of the water table which was corrected by restricting the extraction of groundwater. Flooding followed and had to be obviated by expensive remedial measures as described earlier in the section. Legget[19] has indicated that the groundwater situation under Long Island shows all phases of the development at the western end that have been noted here. However at the eastern end the natural situation still prevails. Long Island is in fact a very interesting location for groundwater studies and probably represents one of the largest usage regions for any single well-defined subterranean reservoir anywhere in the world. Plans have been made for future conservation and these include using artificial recharge wells, reclaiming water from sewage and desalination of sea water. In fact an atlas of the water resources of Long Island has been compiled by the US Geological Survey and the New York State Water Resources Commission.[21] Current consumption is not exactly known, but in 1963 it amounted to 1438 million litres per day from wells. Construction activities throughout the island are closely related to groundwater supplies and may be categorized as, from east to west: few (so that natural conditions are maintained); settlement of individual dwellings and usage of public wells; urban zones with sewerage deriving water by pumping accompanied by depression of the water table; and finally a well-developed region ending in Brooklyn (obtaining water from the New York system, the water table slowly regaining its natural level and some sea water contamination undoubtedly taking place).

The largest hydraulic structures constructed by Man are dams and these are considered in detail in Chapter 9, but one instance can be given here in relation to groundwater and its effects on foundations. This is a tailings dam located upon a very loose, saturated, sandy silt soil at Kimberley, British Columbia, the investigation and design for which were completed in 1974. It was intended to replace an existing embankment of marginal stability and allow for the extension of a waste pond to a total height of 30 m or so. The design was such that it permitted retention of iron tailings to the north in a first phase and gypsum tailings to the south later (after recovery of the iron tailings in the future). Especially important was the presence of a loose, saturated layer of iron tailings which in the design stage was envisaged as forming the foundation for some 240 m of the dam length. Iron and silica tailings waste had been stored in an area to the north of the dam site for over half a century. South of the site recent gypsum pond construction is expected to result in gypsum tailings contacting the southern edge of the dam. Robinson[21] has referred to the fact that over most of the length, gravel or glacial till make up the foundation. Several embankment failures have occurred in the old iron tailings disposal area where embankments are about 0·8 km north of the new dam. The worst was a shear failure in the retention dyke which in 1948 caused liquefaction of the contained tailings, about a million tons of these flowing down the valley towards the new dam site (a part of this is situated on up to 6 m of loose, saturated tailings). It was found that gypsum tailings were accumulating to the south much faster than iron ones. The planned embankment had to satisfy several requirements including adequate storage for iron tailings, containment of gypsum tailings and allowance for the future removal of the former. Stability under static and dynamic loading conditions also had to be attained. The site area is flat and grassy with occasional peaty or boggy ground and two groundwater zones in the foundation materials. Immediately above or at the surface of the bedrock, there is an artesian aquifer with a piezometric level exceeding 3 m above ground level. Groundwater within the iron tailings is within 0·3–0·6 m of the surface. Glacial till provides a rather impervious barrier between these two zones. There exists a very high water table and very loose tailings from which a 'quick' condition can easily develop if anyone so much as walks over these. During probe tests the tailings liquefied underfoot where a man stood too long in one spot. Previous liquefaction failures occurred as a consequence of dyke displacements when the phreatic level rose to critical height in late winter and the safety factor fell below unity. Resultant high pore pressures induced overstressing of tailings and the resultant displacements were sufficient to liquefy saturated sandy silt. Vibration-induced liquefaction of impounded tailings was considered and various concentrations of saturated materials at low relative density modelled at different positions relative to the dam face. Various positions of the phreatic surface within the tailings with regard to toe drains and the embankment were simulated in a wedge analysis. Foundation liquefaction was considered as well. The dam design had to be such that long-term stability conditions could be satisfied and indeed stability was obtained during construction. As a result of pore pressures build up during initial filling over the tailings foundations, it was thought advisable to monitor activities using piezometers and settlement plates. Normally the former were installed to depths equal to 75% of the thickness of tailings and daily measurements were made. As quick response was anticipated from the tailings, simple standpipe piezometers were used and comprised 1·3 cm PVC-coated conduits with slots cut in the lower 0·3 m, the slotted part being protected by a 200-mesh brass screen surrounding the pipe. Additionally the interior was sand-filled to a depth of almost $\frac{1}{2}$ m. Settlement plates were located over the piezometer pipes. There occurred a response in the piezometers as the initial lift of fill neared a few metres of each piezometer location and pre-pressure build up sufficed to cause overflow of the PVC tubes. Initial settlements were rather high. Groundwater may have to be drained away from soil entirely in some operations, e.g. the reclamation of marshes or slope stabilization. The problem is complicated by the fact that replenishment may take place through infiltration or by subsurface flow into the zone of saturation.

Where boggy ground with seepage is involved, a drainage system must drain excess water off and also lower the groundwater table so as to prevent further water incursion. This is done by well points and the approach is applicable to dewatering of foundations also. In large and open excavations, lines of well points can be utilized, wells being set in one or more lines at the edge of these with spacings ranging from 1 to 6 m according to the circumstances. The dewatering process is very slow in clays and here the drainage can be accelerated by electro-osmosis. This involves placement of two electrodes in a saturated soil with a colloidal content in order to increase the flow of water molecules towards the cathode. This is feasible because the electro-osmotic permeability is higher than K and using a perforated tube, it is possible to pump water directly away from this, the cathode. It may be added that although clays can be drained in this manner, it is not possible to use the technique with sands, no matter how fine-grained they are.

The subject of the disposal of toxic wastes is now discussed as this matter today constitutes a crisis. In the USA toxic chemical wastes occupy about 50 000 dumps nationwide and there are also some 180 000 open pits, ponds and lagoons at industrial parks. Environmental Protection Agency personnel point out that at least 14 000 of these are highly dangerous. They are fire hazards, sometimes emit noxious fumes and frequently threaten groundwater. In total, industry is believed to produce about 40×10^9 kg (88×10^9 lb) toxic wastes annually and at least 90% of this material is improperly disposed of.

Cleaning up such chemical dumps is a very expensive business and in the USA costs could reach US$260 billion. However, the urgency of the job must be stressed because, as Dr Samuel Epstein (personal communication) has indicated, the pollution is a public danger second only to nuclear war. Ths most recent and dramatic evidence for this is the case of Times Beach, Missouri, which, due to dioxin pollution, has had to be evacuated and bought by the US Federal Government at a cost of US$36·7 million as was announced by EPA Administrator Anne Burford in February 1983.

One of the major consequences of improper toxic wastes disposal is the threat to groundwater with aquifers and wells increasingly at risk from poisonous materials leaching out of dumps and other sources. Unfortunately there are no Federal US standards for organic-chemical contamination in groundwater and none for verification of the safety of private wells. Referring to Long Island again, this is a particularly vulnerable area and in fact 28 of 428 *public* wells in Nassau County have been closed already due to chemical contamination. Elsewhere similar action has become necessary, e.g. in Atlantic City, New Jersey, 4 of the 12 wells supplying water to the city have closed down because of contaminants seeping towards them from Price's Landfill, a notorious dump, at the rate of almost 0·3 m day^{-1}. Links between such contaminants and human health are not clear, but some of the substances involved are definitely dangerous; for instance dioxin causes headaches, with severe acne and also weight and hair loss.

Most symptoms of toxic poisoning resemble those of common ailments and it must be remembered that diseases such as cancer take an entire generation to develop. Hence illegal dumping and flagrant polluting needs to be stopped; many US states are increasing penalties for toxic wastes crimes with prosecutors seeking jail sentences wherever possible. It might be argued that underground disposal is acceptable if the local hydrogeology is appropriate and engineering and legal controls strict. The geological requirement is that there is a negligible rate of groundwater movement near to the storage region. Disposal wells would have to involve permeable beds overlaid by impermeable capping rocks such as shale. This would be far preferable to dumping tainted soil because many people believe that such landfill can never be made safe. Another alternative is to incinerate toxic materials at high temperature and it is also possible to detoxify some toxic substances using chemical neutralizers or physical pressure or even mutant bacteria (superbugs). In the Federal Republic of Germany, detoxification is a widely utilized technique and makes safe approximately 85% of all hazardous wastes. Perhaps the optimum method of disposal however is resource trading, i.e. supplying a company's unwanted toxic byproduct as raw material to another company. As regards subterranean storage, oil traps can be employed to store natural gas and artificial caverns have been used to store oil and liquefied frozen gas, environmental groundwater being frozen by refrigeration to render cavern walls impermeable. In the Federal Republic of Germany salt domes furnish good storage sites because salt is easily excavated and also impermeable. In Scandinavia unlined caverns in crystalline rock are utilized for oil storage. Fracturation may occur, but the immiscibility of oil with water obviates leakage by outward flow against the hydraulic gradient of the groundwater. Faulting in underground storage locations is really unacceptable because of the high possibility of leakage of fluid along fault zones. Also changes in pore pressures along these may induce local movement. In fact the disposal of waste down a 3 600 m deep well in Colorado stimulated several small earth tremors.[22]

Radioactive waste disposal is ever more important today and the solutions applied to date involve three approaches, namely dilution and dispersal, delaying and decaying and concentrating and containing. This last idea is probably the safest method and there are various ways of effecting it. One is the use of underground chambers, e.g. such a chamber at 300 m depth is used in Kansas and the relevant geology is bedded, dry, impervious salt which is able to flow under pressure and hence close any fissures which may arise. Another way is to place liquid waste in a cement mixture and then inject it under high pressure into deep shale layers. The high pressure makes for local fracturing and facilitates injection into shale.

Presently the UK is participating in an European programme aimed at assessing optimum means for the disposal of small quantities of highly radioactive waste from nuclear power plants and in fact as long ago as 1976 a study of relevant geological criteria for introducing solid radioactive wastes into rocks was made by Gray and others.[23] Impervious crystalline rocks were investigated and in Europe it is proposed to use salt and clay for the purpose. Clearly other factors enter into the matter such as the density of population in an affected area, the existing usage of land and the actual accessibility. However, the actual geological criteria may be listed:

(a) type of rock;
(b) physico-chemical properties of the rock;
(c) seismic stability of the area;
(d) permeability of the rock;
(e) proximity of other permeable materials;
(f) susceptibility of the rock to weathering and erosion;
(g) susceptibility of the rock to climatic changes, e.g. glaciation;
(h) effects of storage on mining and other human activities in the region such as mineral exploration or water impoundment in reservoirs.

The above can be satisfied to a greater or lesser extent in many locations. Crystalline rocks in the Highlands of Scotland are appropriate as also are rocks in North Wales and parts of England such as Cheshire. Of course extensive trial drilling and testing would be necessary in any selected area in order to determine the structural characteristics at depth and also to obtain samples for laboratory studies on such parameters as porosity, heat capacity, strength and resilience. Such an exhaustive programme could entail work over a number of years prior to implementation of a site for waste disposal. Even after their successful conclusion, it would appear to be advisable to make a trial of the site in question during a first phase of exploitation. The waste would have to be sealed in special containers and after emplacement, the disposal storage facility be completely and permanently closed off from the environment.

REFERENCES

1. Bowen, R. 1982. *Surface Water,* Applied Science Publishers Ltd, London; Wiley–Interscience, New York.
2. Chow, V. T., 1964. Hydrology and its development. In: *Handbook of Applied Hydrology,* McGraw-Hill Book Company, New York.
3. Meinzer, O. E., 1923. The occurrence of groundwater in the United States with a discussion of principles. *US Geological Survey, Water Supply Paper,* **189**.
4. Rushton, K. R. and Tomlinson, L. M., 1975. Numerical analysis of confined–unconfined aquifers. *J. Hydrol.,* **25,** 259–74.
5. Theis, C. V., 1935. The relation between the lowering of the piezometric surface and the rate and duration of discharge of a well using ground water storage. *Trans. Am. Geophys. Union,* **16,** 519–24.
6. Rushton, K. R. and Booth, S. J., 1975. Pumping test analysis using a discrete time-discrete space numerical method. *J. Hydrol.,* **28,** 13–27.
7. Tomlinson, L. M. and Rushton, K. R., 1975. The alternating direction explicit (ADE) method for analyzing groundwater flow. *J. Hydrol.,* **27,** 267–74.
8. Ineson, J., 1963. Ground water: geology and engineering aspects. *Symp. Joint Geol. Soc. and Instn of Water Engrs Meeting, 27 February 1963.* (*J. Inst. Water Engineers,* **17**(3) 283–4.)
9. Jacob, C. E., 1950. Flow of ground water. In: *Engineering Hydraulics,* ed. H. Rouse, John Wiley and Son, New York.
10. Conkling, H., 1946. Utilization of groundwater storage in stream systems development. *Trans. Am. Soc. Civ. Engrs,* **111,** 275–354.
11. Harding, S. T., 1927. Groundwater resources of southern San Joaquin valley, California. *Calif. Divn Eng. and Irrig., Sacramento, Bull. 11.*
12. Meinzer, O. E., 1932. Outline of methods for estimating ground water supplies. *US Geological Survey, Water Supply Paper,* **638-C,** pp. 99–144.
13. Dansgaard, W. F., 1964. Stable isotopes in precipitation. *Tellus,* **16,** 436.

14. Craig, H., 1961. Standard for reporting concentrations of deuterium and oxygen-18 in natural waters. *Science*, **133,** 1833.
15. Fritz, P., Suzuki, O., Silva, C. and Salati, E., 1981. Isotope hydrology of groundwaters in the Pampa del Tamarugal, Chile. *J. Hydrol.*, **53,** 161–84.
16. Clayton, R. N., Friedman, I., Graf, D. L., Mayeda, T. K., Meents, W. F. and Shimp, N. F., 1966. The origin of saline formation waters: 1: Isotopic composition. *J. Geophys. Res.*, **71**(16), 3869–82.
17. Lizzi, F., 1982. *The Static Restoration of Monuments*, Sagep Publishers, Genoa.
18. Anon., 1923. Modern engineering to save medieval tower. *Engineering News-Record*, **91,** 505.
19. Legget, R. F., 1973. *Cities and Geology*, McGraw-Hill Book Company, New York.
20. Anon., 1965. Foundation reinstatement without interference to services. *The Engineer*, **220,** 791.
21. Robinson, K. E., 1977. Tailings dam construction on very loose saturated sandy silt. *Can. Geotech. J.*, **14,** 300–407.
22. McLean, A. C. and Gribble, C. D., 1979. *Geology for Civil Engineers*, George Allen and Unwin, London.
23. Gray, D. A., *et al.*, 1976. Disposal of highly active solid radioactive wastes into geological formations—relevant geological criteria for the United Kingdom. *Rep. Inst. Geol. Sci., 76/12.*
24. Bowen, R., 1980. *Ground Water*, Applied Science Publishers Ltd, London.

CHAPTER 5

Permafrost and Ice

5.1. PERMAFROST

This is permanently frozen ground, i.e. ground frozen for at least several years and it can be defined on the basis of temperature which clearly is below 0°C (consequently the presence of ice is not mandatory and where this is the case dry permafrost is present).

Ice is normally associated with permafrost. The upper limit of such perennially frozen ground is termed the permafrost table and it represents an almost impermeable surface. Continuous permafrost constitutes material with an annual temperature at 10 m depth of −5°C. The permafrost table is usually $\frac{1}{2}$ m or so below the surface, but in granular materials it may be deeper. Discontinuous permafrost is much thinner and may be interrupted by thawed areas; it possesses a lower permafrost table and the mean annual temperature at 10–15 m depth is between −5 and −1·5°C.

Above the permafrost table there is an active layer, a zone subjected to alternate freezing and thawing and one in which freezing may occur from the bottom upwards as well as from the top downwards, which is often a seasonal matter; this active layer is the zone with the maximum fluctuation of temperature and may be termed the frost zone, its irregular lower surface constituting the frost table. Occasionally the active layer and the frost zone do not coincide so that a residual thawed ground region lies between the frost table and the permafrost table. The term talik applies to such ground and indeed to any thawed area in and under the permafrost. It may act as a viscous liquid and flow. Finally it may be mentioned that the lower limit of the active layer varies from one year to another.

In 1969, Corte[1] indicated that outside permafrost areas freezing occurs from the top downwards and thaw takes place both from the top and bottom of frozen ground with freezing and thawing being seasonal or sporadic. The commonest phrase used in respect of permafrost phenomena is a periglacial environment which covers that environment in which frost processes predominate and permafrost is found.

The importance of this subject should be noted—about 20% of the entire land surface of the Earth is underlain by various types of frozen ground. As a result of the fact that there is much more land in the high latitudes of the northern hemisphere than in the southern hemisphere, the permafrost areas of the north are much larger than those in the south—22·4 million km^2 compared with 13·1 million km^2. In terms of countries, more than 80% of Alaska, 50% of Canada and 47% of the USSR are underlain by permafrost. Usually permafrost extends further south on the eastern and more continental land areas. Outside the limits of permafrost, most land north of 30°N is influenced by seasonal or sporadic freezing and thawing. The vertical profile of permafrost is also variable and the maximum recorded thickness is in Siberia at 1500 m. Throughout Europe and Asia it averages much less than this however, the range being 300–460 m and in North America, according to Black,[2] the range is 245–365 m. It has been found that the layer of permafrost actually thins towards the north as well as southwards. Discontinuous permafrost usually does not exceed 60 m in thickness and the active layer is of variable thickness, from a few centimetres to 4 m (the former over continuous, the latter over discontinuous permafrost). Clearly these matters are very important in engineering work and the various techniques of geophysics are applied to the determination of the thicknesses of the various zones. Drilling can be employed, but electrical resistivity and seismic techniques have also proved useful.[3] In order to investigate the thermal profile occurring in a frozen region, glass thermometers, thermocouple resistance thermometers and thermistors may be utilized as for instance Hansen[4] has indicated. It has been found that temperatures show their widest fluctuation at the surface and in the active zone (mainly responding to daily and seasonal changes in atmospheric temperature). Fluctuation decreases with depth until a level of zero annual amplitude is reached and below this the temperature rises with depth. Of course wherever temperatures exceed 0°C, there can be no permafrost.

Now it is apparent that the planetary distribution of frozen ground must reflect some sort of thermal equilibrium which in turn must spring from the relations between the climatic regimen, heat flow from the interior of the planet, the properties of soils and rocks and the presence of water. Probably the climatic conditions are the most important of all of these factors and according to Muller[5] the most favourable for the development of permafrost are cold and long winters with little snow followed by short, dry and rather cool summers with low precipitation throughout the year.

Attempts have been made to predict the depth of freezing and thawing using climatic indices one of which, the freezing index, is founded upon the cumulative total of degree-days, and is the number of these between the maximum and minimum points on a cumulative curve constructed from appropriate data plotted against time in a freezing season. There is a satisfactory correlation between the freezing index and the penetration of frost into the ground when dry unit weight and moisture contents of the soils are taken into account.

In a climatic regimen, the distribution of permafrost is influenced by variables such as vegetation, surface water, terrain and surface materials. Most permafrost occurs either beneath northern boreal forest or beneath tundra vegetation and in fact the boundary between continuous and discontinuous permafrost in some regions coincides roughly with the southern limit of the tundra. As Brown has indicated,[6] the significance of vegetation is that its thermal properties determine the movement of heat into and out of the ground. Clearly the depth of thaw increases if vegetational cover is removed and in the zone of sporadic permafrost ice may occur preferentially beneath a blanket of insulating peat. Surface water bodies also affect the occurrence of ground ice so that, for example, under lakes and rivers which do not freeze completely in winter, permafrost is either thinner or absent.

The important properties of surface materials in terms of frozen ground have been given by Corte.[1] They are as follows:

(a) thermal conductivity;
(b) specific heat;
(c) volumetric heat capacity;
(d) thermal diffusivity.

These relate to fundamental characteristics such as the packing of particles and the mineral composition. For instance, thermal conductivity is greater by a factor of four for ice as opposed to water; hence frozen ground has a much greater thermal conductivity than non-frozen ground. Thermal conductivity varies with the type of material so that, for example, sand has a thermal conductivity of about half that of clay.

It must be emphasized that although the thermal equilibrium now in existence seems to relate to prevailing conditions, there can be no doubt that it has actually arisen over thousands of years and its general pattern is probably under modification by secular changes in climate and other controls. From the engineering standpoint, the relevant matter is that it can be altered quickly so that permafrost may be degraded or aggraded by manmade works, i.e. human intervention. These can be summarized as:

(a) removal of vegetational cover (e.g. land clearance for agriculture);
(b) modification of drainage conditions (e.g. by river diversion);
(c) construction (of buildings or pipelines or roads).

All of these activities result in permafrost degradation, involve usually irreversible changes and may initiate a new thermal equilibrium in the ground. This latter state can be established in a few years, but where extensive land usage modifications are involved it may be decades or even centuries before this takes place.

The most destructive processes in permafrost regions relate to the formation and thawing of ground ice, either in the permafrost zone or in the overlying active zone. The two main groups of processes are frost action and solifluction. Ground ice is important in both and the characteristics of the main ice forms are as follows:

(a) open cavity ice—forms by sublimation of ice crystals directly from atmospheric water vapour;
(b) single vein ice—forms when water enters and freezes in an open crack which penetrates permafrost;
(c) ice wedges—form through the thermally induced cracking of frozen ground;
(d) tension crack ice—grows in cracks resulting from mechanical rupture of the ground;
(e) closed cavity ice—forms by vapour diffusion into enclosed cavities in permafrost;
(f) segregated ice—grows in lenses in material older than these (epigenetic ice) or develops when the permafrost table gradually rises (aggradational ice);

(g) intrusive ice—forms by the intrusion of water under pressure and, on freezing, it causes ground heave;
(h) pingo ice—found when intrusive ice domes the overlying surface;
(i) pore ice—holds soil grains together.

The processes and control of frost action and solifluction entail investigation of frost heave associated with segregated ice in supersaturated materials and here there are two aspects to be considered. One is vertical displacement of the ground surface arising from the growth of ground ice. Consequent surface disruption may be as much as many centimetres and have serious consequences for engineering structures. Distinctive landforms are the result. The second aspect is the vertical sorting of particles in which large stones in mixed sediments migrate upwards and may be forced out of the ground to form stone circles or nets. Ground hummocks (or palsas or mima mounds) may result from ground heaving and can be as much as 10 m high. Vertical upwards migration of large objects in soils has been investigated and two types of mechanism recognized:

(a) frost-pull which depends upon the ground expanding during freezing and carrying the large objects upwards—thawing causes fines to move under them and hold them in place;
(b) frost-push which depends upon the fact that the thermal conductivity of stones is greater than that of soils; consequently, stones heat and cool much more rapidly than soil, and ice forms first beneath them, forcing them upwards. During thaw, fines move beneath the stones and prevent them from regaining their original positions.

Controlling frost heave requires a knowledge of climatic, material and water conditions and also the nature of freezing and thawing. There are three approaches to the reduction of frost heaving.[1] They are as follows:

(i) chemical treatment of the soil, e.g. injection of calcium chloride reduces the freezing temperatures of water in soil and minimizes the loss of strength of materials caused by freeze–thaw cycles; also, soil voids may be infilled with Portland cement or other grout;
(ii) grouting, as stated above, or by using polymers or resins;
(iii) structural design precautionary measures such as increasing the thickness of pavement and sub-grade materials in order to

prevent frost penetration, insulating the surface water with such natural materials as straw or peat or artificial materials such as cellular glass blocks, etc.

Permafrost areas are often marked by polygonal ground patterns which result from cracks indicative of the location of ice wedges. The importance of such features is that they furnish evidence of ground surface conditions. If the ice melts, they act as the centres of subsidence (see ref. 7). Of course this may seriously disrupt any superincumbent engineering structures which may be present. Cracks may be up to 10 m deep and the diameters of such polygons can exceed 100 m. Centres of polygons may be higher or lower than their peripheries.

Once thaw has set in, the properties of the ground alter from those associated with its freezing. Thus the mechanical properties of frozen ground are rather similar to those of ice and hence have higher shear strength and thermal conductivity than similar but non-frozen material. When sediments in the active zone thaw, therefore, they show a behaviour different from that of the frozen ground and may subside and behave as a viscous fluid. Some materials are particularly susceptible to settlement, e.g. gravel embedded in large masses of ice is inherently unstable when thawed and silty frozen soil with a high proportion of ice usually becomes waterlogged when thawed. Where the ground surface is inclined, flowage of thawed material may occur, a process termed solifluction.

Those features which result from the degradation of ground ice in permafrost areas are termed thermokarst landforms. Among them are included types of pit, dry valleys and lakes. All of these are subsidence features and result from the thawing of supersaturated and icy soils at the top of permafrost.[7] They become most pronounced when melting affects abundant ground ice in the upper permafrost. The cause of thawing is an important consideration and may be due to climatic change, as for example in the southern permafrost regions of the USSR where thermokarst is most extensive, as indicated by Kachurin.[8] Seasonal thaw may also lead to thermokarst development and local thawing may be induced by the activities of Man such as peat compaction, bulldozing, ploughing, ditch-digging, etc.

Permafrost degradation may proceed in a lateral sense, usually because of the horizontal erosional effects of water and indeed the extension of some lakes (thaw lakes) is believed to be due to this. However, vertical degradation is commoner and is illustrated in Siberia by the development of alases (large and flat-floored, vegetated basins) which result from melting of ice wedges which become the loci for lakes. As the lakes deepen and merge, they attain a depth too great for complete freezing in winter and the rate of permafrost degradation beneath them is increased. Ultimately the large lakes may be filled or drained, vegetation may cover their floors and the creation of alases is completed (see for instance ref. 9).

The significance of solifluction to engineers is its ability to disrupt structures at the surface and it may be regarded as comprising two major processes:

(i) the flow of waterlogged debris which involves the thawing of ground in the active zone or the adding of water to a thawed surface through snow melt—as the ground thaws or water is added to it, there is a concomitant increase in the weight of material and its shear strength diminishes through the action of water in reducing internal friction and cohesion; the nature of the resultant flow depends upon the properties of the material involved, availability of water, depth of thaw, ground inclination and vegetative cover; high silt content has been found to facilitate solifluction as also does an abundant supply of water;

(ii) the creep of surface material by freeze–thaw action—this is initiated by the heaving of particles on a slope upwards (normal to the surface of the ground) and lowering of particles vertically on thawing under gravity.

Solifluction rates have been measured by recording the translocation and vertical deformation of linear objects such as pipes and cables which are inserted through the mobile layer. As might be expected, the majority of movements take place during spring thaws and they have been shown to range from under 1 cm to more than 30 cm annually. Solifluction produces a number of characteristic landforms such as terraces which may be delimited by stone banks covered by vegetation and occur on slopes exceeding 5°.

5.2. THE RESOLUTION OF PROBLEMS RELATED TO PERMAFROST

Usually, areas of permafrost are inaccessible and remote sensing imagery may be useful in reconnaissance surveying which is followed by more detailed site investigations. This work should result in accumulation of data apropos the nature of the terrain, the relevant materials, the ground ice conditions and the sources of materials of construction. Thereafter, adequate planning may be effected. Appropriate engineering reaction to various permafrost problems may be outlined: it consists of neglecting, eliminating or preserving frozen ground or alternatively structures may be so designed as to take into account expected movements.

Permafrost may be ignored where good drainage conditions exist and ground materials are not susceptible to severe frost action and solifluction. Where the permafrost is thin, sporadic or discontinuous, and in the case where thawed ground possesses adequate bearing capacity, there is a possibility of removing frozen ground by, for example, stripping away surficial vegetation or excavation or treating the ground or thawing the ice using steam. Material which is not susceptible to frost action may be placed on the surface. This technique is referred to as the active method as compared with another approach, the passive method in which the permafrost may be preserved, e.g. by insulating the surface using mats of vegetation or gravel blankets.

In locating buildings, certain optimum conditions must be looked for, including a thin active layer, bedrock near to the surface, thawed material with satisfactory bearing properties, good drainage, absence of frost-susceptible materials and a stable site removed from potential water seepage areas and icing. In fact most building construction problems arise from the active zone with the foundations fixed in the underlying permafrost zone.

A very important factor in building construction is the adfreezing strength of frozen ground, i.e. the resistance to the force necessary to pull the frozen ground apart from objects such as foundations, to which it is frozen. Generally this strength is higher in sands than in clays. Pilings are commonly utilized in building foundations and the adfreezing strength is taken into account in emplacing them. The aim is to freeze the pilings into the ground to a depth of about twice the thickness of the active zone and to avoid adfreezing in the active zone by lubricating, insulating or collaring the piles in that zone. Thus the downward acting force in the permafrost zone plus the superincumbent load are enough to overcome the upward heave force acting on the protected piling in the active zone and this prevents pile movement.

Other techniques in building construction include making a trench around the building so as to eliminate possible horizontal stresses, elevating the floor by a metre or so in order to allow air circulation, insulating the floor to restrict temperature rise under it and providing skirting insulation between the ground and the floor with air vents which may be closed in summer and opened in winter so as to permit free air movement. Sometimes jacks may be employed so as to adjust the building level if there is any movement. Pads of concrete, wood or gravel may be sited on natural vegetation so as to provide an insulating foundation for small, temporary buildings. Similar problems may occur in constructing roads or runways and in such cases the passive technique requires the creation of adequate natural or artificial insulating layers whereas the active technique necessitates the replacement of frost-susceptible surface material with surface material which is not frost-susceptible.

Always in construction, the objective is to create a stable in place of an unstable surface. The case of Aklavik is now cited: The expansion of this town, which was founded in 1912 on the Mackenzie River delta by the Hudson's Bay Company, proved to be impracticable and so a new site was selected at a distance of 48 km but this was again in a permafrost area. Initial planning requirements were satisfied because the new site was on a terrain near the Mackenzie River; hence it was near navigable water which was not susceptible to flooding. The permafrost was up to 90 m thick and it was necessary to effect the construction of the town with minimum disturbance to ground conditions. It was decided to leave the natural moss cover intact so as to preserve its insulation value and piles were used to found all permanent structures securely in the permafrost. In addition road cuts and ditches were not permitted in order to obviate permafrost degradation and all culverts were installed in gravel fill so as to accommodate the surface runoff. All traffic routes were based on gravel pads surmounting natural vegetation. The work entailed the use of over 20 000 piles, mostly made of local spruce although some concrete and steel piles were brought into the site. Emplacement was achieved by steam thawing and usually effected to depths of 4·5 m. Extra care was taken to insulate the ground from the effects of new structures; for instance all the main buildings had air space beneath them of at least 1 m. The result was a great success and in fact the relocation of Aklavik at Inuvik, the new site, was justified when in June 1961 an ice jam on the Mackenzie River caused serious flooding of the old town.

5.3. GLACIATION

Glaciers comprise ice bodies consisting principally of recrystallized snow and they flow on land surfaces. They play a critical role in the water cycle, erode land, transport rock material so produced and deposit it elsewhere; there are several types.

The most common type is the cirque glacier. These are very small glaciers in cirques, steep-walled niches (bowl-shaped) in mountain sides excavated by frost wedging and glacial packing. In the continental USA there are at least 950 such glaciers and there are over 1500 in the Alps.

A valley glacier is one flowing downwards through a valley and it may vary in size from a few acres to a tongue-shaped form tens of kilometres long. In the conterminous United States there are at least 50 of this sort and even more in the Alps.

A piedmont glacier is one on a lowland at the base of a mountain and fed by one or more valley glaciers.

Finally, an ice sheet is a very large glacier of irregular shape which usually blankets a large land area. If of very small size, an ice sheet constitutes an ice cap. The most famous ice sheet is that of Greenland which has an area of $1{\cdot}726 \times 10^6$ km^2 with a maximum elevation of over 3 km above mean sea level. The Antarctic ice sheet is much larger, however, and its area is approximately 1·4 times that of the USA. Its highest altitude is over 4 km and its thickness exceeds 4 km in places, the total volume exceeding 24 million km^3. 89% of the world's ice and two-thirds of all the fresh water on the planet are contained in the Antarctic.

The basic requirements for the existence of glaciers are snowfall and low temperature, conditions which are fulfilled at high latitudes and high altitudes and which occur more often in wet coastal areas than in dry continental interiors. Glaciers interrelate through the snowline, the lower limit of perennial snow. Above this are snowfields and the snowline ascends from near mean sea level in the polar areas to as much as 6000 m in tropical mountains. During snowfall, in temperatures below 0°C some atmospheric water vapour alters to the solid state and forms hexagonal ice crystals (snowflakes). When newly fallen, snow is light and porous and air penetrates it. Ice evaporates inside it and water vapour condenses. Hence snowflakes gradually become smaller as voids disappear. When the specific gravity of granular snow attains a value of 0·8, the snow has been compressed to a point of compaction at which it is impermeable to air and in fact is converted into ice. In a sense, ice can be regarded as a metamorphic material, comprising interlocking crystalline grains, which is transformed by pressure exerted by overlying ice. The specific gravity of glacier ice is approximately 0·9; hence it can float and form icebergs. Glaciers can develop tensional cracks known as crevasses and these rarely exceed 50 m depth. The surface velocity of a valley glacier is measured by surveying from the sides of the valley a line of markers extending from side to side. This is repeated at regular intervals and the results show that the central part of a glacier surface moves more rapidly than its sides (this is also true for a river). The velocities recorded range from a few millimetres to several metres per day, i.e. approximately the same rate as water percolation in the unsaturated zone. If the climate changes, related changes occur in glaciers. For instance, a change to cooler conditions will cause glaciers to expand if more moisture is available whereas under warmer dryer conditions, shrinkage of glaciers will take place. Although the end of the glacier may melt back faster than ice can be supplied to it by flow and the glacier retreats, ice flow continues in the same direction. Occasionally there are violent phases of change termed surges during which very high velocities may be attained—in one case over 6 km annually. The reason for such surges is not yet known.

Glaciation, the alteration of land surfaces by massive movement of glacier ice over them, includes erosion, transport and deposition. A glacier may be regarded as exerting the actions of a plough, a file and a sled. In the first case it gouges out rock and 'plucks' or scrapes up regolith (loose soil); in the second it erodes solid rock. In the third case it removes materials. Some associated phenomena are now considered and include the following:

(a) frost wedging;
(b) cirques;
(c) glaciated valleys with U-shaped cross-sections (as opposed to the V-shaped ones found in young river valleys) and floors below those of tributaries which therefore 'hang' above main valleys;
(d) mountain features such as aretes (knife-edge ridges created when two groups of cirque glaciers eat into them from both sides) and horns (bare pyramidal peaks remaining where glacial action in cirques has occurred from three or more sides, e.g. on the Matterhorn).

(e) fjords which are glaciated troughs partly submerged by the sea; examination of the fjords in Norway, Alaska, Chile and elsewhere demonstrates that depths of glacial erosion of 300 m or more may be involved.

As regards transport by glaciers, this differs from stream transport in that the load can be carried in their sides and tops and much larger rocks may be moved. Deposits laid down by glaciers are unstratified and unsorted and such boulder clays are most unsatisfactory from an engineering standpoint. As well as large fragments, rock flour may result from internal crushing and grinding in a glacier. The main glacial deposits are as follows:

(a) Glacial drift or till—These are sediments deposited directly by glaciers or indirectly in glacial streams, lakes and the sea. There are two varieties, namely till and stratified drift, which may grade into each other. As noted above, deposits laid down by glaciers are unstratified; hence stratified drift is not deposited in this way, but is laid down by glacial meltwater.
(b) Moraine—This is widespread thin drift with a smooth surface and it comprises gently sloping knolls and shallow closed depressions as ground moraine. Terminal moraine is a ridgelike accumulation of drift deposited by a glacier along its margin and it may range in height from a few metres to hundreds of metres. In a valley glacier this type of moraine is built not only at the terminus but also along the sides of the glacier and for some distance upstream. The side parts constitute lateral moraines.
(c) Erratics—These are glacially deposited rock particles with compositions quite different from that of the underlying bedrock. A group of erratics spread out in a fan comprises a boulder train.
(d) Outwash sediment—This is stratified drift deposited on the downstream sides of most terminal moraines by streams of meltwater flowing away from the glacier. A body of outwash forming a broad plain beyond the moraine is termed an outwash plain. In contrast a valley train is a body of outwash which partly fills a valley (see Fig. 5.1).
(e) Ice-contact stratified drift—When rapid melting and evaporation reduce the thickness of the terminal part of a glacier to 100 m or less, movement practically ceases. Meltwater then deposits stratified drift which constitutes ice-contact stratified drift where

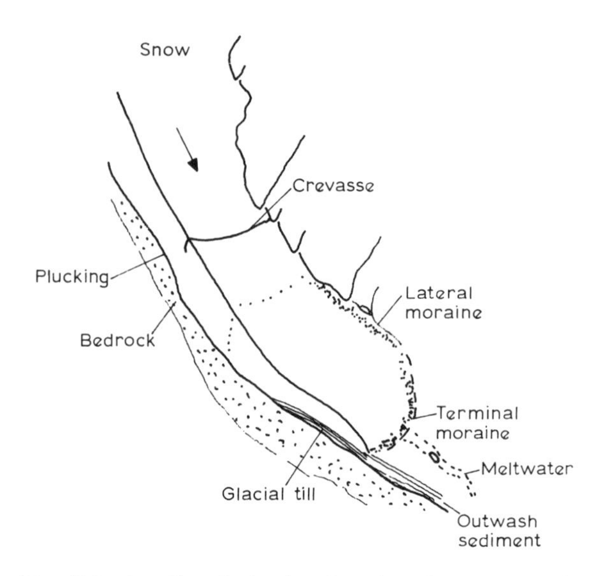

FIG. 5.1. A valley glacier (sectioned to show only half).

it contacts supporting ice. Bodies of such material may be classified according to their shapes.

Thus short, steep-sided knolls and hummocks constitute kames while terrace-like forms along valley sides comprise kame terraces. Long and narrow ridges usually sinuous in shape form eskers while basins in drift are termed kettles. By contrast drumlins are streamline hills constituted of drift and are elongated parallel to the direction of glacier movement. As regards this latter, the under surface of ice which is impregnated with rock fragments of various sizes abrades bedrock over which it moves and makes long scratches (glacial striae) and grooves. These are important indicators of the direction of movement of the relevant glacier.

It has been established that many glacial ages have occurred during the Earth's long history, the most recent being the multistaged Pleistocene. When the continental ice sheets (which were about 13 times larger on non-Antarctic lands than is now the case) receded, temporary lakes were formed and these had material deposited into them. This was derived from seasonal meltwaters and hence is layered. Each layer constitutes a varve and counting varves together with subsequent correlation has enabled a Pleistocene chronology to be developed for the Northern Hemisphere. A major effect of the Pleistocene ice stages was the spreading of permafrost which now extends over about one-fifth of the land area of the Northern Hemisphere. In places it has reached thicknesses of as much as 1·5 km.

The cause of these glaciations is not fully understood, but certain data are very important to attaining an explanation. These are as follows:

(a) climatic fluctuation seems to have been worldwide and simultaneous;
(b) glaciations commence on land and usually in the highest locations;
(c) glaciers diffuse until stopped by warmer conditions;
(d) as regards the most recent ice episode, the Pleistocene, glaciers reappeared in the same places as their predecessors occupied.

The following suggestions have been made as to the reasons for initiation of colder climatic conditions at particular localities on a continent:

(i) slow movement of a crustal plate from a lower to a higher latitude may carry a continent into a colder climatic regimen;
(ii) plate-related uplifts of parts of continents may create mountain ranges and elevate them sufficiently to intersect the snowline thus permitting glaciers to form;
(iii) solar radiant energy may vary chronologically.

It must be pointed out that, although the first two could explain glaciation of particular continents, the mechanisms involved are too slow and gradual to account for the disappearance and reappearance of glaciers and short-term glacial fluctuation on two or more continents simultaneously. The third item could explain the fluctuation, but of course there is no evidence suggesting it.

The Pleistocene ice age ended about 10 000 years ago, having spanned 2 000 000 years, and in the UK and USA vast amounts of glacial drift were deposited and landscapes were modified by erosion so that such features as roche moutonnees and crag and tail resulted. The former resulted from a combined scraping and plucking action of the ice while the latter is an excellent example of the occurrence of erosion and deposition in close proximity. A small hill formed by an igneous intrusion may be ice-eroded until a steep, craggy face (the crag) is left to face the ice assault. Behind this and protected by it is a streamlined soft rock or till tail. There were also numerous fluctuations in sea level during the Pleistocene and these have left coastal deposits lying far above the present high water mark. Of course the planetary temperature dropped during the glacial periods and also periglacial phenomena arose beyond the ice front and produced effects comparable to those recognized in contemporary tundra. During the interglacial periods, the climate ameliorated and in the UK for example was often warmer than it is today. The optimum climate with maximum worldwide temperatures was around 5000 BC. While there have been short-lived temperature fluctuations since the last glacial period, there appears to have been no sustained increase in planetary temperature since then.

5.4. ENGINEERING PROBLEMS

Observations show that glacial deposits comprise a number of sand and gravel beds which are rather pervious and there may be contents of impervious clay or clayey silt present as well. As well as comprising large deposits, these sands and gravels may be covered and lenslike in form, occurring in isolated situations or interconnected by stringers of pervious materials. There may be a mixture of pockets of impervious, soft, compressible clay with pervious, competent sands and gravels. Krynine and Judd[10] have cited a geologist's remark on optimum foundation materials for a structure to be built in a glacial drift area—he compared them with the contents of a dustbin, i.e. a wide range of materials may occur in a glacial deposit. As noted, many of these possess a high permeability and this is a characteristic potentially useful to the engineer. For instance in the case of roads and runways, such foundation materials obviate the necessity to install costly artificial drainage. On the other hand, where water-impounding structures are involved such pervious beds are dangerous and unfortunately often hidden as well. As it may be prohibitively expensive to make a complete exploration of such an area, the best course is to assume that, for example, if construction is to proceed on a moraine, perviousness will be encountered so that appropriate design precautions will have to be taken.

The design of foundations for light superincumbent structures in glacial zones usually does not cause problems. However heavier structures in glacial zones may undergo differential settlement. This can arise because of the variety of materials encountered. While sands and gravels may be competent, clay pockets will not be and if they are undetected, damage can result.

REFERENCES

1. Corte, A. E., 1969. Geocryology and engineering. *Reviews in Engineering Geology*, **2,** 119–85.
2. Black, R. F., 1954. Permafrost—a review. *Bull. Geol. Soc. Am.*, **65,** 839–55.
3. Barnes, D. F., 1966. Geophysical methods of delineating permafrost. *Proc. Permafrost Internat. Conf.*, NAS/NRC Publn. 1287, pp. 349–55.
4. Hansen, B. L., 1966. Instruments for temperature measurements in permafrost. *Proc. Permafrost Internat. Conf.*, NAS/NRC Publn. 1287, pp. 356–8.
5. Muller, S. W., 1947. *Permafrost or Permanently Frozen Ground and Related Engineering Problems*, Edwards, Ann Arbor, Michigan.
6. Brown, R. J. E., 1966. Influence of vegetation on permafrost. *Proc. Permafrost Internat. Conf.*, NAS/NRC Publn. 1287, pp. 20–5.
7. Mackay, J. R., 1970. Disturbances to the tundra and forest environment of the western Arctic. *Can. Geotech J.*, **7,** 420–32.
8. Kachurin, S. P., 1962. Thermokarst within the territory of the USSR. *Biul. Peryglac.*, **11,** 49–55.
9. Cydek, T. and Demek, J., 1970. Thermokarst in Siberia and its influence on the development of lowland relief. *Quaternary Res.*, **1,** 103–20.
10. Krynine, D. P. and Judd, W. R. 1957. *Principles of Engineering Geology and Geotechnics*, McGraw-Hill, New York.

CHAPTER 6

Foundations

6.1. FOUNDATIONS

The foundations of buildings are greatly influenced by geological considerations and so the site engineer and the engineering geologist should work closely together in order that designers of construction projects may be advised properly as regards the mechanical behaviour to be anticipated in the soils and rocks of the substrate which has to be dealt with during the execution of such works. Consequently every relevant geological feature has to be examined on site and the results submitted so that the best design can be produced. By this is meant a design which is both efficient structurally and economic. In clarifying problems faced prior to building, it is useful to list the characteristics desirable for a good foundation.

6.1.1. Ground Stability

The absence of any tendency to slide is the concern here. Slope movements are looked into at length in Chapter 7, but it may be stated here that they can cause much damage and large scale landslides are capable of destroying towns and killing many people as well as disrupting roads and railways. The causes of instability in slopes are well known and include changes in the gradient, surplus loading, vibrations, changes in the moisture contents of soils, groundwater pressure, frost weathering and vegetation. Hence it is feasible to recognize potentially unstable ground in projected foundation areas before any actual work is done in them. For instance, topographical features may indicate instability in the ground as also may irregular waving of contours in maps of a sloped region. It is regrettable but true that building may have to proceed even in such unsuitable areas and then, attempts must be made to stabilize the ground. First, the potential hazard must be evaluated and then a stability analysis effected after which the slope form may be altered and defensive drainage measures undertaken.

6.1.2. Solid Ground

This is ground devoid of subterranean cavities (either natural or manmade). Very large cavities may be concealed by a drift cover, but subsidence will begin as soon as a massive structure is erected. Artificial and often ancient cavities may occur in many old cities and they are costly to stabilize. Abandoned mine works leave numerous unfilled cavities and even plotting these requires a rather complex series of prospecting holes.

6.1.3. Favourable Foundation Material and Groundwater Properties

Foundation materials should have favourable mechanical properties, e.g. high strengths and durability. Groundwater conditions should be such that corrosive chemicals, such as sulphates, are absent.

6.1.4. Absence of Materials Liable to Volume Changes

These can arise with unsuitable soils or rocks through freezing or desiccation. Unsuitable soils or rocks are those containing appreciable quantities of clay particles. In low temperature conditions, water in clayey soils freezes. If such soils are analysed, it will be noted that they have the capacity to absorb water from underground layers and transmit it to such thin ice layers. In consequence the ground surface can heave by tens of centimetres and this phenomenon affects light buildings or the less rigid parts of buildings such as ground floor window parapets. Of course building standards provide for satisfactory methods of prevention of this hazard; they usually require a satisfactory depth of foundation. Desiccation is another source of serious damage to buildings and damage can appear on walls exposed to the sun. The foundation soil has a water content which diminishes beneath the outer boundary of the wall as compared with the inner boundary so that the foundation acquires a tilt away from the building. This can cause vertical cracking or, if the ceiling is anchored, oblique cracking. Artificial desiccation may arise from shrinkage of clayey soils and occurs around kilns in brick factories—differential settlement results.

6.1.5. Vibration-Free Ground

This is very hard to find in urban areas nowadays because of extensive machine operations and heavy traffic. The bases of such machines ought to damp such vibrations and thus obviate their propagation into the surroundings. As most machinery foundations do not vibrate excessively, there is a paucity of investigations that have been carried out, especially because even where vibrations occur they have a very low natural frequency if the machinery in question is massive enough.

Problems of resonance do occur, however, and remedial grouting has been applied to solve them. The natural frequency referred to is that at which the entire system (soil–foundation–machine) vibrates at maximum amplitude when forced vibrations in any of the six modes of vibration are applied. Earlier studies on the natural frequency of machine foundations assumed that a mass was supported on a spring so that the natural frequency was proportional to the square root of the spring constant and inversely proportional to the square root of the mass. Later a damping constant was introduced and this reduced the natural frequency from that calculated for the freely vibrating linear motion of the mass on the spring. Unfortunately the concept is not particularly valuable because damping constants for all possible soils are not known and may not even be constant for any one soil. The soil under the foundation may be assumed to vibrate with this when vibration takes place so that the natural frequency can be considered as inversely proportional to the square root of the weight of the machine which is vibrating, its foundation and a mass of soil, the size of which unfortunately is unclear.

The natural frequencies of various soil types, according to Tschebotarioff,[1] are within the range 500 cycles min^{-1} for soft organic soils through 1100–1700 cycles min^{-1} for clays and sands to 2000 cycles min^{-1} and above for rock. From this it should be noted that the variation in natural frequencies is much less than other characteristics such as bearing capacities. Hence it may be inferred that the natural frequency cannot be directly related to the bearing capacity of a soil. Probably, the natural frequency of a soil–foundation system varies with the soil or rock characteristics (especially bedding), the total weight of equipment which vibrates, with the foundation and the area and also the nature of the vibrations and the effects of adjacent excavations.

Resonance has been mentioned and may be a troublesome factor with vibrating machinery foundations and another deleterious factor is densification—granular soils have been observed to undergo accidental densification due to the effects of vibrating machinery and such effects must be anticipated if the geological circumstances are such as to permit them. Cohesive soils will not normally permit densification. Grouting is a useful corrective treatment and chemical grout can alter the elastic properties of a soil so that the natural frequency of the foundation is changed and the amplitude of the vibrations reduced to acceptable levels. This matter is treated by the author in greater detail in *Grouting in Engineering Practice*.[2]

6.1.6. Absence of Seismic Shock Hazards

This entails building in earthquake-free areas of the world, not a major problem in Western Europe, but a considerable difficulty in many countries in active seismic regions, e.g. the USA, Japan, Italy, Iran, etc. Often, seismically active regions include tectonically active faults and where these occur slip movements are to be expected. However, great seismic activity with accompanying damage to buildings, etc., is not necessarily linked with faults for these can be found kilometres away from earthquakes.

In seismic areas the type of foundation soil is especially significant because a weak soil almost doubles the amplitude and lengthens the time of swing. Hence normal buildings which are not of the skyscraper type and possess a short time of natural vibration (say 0·333 s) tolerate earthquakes quite well, the more so if the foundation is stiffened with piles. If founded on rock, such buildings suffer from resonance. Underpinning of buildings with conventional drilled or driven piles is possible as is also underpinning by means of segmental piles jacked in the soil using static pressure. However, this latter technique cannot be utilized if rocks or masonry ruins occur and it requires laborious excavation work to permit the operation of jacks.

In 1952 the concept of the palo radice or root pile was introduced by Lizzi[3] and as noted earlier in this book, this entirely altered the field of underpinning. The main features of this technique are rotary drilling into the soil, bonding of the pile and concrete casing to the superincumbent structure with the concrete composed of 600–800 kg cement m^{-3} sieved sand (a high strength product results) and reinforcement comprising a single bar for the piles with the smallest diameter (10 cm) and a multi-bar frame or tubes for larger diameters. Smaller diameters are generally preferable. Load testing has established that except for very soft soils, a palo radice develops maximum bearing capacity over lengths not exceeding 30 m and an empiric relation is available for the limit load, *P*, namely:

$$P = \pi DLKI$$

where *D* is the nominal diameter of the pile, *L* is its length, *K* is a coefficient representing the average interaction between the pile and the soil for the entire length (in kg cm^{-2}) and *I* is a non-dimensional coefficient the form of which depends upon the nominal diameter of the pile. Values of *K* are given as (a) 1 for loose soil, (b) 1·5 for average compactness and (c) 2 for stiff gravels and sands. In all cases there can be no optimum resistance to earthquakes where a poor foundation exists.

6.2. FOUNDATIONS AND THE MECHANICAL PROPERTIES OF ROCKS

It is interesting to observe how much these are taken for granted in everyday life, most particularly with regard to strength and durability. Consequently, though people are aware of frost damage and landsliding, dam disasters and tunnel collapses the nature of these occurrences is not generally understood. The strength of all natural materials varies systematically and can be predicted; it is much influenced by the environment of formation of the soil or rock as well as by the subsequent history, especially as regards stress. The strength of a rock or soil can be defined as that resistance which is opposed to any applied stress.

Significantly, this mechanical strength may not be matched by chemical resistance to weathering agents, i.e. some strong rocks are broken down quite easily chemically. This reflects the different features constituting the two types of resistance. Mechanical resistance depends upon the fabric of the rock, the degree to which its constituent crystals are interlocked. Chemical resistance depends entirely upon the mineral contents and the effect on them of the atmosphere and soil solutions. However, it must be emphasized that the strength of rocks refers also to their ability to resist a wide variety of stresses, particularly those arising from crushing, flexure, shear, impact and abrasion as well as compression and tension. As seen earlier, igneous rocks are usually the strongest although they can be chemically weathered rather easily. An exception is granite which is moderately strong, but very resistant to chemical attack. This is because of its abundance of quartz and paucity of ferromagnesian minerals. Structural features in igneous rocks may reduce their strength, e.g. volcanic basalt may be vesicular with gas bubbles and flow characteristics, which will lower its strength. Rift and grain in granite represent planes of weakness as do bedding planes in sedimentary rocks. In metamorphic rocks, foliation and schistosity constitute the potential planes of weakness.

It is probably true to say of all rocks that their strength is reduced by any discontinuities. These cause values for strength which vary according to the direction in which the rock is stressed. Table 6.1 gives some comparative data taken from ref. 4. Of course there is a wide spectrum of strengths in nature and Table 6.2 shows variations that may occur.

TABLE 6.1

THE COMPRESSIVE STRENGTHS OF COMMON ROCKS (UNIAXIAL COMPRESSIVE STRENGTH, kg cm^{-2})

	0, very low . . . 28, low . . . . . . .56, medium. . . . . 1 125, high. . . 2 250, very high . . 4 500
Igneous	
Dolerite	
Basalt	
Granite	
Metamorphic	
Quartzite	
Gneiss	
Marble	
Schist 1	
Schist 2	
Sedimentary	
Limestone	
Sandstone	
Shale	

TABLE 6.2

COMPRESSIVE STRENGTHS OF COMMON ROCKS AND THEIR WEIGHTS PER CUBIC FOOT

Rock/material	*Compressive strength, kN m^{-2}*	*Weight per cubic foot*
Granite	34 500–414 000	162–172
Marble	55 200–186 300	165–179
Limestone	17 940–193 200	115–175
Slate	—	168–180
Quartzite	110 400–310 500	165–170
Sandstone	34 500–138 000	119–168
Brick	6 900–138 000	—
Concrete	17 250–27 600	—

It may be observed that the values in Table 6.2 differ from those of Table 1.12 and this discrepancy is due to their derivation from different sources. It illustrates the variability of such measurements which nevertheless offer a good guide to the value of such rocks from an engineering standpoint. The strengths represented are really *crushing* strengths and usually exceed requirements for manmade loads by far. For instance the Washington monument in Washington DC, exerts a pressure of only 2173·5 kN m^{-2} on its foundations.

Turning to flexural strength, i.e. resistance to bending, R, this is measured in terms of the modulus of rupture computed from

$$R = \frac{3wL}{2bt^2}$$

where W = load, at the middle of the span, required to produce a rupture; L = distance between supports; b = width of specimen; t = thickness of specimen. The moduli of rupture of various rocks are listed in Table 6.3. Apropos shearing strength (see Table 6.4), this may become important as regards stones utilized in large structures because of the possibility that improper masonry installation may concentrate loads near edges.

TABLE 6.3

FLEXURAL STRENGTHS OF COMMON ROCKS

Rock/material	*Modulus of rupture, kN m^{-2}*
Granite	9 522–38 295
Marble	4 140–27 600
Limestone	3 450–13 800
Slate	41 400–103 500
Sandstone	4 830–15 870
Concrete	4 140–6 210

TABLE 6.4

SHEARING STRENGTHS OF COMMON ROCKS

Rock	*Shearing strength, kN m^{-2}*
Granite	25 530–55 200
Marble	8 970–44 850
Limestone	5 520–24 840
Slate	13 800–24 840
Sandstone	2 070–20 700

Of course other rock parameters are significant in foundation work and these include porosity, permeability, resistance to heat and durability. Porosity, the percentage of the total volume occupied by pores, is not directly related to the absorption of water but it is true to say that rocks with low porosity usually possess low absorptive capacities. Usually igneous and metamorphic rocks have low porosities whereas sedimentary rocks may have very high porosities (see Table 6.5). Porosity has a bearing on foundation work because the more porous a rock or soil is, the more compressible it becomes. Again the above data are not exact because there is great variability according to components in soils and rocks. For instance in other sources sandstone is restricted to a porosity range 10–15% and limestone (with marble) to a mere 5%. Hence figures given must be regarded as a guide only. Every rock and soil sample must be evaluated individually.

TABLE 6.5
POROSITIES FOR SOME COMMON SOILS AND ROCKS

Soils:	
Average	50
Sands and gravels	20–47%
Cemented sands	5–25%
Clay	49%
Chalk	Up to 50%
Rocks:	
Granite	0·4–3·84%
Marble	0·4–2·1%
Slate	0·1–1·7%
Quartzite	1·5–2·9%
Sandstone	1·7–27·3%
Limestone	1·1–31%

As to heat resistance, at temperatures of 850°C and above, rocks are seriously damaged. However sandstones show the least damage and hence the greatest thermal resistance (see Table 6.6). It is important to note that chilling with water increases fire damage.

Turning to durability, this is in part determined by the mineral content. Some rocks contain minerals such as pyrite which, when exposed to the atmosphere, deteriorate leaving pits or causing stains. In order to forecast durability, field exposures must be observed or

TABLE 6.6
THERMAL RESISTANCE OF COMMON ROCKS

Most resistant (850°C and up)	Sandstone Fine-textured granites Coarse-textured granites Gneiss
Least resistant (calcining between 600°C and 800°C, ruined at 900°C)	Limestone Marble

alternatively buildings of fair age embodying blocks of the desired rock.

In the field the measurement of some of these rock index properties can be satisfactorily effected for building purposes using simple equipment. For instance the *in situ* strength of rocks can be determined utilizing a Schmidt concrete test hammer type N in order to measure the rebound number of the rock (R). R is plotted against the unconfined compressive strength of various common rocks in Fig. 6.1. From this it may be seen that an indication is given of the technique required for excavation, i.e. ripping or blasting. Comparison should be made with Fig. 6.2. These figures can be related to an empirical engineering rock property termed 'rippability' which determines whether excavation shall proceed by mechanical means using a tractor-mounted ripper or by blasting. The upper limit for rippable rocks is about 2 km s^{-1} for seismic velocity and 70 MN m^{-2} for unconfined compressive strength. In studying foundation soils and their behaviour, soil mechanics is employed and the usual classification has been given in Table 1.4. Rocks cannot be classified in a similar manner because they possess a complex macrostructure. Appropriate engineering classifications for rocks have been proposed and these have been discussed in Chapter 1.

6.3. SETTLEMENT

This is one of the most serious problems which can develop after the construction of a building and it is caused by the deformation of

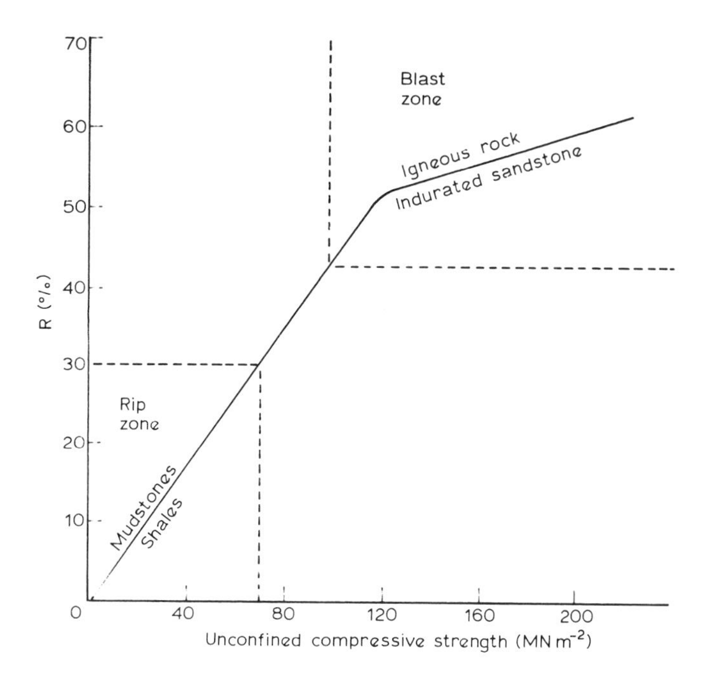
Blast
zone
Igneous rock
Indurated sandstone
Rip
zone
Mudstones
Shales
R (%)
0
10
20
30
40
50
60
70
40
80
120
160
200
Unconfined compressive strength (MN m⁻²)

Fig. 6.1.

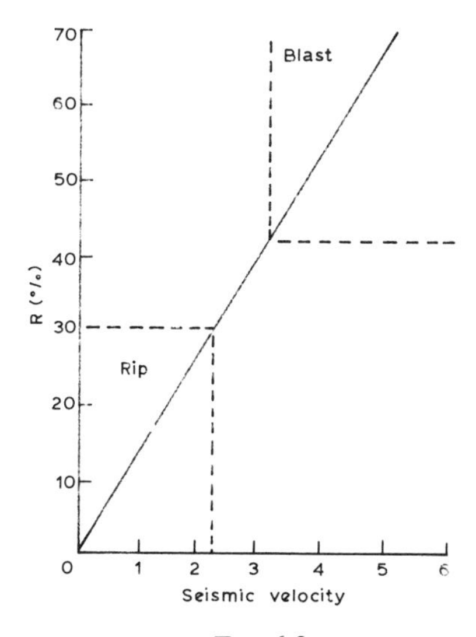
Blast
Rip
R (%)
0
10
20
30
40
50
60
70
1
2
3
4
5
6
Seismic velocity

Fig. 6.2.

superficial layers underground, often due to groundwater ascending to these levels. The amount of settlement depends upon the contact pressure exerted by the base of the foundation in the soil. Almost all soils are compressible but some are more so than others. More compressible soils include residual soils, detrital material, layers of loess and fills. Implicit in the above remarks is the idea that settlement may be uniform if a foundation area is uniformly loaded. However this is almost never the case—partly because of the very nature of man-made structures with their inherent variations; e.g. corners will be heavier than other parts. Consequently, differential settlement is to be anticipated and from this cracks may arise.

Settlement may be estimated from laboratory and field tests. When the latter are executed with loads distributed over relatively small areas, the effects of weak soil layers or saturated clays at some distance below grade may not be disclosed. Preferably the design should be based upon pressure–settlement curves determined from confined compression tests and consolidation tests on soil samples. It is possible to render the data in terms of void ratio, e, where

$$e = \frac{n}{1-n}$$

n being porosity, and then they can be converted to settlement using the expression

$$h = \frac{(e_i - e_p)12h}{1 + e_i}$$

where h = total settlement expected under load, e_i = initial void ratio, e_p = void ratio under pressure p, value h referring to the thickness of the soil layer in feet. If settlement is plotted for a layer of soil of say 10 ft in thickness, settlement on a soil layer located much deeper may be estimated by dividing the layer into 10 ft thick sections and then calculating the pressure in the middle of each section determining from pressure–settlement curves, the settlement for each section and then adding the settlements. When computing pressures, these are assumed to spread from a footing on a 30° or 1 in 2 slope.

Alternatively the pressure at the centre of the thick layer may be computed. Then the settlement for a 10 ft layer with this pressure may be obtained from pressure–settlement curves. Lastly the total settlement may be derived by multiplying the settlement by the ratio of the layer thickness (in feet), to 10 ft.

The results obtained by these methods are not quite concordant, but they are usually sufficiently close for practical purposes. Each may be adapted to computation of settlements where different soil layers underlie a footing if the pressure–settlement curves for each type of soil are available. Limitations of settlement have been calculated for buildings; for instance for brick buildings founded on sand, these are of the order of 4·8 cm (ref. 4).

Measures can be taken to supply optimum foundations for particular types of structure and circumstances and these are referred to below.

6.4. FOUNDATION TYPES

6.4.1. Spread Footing and Mats

These enable loads to be distributed over areas large enough for soils to support them safely without any excessive settlement. Normally such foundations are made of concrete, sometimes simple and sometimes reinforced. The compressive strengths of concretes made with various crushed rock aggregates have been cited by McLean and Gribble (in 1979).

Concrete for footings should be placed upon undisturbed soil. Where practicable the last few centimetres of soil should be excavated immediately prior to commencement of the concreting. As regards isolated footings these are usually placed under concentrated loads such as columns and they are generally square or rectangular. In order to prevent overturning and unequal distribution of stress in the soil, the centroid of a footing must be placed as near as possible to the resultant point of the imposed loads. The area of the footing must be sufficiently large for the bearing capacity of the soil not to be exceeded and in order to ensure that the maximum settlement is within acceptable limits. The dimensions of the footings must be such that no differential settlement will take place—to ensure this the footings are proportional for a particular structure so that the unit pressure for working loads is the same under each one (working load equals dead load plus normal live load) (see Fig. 6.3(a)).

Another approach is to employ combined footing, i.e. a spread footing supporting two or more concentrated loads.

6.4.2. Combined Footings

These may be rectangular or trapezoidal in shape or even an inverted T-beam symmetric about a line between the two loads. The centroid of

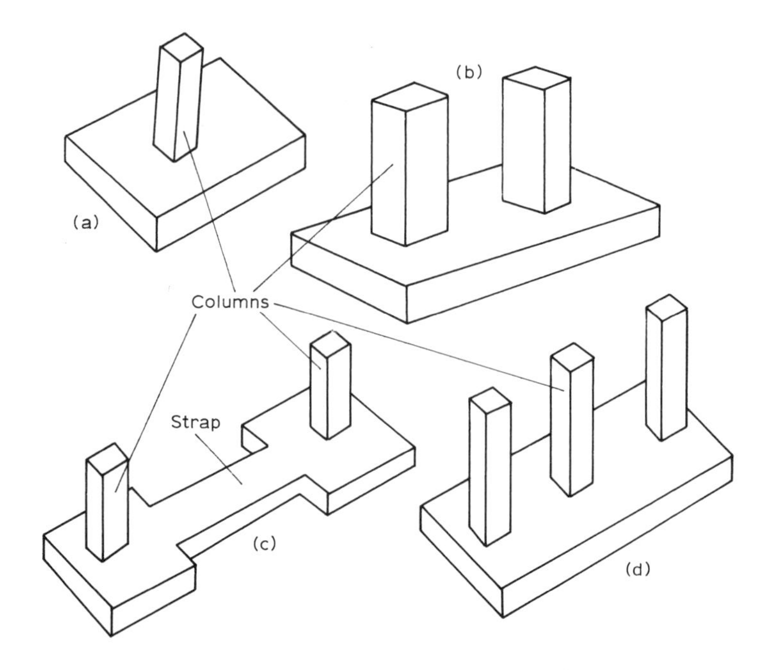

FIG. 6.3. Types of foundation. (a) Spread footing; (b) combined footing; (c) cantilever footing; (d) continuous footing.

the footing should be set as close as is feasible under the resultant of the loads. The footing must be so designed that the soil pressure is kept within desirable limits (assuming linear variation in pressure under eccentric loading). The design must maintain bearing pressure under the entire area of the footing under all conditions of loading (see Fig. 6.3(b)). Sometimes this type of footing has certain advantages over the spread type, e.g. where two piers are very close, two (isolated) spread footings would overlap. An adverse type of soil as regards combined footings is that sensitive to variations in load.

6.4.3. Cantilever Footings

These are shown in Fig. 6.3(c) and may be utilized in supporting two columns. They comprise a pad under each column with a lever or strap between them and they are more economical than combined footings because they require less concrete and reinforcing steel. Normally the pressure under the pad is assumed to be uniform and they are made and placed so that their centroid lies under the resultant of the loads.

The strap is designed as a rigid beam and ignores soil pressure underneath.

6.4.4. Continuous Footings

These are shown in Fig. 6.3(d) and are used under walls or in supporting a row of columns or piers.

In designing a wall footing it is customary to select a typical 1-ft long section of footing and design it so that it transmits the soil pressure transversely to the wall. The resulting design is applied to the remainder of the wall unless very different conditions are encountered such as a marked change in load or a variation in the size of the wall. Of course basement walls for buildings must be so designed that they resist active lateral earth pressures.

6.4.5. Rafts

These are continuous under a structure (like mats) and can be used to ensure watertightness or because the bearing capacity of a soil is low with the consequence that isolated spread footings would be so big that a set of them would cost more than a joined mat. There are many types of such foundations, e.g. a thick reinforced concrete flat plate, a thick inverted flat slab, an inverted beam and girder construction or a flat plate with depressed footings under columns.

In the design of continuous linear footings, if uniform pressure is assumed under the mat, an unduly conservative design may result; a more economic design is obtained by allotting column loads to areas symmetrical about the column. Each area is sized so that the permissible soil pressure develops under the column load, such loaded areas sometimes taking the form of continuous column strips. Portions of the footing between such areas are taken to be unloaded. The resulting bending moments are smaller than those for the assumption of uniform loading on the mat. In addition, it is possible to make the unloaded portions thinner than the rest of the mat with less reinforcing steel.

Floating foundations comprise mats constructed on poor soil and of course designed for low soil pressures. Such foundations are to be used cautiously because the ground may be settling under its own weight and fills around the structure may cause even more settlement. Mats have to be made watertight because they are subject to uplift from water pressure. Even where there is no uplift, a waterproofed layer is advisable in order to obviate loss of moisture from the soil through the concrete. Such movement of water out of the soil may well promote an increased rate of settlement. Footings subjected to overturning forces are designed assuming linear variations of soil pressure.

6.4.6. Pile Foundations

Piles are slender columns positioned underground and are usually placed in groups; their loads are supported through bearing at their tips, friction along their sides, adhesion to soil or a combination of all these. Clearly the behaviour of a pile foundation depends upon several factors:

(a) strength of the piles;
(b) bearing and shearing capacities of the soil;
(c) how close the field conditions are to the design assumptions.

As regards item (c), it may, for instance, be assumed in the design that the load is shared equally, by each pile in a group, but to realize this it is necessary to ensure that the tops of the piles are embedded in a rigid reinforced concrete cap, their centroid being under the resultant of the loads.

Deviation of piles from the vertical is measured using a plumb bob inside a pile casing. However, if solid piles are utilized this cannot be done; the job thus becomes more difficult, the more so because the bending of a pile may not be apparent at ground level. The permitted tolerance is 2% of the length of the pile. Pile foundations are expensive; hence they are only utilized where the soil conditions are such that mat foundations are not feasible. They are employed in such situations because they can penetrate weak soils and transmit loads to a deeper stratum with satisfactory results as regards bearing. Additionally piles can distribute loads over a sufficiently large vertical area of relati ely weak soil to enable it to bear a required load safely. They may be sloped or battered in order to resist horizontal forces.

Piles are either of the end-bearing or friction type depending on the means of transmission of loads to the soil. Normally, driven piles transmit loads through both end-bearing and friction, displacing soil downwards at the tip and laterally at the sides as they are driven. They must be driven into a good bearing bed and relatively weak soils cannot be relied upon to be sufficiently compacted below the tip to provide any appreciable bearing capacity. The taper of a pile may provide some bearing, but only in good soils. In soils such as silts, clays, sands or gravels, there may arise a significant resistance to penetration. The reticulated root piles (palo radice) of Lizzi,[3] mentioned earlier, act through pile–soil interaction and vice versa to establish a 'knot' effect if placed fairly near, this unfortunately entailing greater expense.

The transmission of loads to deeper soil layers implies that these exist on site. In some cases they may be deformable and hence deep-seated deformation may take place. An instance is afforded by Neogene or London Clay (Eocene) as well as loess. The larger the structure built, the more involved the deep layers become in the settlement. Where a deformable foundation soil extends to considerable depth, the settlement of a uniformly loaded area is non-uniform. The settlement is caused not only by the loading at one point, but in addition by the loading of its surroundings. The reason for this is that stressing does not only propagate vertically, but also in a lateral sense (see Fig. 6.4). Isobars are plotted in the figure and indicate the magnitudes of vertical stresses beneath a foundation beam and a foundation plate. The magnitude of the vertical pressure is 10% of the loading pressure at a point located three widths of the beam below the surface and one and a half widths away from the edge. At depth the pressure below the axis is greater than that below the edge of the beam.

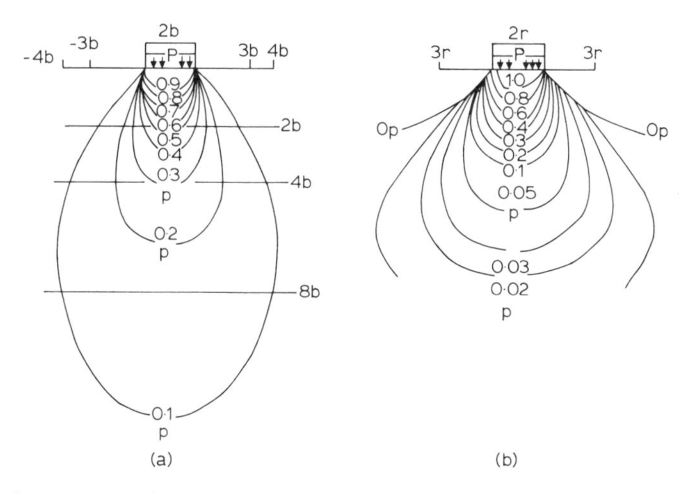

FIG. 6.4. Isobars (curves of equal stress) illustrating lateral spreading of pressure at depth. They are in an idealized soil with elastic properties: (a) beneath a uniformly loaded strip $2b$ in width; (b) beneath a uniformly loaded circular area.

From these considerations it is clear that as stated the settlement of a uniformly loaded area is not itself uniform because it is greater centrally than peripherally. The lateral spreading of the pressure at depth causes areas around the loaded area to exhibit settlement. Deviation from the theoretical uniform settlement of a uniformly loaded area is also due to the fact that an innate stiffness may occur and the greater this is, the less will be the differential settlement centrally and the greater the transmitted contact pressure, i.e. it will be transferred from the structure and the soil to the periphery of the area of the foundation. The result is to set up isobars of vertical pressure of the type shown in Fig. 6.5. As the modulus of deformation of soils increases with depth and the pressure of a structure spreads out laterally below the surface, then the settlement will not be as large as would be expected and of course the shape of the settlement curve will differ from that of an ideal soil. If the soil contains clay then it is necessary to consider settlement in terms of time. This results from the fact that where such clay occurs, consolidation, and therefore settlement, are slow. In fact the more clay there is, the slower will be the settlement.

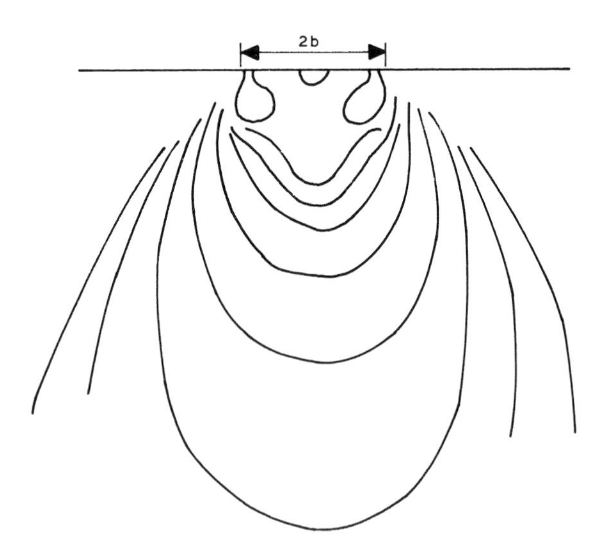

FIG. 6.5. Isobars in an ideal soil having elastic properties beneath a stiff foundation beam of width $2b$.

Turning to the subject of really unstable soils for foundations, these of course produce very undesirable results, in the form of large settlements, with their concomitant effects. There are several types of such soil. Loose sand is one type, sometimes with silt; another is fill material which is poorly compacted. A third type is composed of loose loess, material deposited by aeolian action which is liable to very abrupt settlement when it becomes saturated.

Fills are unsuitable because they not only contain pores but also largish voids and cavities. Consequently there can be quite large settlements with fills. If they have to be utilized as foundation materials, an analysis of their porosity must be effected. If the porosity exceeds 30% of voids or thereabouts, it will be unsafe to utilize them.

Sands may cause trouble if loose and subjected to shock such as vibration. The critical parameter is density which can be measured or inferred from porosity measurements.

In order to cope with unstable soils, it is essential to be well acquainted with the local geological conditions. It is interesting that in an attempt to quantify these matters the Czechoslovakian soil mechanics specialists have introduced the concept of 'number of collapsibility' apropos loess. The appropriate standard (No. 731001) defines it as the percentage contraction of an undisturbed soil sample loaded to simulate the anticipated structure and saturated. If the number exceeds 1% the loess is taken as collapsible.

Of course bearing capacity must be examined also and if this is attained for a soil, contact pressure under the foundation becomes large enough to exceed the shear strength of the soil so that soil is squeezed out causing an accelerating subsidence of the building. The development of a compressed wedge of material below the foundation may be envisaged and the soil is pushed outwards and resisted by the actual soil strength. The relation may be expressed thus:

$$q = \gamma B N_\gamma + \gamma D N_q + c N_e$$

where q = bearing capacity, γ = unit weight of soil, c = cohesion of soil, the magnitudes of the other components (all coefficients) depending upon the magnitude of the shear strength angle.

Taking the above expression into consideration and approaching the matter pragmatically it is possible to delimit a number of soil types. The first will include non-cohesive soils, i.e. sands of various grain sizes, etc., and these will have a negligible value for c, the apparent cohesion. However when they are subjected to pressure they increase in strength and develop a high value for the angle of shear strength. As foundation depth and width increase, so does the bearing capacity of such soils. As regards ϕ, this angle, it will be of variable magnitude in various soils. In fine-grained silty sands it has a value of about 35° and determines the values of N_γ (=18) and N_q (=34). In normal site work a safety factor of approximately 2 is adopted.

Another soil type is fine-grained and rapidly consolidating, i.e. loam-type soils with a small clay content. These are characterized by rapid rates of consolidation and the strengths increase by loading. In this type and alluvial deposits generally, consolidation sheets advance water drainage from the soil when placed under the foundation. The value of ϕ here will be about 25° with $N_\gamma = 3{\cdot}8$ and $N_q = 12$.

In the case of cohesive soils which consolidate slowly, e.g. Neogene clays, the value of ϕ is zero and $N_\gamma = 0$ with $N_q = 5{\cdot}4$.

6.5. GROUNDWATER

This is one of the most important factors in foundation engineering and if it affects the construction work adversely, a change of approach may become essential. Perhaps the worst case is that in which the foundation excavation penetrates below the water table. In principle this should never be allowed to happen, but it often does—sometimes due to a miscalculation of the depth-to-water. Even where the water table is not involved so that groundwater is not involved, ordinary soil water is a sufficient hazard in itself and the presence of this (in almost all soils) requires waterproofing. The technique cannot support a head of water of appreciable size unless a very heavy waterproofing is resorted to; this is costly and on occasions unsafe. The optimum situation is one with minimum soil water located above the water table. Of course well pointing may be used to lower the water table from the foundation zone during excavation and construction, but thereafter it may rise; hence appropriate measures have to be taken in the course of construction. Occasionally, in inclined sites, the water level on site can be depressed by means of horizontal boreholes acting as drains. Such gravitational drainage is only feasible where a suitable gradient occurs in the relevant area and if this is not present, drains can be made and led into pumping wells. Here only a light waterproofing is required which makes for economies in job costs. After construction, there are two potential dangers: one is elevation of the water table which may reduce the bearing capacity of the soil; the other is the reverse, i.e. depression of the water table. The latter promotes settlement and may initiate cracking in buildings. Perhaps the best known case of the deleterious effect of such a lowering is that which occurred in the City of Venice where, as a result of the increased consumption of artesian aquifer water through pumping to supply industry in Mestre, the rate of settlement rose from about 1·5 mm to as much as 5 mm annually.

If any corrosive materials occur on site they have to be investigated. Sulphates are some of the most damaging agents and are found almost everywhere in industrial countries. Chemically, water can be described according to its degree of softness or hardness, depending on its mineral content; there are three major types, namely water of low mineral content, water with dissolved CO_2 and acidic waters. If pure, soft water will dissolve calcium bicarbonate and, though more slowly, silicates and even aluminates of cement binder. Such water occurs in igneous and metamorphic rock areas in which the mineral constituents do not easily contribute to water. Highly mineralized water is hard, containing an excess amount of chemical compounds. Sulphate-bearing water can form gypsum, $CaSO_4.2H_2O$, in contact with calcium hydroxide and this can cause the (undesirable) expansion of concrete. Even worse is the volume increase triggered by calcium aluminium sulphate on crystallization with 31 molecules of water. This is known as Candlot salt and the effect is so devastating that it can completely disintegrate concrete. Sulphates derive from rocks which contain sulphate compounds such as pyrite.

Water containing carbon dioxide is very corrosive, especially to concrete and iron. Hence, if they are to be transported, relevant pipe installations must be of specially resistant material. In any area with sulphates, the cement to be used in construction should have a low lime content.

6.6. EXCAVATIONS

These may involve soils or rocks in the engineering sense and can be effected in a number of ways according to the nature of the material. Drilling is employed frequently if hard rock is involved and normally this is done either by rotary or percussion drilling instruments. In the former the material is crushed by rotation of the bit against the rock under load; in the latter a chisel-shaped bit is struck repeatedly against the rock so as to pulverize and fracture it. The fine debris so formed is flushed out of the hole by a returning air current. In 1967, McGregor[5] grouped various rock types according to the concept of drillability and his schema is shown in Table 6.7. As regards the relative hardnesses of the various types of bit, these are listed and compared with some geological minerals and rocks as well as with the Mohs' scale of hardness in Table 6.8.

TABLE 6.7

THE DRILLING CHARACTERISTICS OF COMMON ROCKS

Igneous:

Abrasive—obsidian, rhyolite, aplite, felsite, fine and coarse granites, pegmatite, quartz–porphyry, welded tuffs

Not so abrasive—basalt, dolerite, gabbro, andesite, diorite, syenite

Weathered—such decomposed materials as kaolinized granite and serpentine which can be dug or ripped (although blasting is required in some cases)

Sedimentary:

Abrasive—siltstone, siliceous limestone, some sandstones, pyroclastics

Friable—grits, some sandstones

Hardish, but not abrasive—limestone, mudstone, shales

Soft—marls, some mudstones, shales, chalk, coal, oolitic limestones

TABLE 6.8

HARDNESSES OF DRILLING BITS AND GEOLOGICAL MINERALS AND ROCKS

Mohs' scale numbers	*Drill bit types*	*Minerals*	*Rock*
1			
2		Coal	Evaporite
3		Mica	
4		Calcite	
5		Pyroxenes	
6	Steel	Amphiboles Feldspar	
7		Olivine	
8	Tungsten carbide	Quartz	
9	Corundum		
10	Diamond		

As well as drilling, excavation may be effected in several other ways according to the nature of the ground and these include:

(a) water-jetting;
(b) electrical current breakage;
(c) thermal lance melting.

In all excavations an important question is whether or not the bottom will be free of water. The absence of such water in the ground (not necessarily groundwater) is a great asset because it renders the excavation stable. Sloped pits are usually dug and the inclines may be at relatively greater angles than in road cuttings—e.g. dry loam may be excavated at an angle as great as 45°. Naturally this will depend upon the rapidity with which the soil strength decreases through dilatancy and the removal of previous load. Failure of slopes is a complex subject discussed elsewhere in this book; here it is sufficient to say that it may occur through fracture-induced collapse, fissuring and/or horizontal residual stresses and occur up to months after excavation. The prediction of such effects is almost impossible using only laboratory samples and of course larger ones become available only after completion of the excavation.

There is always the danger that water may infiltrate an excavation and this will strongly promote slope failure. In urban areas braced excavations must be utilized in place of the sloping walled pits alluded to above and these may use concrete diaphragm walls or pile walls. Sometimes anchors are employed to stabilize the walls. While the safe resistance of anchors is low in clayey shales, it increases considerably in hard rocks where grouting may be effected. Care is necessary when using anchors in built-up areas with existing buildings because of residual stresses which may increase tension in the anchors so loading them above expectation. Cable anchorages are more economical than retaining walls or strutting and timbering; in addition they do not reduce the working space available to such a great extent. Another reason for the preference for stressed cables as opposed to retaining walls is that slight deformation must occur in the latter to mobilize earth pressures. This is not the case with cables. The actual length of cables to be used is determined by the most probable slip surface—the anchorages are extended beyond this. If a permanent excavation is required then the cable heads must be left accessible so that any loss of stress due to deformation of the cable, the anchorage or indeed the rock itself may be checked and restressing carried out as necessary.

The case of the Hallamshire hospital in Sheffield, England provides an excellent example of a temporary use of cables. The north face of the excavation for this new hospital was 12 m high and excavated in mudstone dipping at 15° towards the face. Several rows of cables were installed in boreholes driven into the rock and these were anchored into the rock at their inner ends with the outer ends stressed against vertical concrete pillars constructed at 3·6 m centres. Horizontal sections of sheet piles made of steel were held in place by the cables against the concrete posts. Together with widely spaced vertical planks, these finalized the reinforcement. With deepening of the excavation, concrete was cast so as to extend the posts and new sets of wallings and cables were placed at each 1·8 m vertical interval. Cables were stressed to 25 tons and altogether 400 cables were installed in the 180 m long face.

Cable anchorages are used also in dam foundations.

The above factors, water in an excavation and the concomitant danger of slope failure, are compounded when the foundation extends below the water table as may happen in a valley floor for example. In such a case, it is necessary to evaluate the anticipated quantity of water influx and this entails a pumping test during construction. Appropriate measures must be taken, such as effective drainage accompanied by the installation of impermeable diaphragms or sheet-pile walls. Pumping wells will lower the water table during construction (well pointing).

6.7. SELECTION OF FOUNDATION METHOD

This is critical to the success of a construction project and usually comes down to a straight choice between spread or deep (e.g. piles, caissons, etc.).

6.7.1. Spread foundations

These are possible wherever the soil or rock has an adequate bearing capacity so that no large differential settlement of the ensuing structure is to be expected. A foundation resists differential settlement more as its stiffness is increased. There are two potential situations, namely shallow and deep.

In the case of surficial layers such as typical Holocene deposits, these may be responsible for unwanted differential settlement. Consequently the widths of foundation structures in such materials are determined in order to create a footings situation adequate to obviate unequal settlement, i.e. to obtain equal settlement of columns or walls. In dealing with big buildings in urban areas and structures of the industrial type, the applied loads transmitted through their columns will be very great and hence the foundation areas concerned must be correspondingly large. This often entails the utilization of piles. However with houses and other such small buildings, this is unnecessary and beam foundations are adequate. Such beams can add to the stiffness of the structure if the soil is compressible with sufficient bearing capacity. A raft foundation may have to be used with such soils occasionally. In any case, if the shallow layer under the foundation is compressible and has a *low* bearing capacity, the beam width required is so large that it becomes indistinguishable from a foundation mat.

When the foundation soil is compressible to a considerable depth, e.g. in deltaic clays, the width of the foundations will be greatly influenced by the calculated settlements. Obviously the greater the load, the larger the required footing—however, the larger the footing, the deeper the transmittance of the load. Consequently a large footing settles more than a small one. In order to derive uniform settlement it is therefore necessary to reduce the contact pressures under large footings (as compared with small footings).

Very large foundation areas will show dish-like settlements so that two proximate structures may actually tilt towards each other. Floor structures are incompressible horizontally so that a tilting process of this type produces bending deformations in the outer columns. Similarly, tensile cracking may occur in the walls of such buildings. In most structures it has been found that the permissible radius of curvature is about $250L/2$ where $L =$ the length of the building. It has proved feasible sometimes to reduce such a dish-like settlement by loading the corners of the relevant buildings as was the case at the University of Moscow.[4] Of course this approach was applied also in castle building. If an intolerable curvature of structure arises, then it is divided into blocks by means of dilatation joints, the lengths of these blocks fulfilling the requirements of allowable curvature. In the case of rigid structures, this technique is particularly important because the permissible curvature is more restricted. Thus the radius of curvature becomes greater than $1000L/2$ with monolithic reinforced concrete bunkers so that the lengths of blocks is smaller. With such a dish-like settlement, the dilatation joints close towards the top of the structure and can be allowed for by permitting the joints to open slightly in order to avoid crushing of blocks at the contact.

Elimination of the tilting of proximate structures has been achieved by making the foundations irregular. Utilizing a varying thickness of loam under a foundation may assist in obtaining a uniform settlement. The method is applied when a new structure is expected to incline away from a nearby older structure.

Great difficulty may arise where the foundation material is more deformable than say a stiff clay because then the tilting of blocks towards one another could become sufficient to promote serious damage. To obviate this happening, a deep pile foundation is to be recommended.

6.7.2 Pile Foundations

These are expensive but assist greatly in obviating problems caused by making foundations on weak soils. This is because they transmit loads to deeper layers; usually these possess a greater bearing capacity coupled with a smaller differential settlement. The base of the actual foundation ought to be placed at a level such that it is not subjected to the effects of groundwater. In the event that a competent layer exists not too far below the surface, the piles are embedded in it and the load transferred directly to it. Such a bed could be bedrock or, e.g., a gravel. Geological investigations are essential here in order to ascertain whether such a layer exists or not and if it does, how far down it is. If this is not done, there can be very adverse results. For instance if a gravel layer adjudged to be competent occurs at depth and is actually thin and underlaid by clay, then piles inserted into it may go right through, penetrate the clay and thus become embedded in a layer with a much *lower* bearing capacity. Occasionally the ratio of penetration depth to number of blows necessary in sinking the pile gives the resistance of the pile, but even then the geology must be examined. Such a relationship is useful in the case of non-cohesive soils which are permeable, but with fine-grained cohesive ones it is not applicable. However geological investigation is necessary with both types.

When all possible analytical work and subsequent evaluation have been effected and the results of all laboratory and field testing of soils and rocks are in, it is necessary to judge the suitability of the selected site for the proposed construction. This is the interest of the engineering geologist and site engineer since such arcane matters as budget considerations are dealt with by the appropriate financial personnel. Of course the conclusions reached must relate to the actual building intended and reflect the fact that some structures made by Man exert a much greater load than others. Thus a soil suitable for the foundation of a house may be totally inappropriate for the foundation of a nuclear power plant.

Zaruba and Mencl[4] have proposed a site classification tailored to the above factors and this is as follows:

(i) suitable if the soil is satisfactory and a small footing width is acceptable;
(ii) suitable if drainage is provided, e.g. with a gravel soil;
(iii) suitable on condition that the design of the proposed structure is adapted to the foundation conditions;
(iv) not very suitable, i.e. with foundation conditions so adverse that high costs would be incurred to cope with them;
(v) unsuitable, i.e. a site in which construction would be feasible only if a very high expenditure can be accepted.

As regards (iii) it may be necessary to eliminate differential settlement by specially designing the foundation or modifying the structure, for example using dilatation joints. Apropos (iv) an excellent instance is provided of thin permeable gravels which underlie Holocene deposits and are themselves underlaid by weak Tertiary clays. As for (v) this state of affairs occurs where unstable ground is involved. The whole schema alluded to in (i) to (v) above has been incorporated into the Czechoslovakian standard for foundation soils.

The regional setting of the particular site and the proposed structure must also be considered. Zaruba and Mencl[4] made some very interesting observations on this matter. For instance, they noted that in peneplains of the Bohemian massif, igneous and metamorphic rocks are characterized by deep and non-uniform weathering which produces residual soils of variable thicknesses. Some are sandy loams with appreciable contents of mica and these are compressible. The problem arises because in such areas one part of a structure may be founded on hard competent rocks while another part requires a pile foundation. Igneous rocks near young depressions are bordered by faults quite frequently and they can be a nuisance as for example with the original building site for the International Hotel in Brno, Czechoslovakia which had to be abandoned. In Prague most of the bedrock is Ordovician clay–shale alternating with coarse-grained quartz sandstones and quartzites, all folded during the Variscan orogeny. Transverse and longitudinal faults were disruptive agents so that the physical and mechanical properties of the rocks are determined partly by tectonic disturbance. Also, the rocks are very deeply weathered as a result of deep frost penetration during the Pleistocene. Several layers of Ordovician rocks contain finely dispersed pyrite so that joint water in them is highly corrosive due to its high sulphate content. As regards Devonian rocks, most of these in Bohemia and Moravia are of karstic type and of course constitute a hazard for dam construction. Mesozoic rocks in Slovakia are tectonically affected because of their involvement in the Carpathian nappe structure and the limestones again are frequently karstic in nature with heavily disintegrated dolomites.

The Carpathian flysch belt is particularly interesting—in the late Cretaceous and early Tertiary, a large foredeep developed at the periphery of the now Inner Carpathians and was completely filled with marine deposits. These took the form of thick complexes of alternating sandstones, claystones and conglomerates and the flysch complexes were folded and overthrust. Naturally such tectonic disruption is reflected in the state of the flysch, its sandstones being fractured and cracked with claystones of reduced strength compacted along bedding planes. Flysch is easily weathered and thus a number of slopes have become covered with debris which is susceptible to sliding.

Later Neogene rocks are variable in their suitable characteristics as building sites with the Miocene sands normally rather deep and excellent as foundation soil. Miocene clays and marls on the other hand are dangerously liable to landsliding and large scale settlement. The Pliocene comprises silts and sands as well as clays and the Quaternary is often present as a foundation soil. Of course many residual soils occur including loess.

If the situation in Great Britain is compared briefly, the geographical variability of potential regional geological situations is revealed. Precambrian occurs widely, but is best represented in Scotland where it includes both metamorphic rocks (mostly gneisses) and sedimentary sandstones and arkoses with shales. Of course metamorphosed sediments also occur, e.g. in the Dalradian which is a set of strongly folded schists including mica schists, quartzite and also marbles. Actually the Scottish together with Greenland and Canadian rocks of this age constitute part of a single Precambrian landmass preceding the opening of the North Atlantic Ocean.

Palaeozoic rocks are present and in the older Palaeozoic they were deposited mostly in an elongate depositional trough (geosyncline) extending from southwest to northeast across the North Sea into Norway. In fact, over Wales the deepest part of this trough, about

12 000 m of sediments, accumulated and comprise marine sandstones and shales with pyroclastic deposits of tuff and agglomerate representing Ordovician vulcanicity. Later, an orogeny, the Caledonian, commenced and caused enormous overthrusts to develop in the Northwest Highlands of Scotland, e.g. the Moine thrust. In the younger Palaeozoic, Devonian, Carboniferous and Permian rocks were deposited. Devonian muds became slates and were laid down under water whereas to the north are Old Red Sandstone deposits of continental origin (aeolian). Both of these sets of rocks were extensively folded during a new orogeny, the Armorican or Hercynian, regarded as a later stage of the Caledonian by some. In the Carboniferous, initial limestone deposits are succeeded by the famous Millstone Grit and then the Coal Measures and towards the end, the Hercynian orogeny started affecting most of northwestern Europe so that rocks in southwest England became folded along an east–west axis. The result was the impressing of cleave on argillaceous rocks. North of the area were other areas having a north–south trend and here the coal measures were preserved in synclines flanked by uplifts, the existing coal fields. Later, in the Permian, lakes were formed and contained deposits of breccias, sands and red marls. In the subsequent Mesozoic, sands were initially deposited in inland basins—they arose from the denudation of mountains in arid climates and were transported by wind and water. At the end of this phase (the Triassic), salt lakes arose together with evaporation pans in which evaporites were laid down (reaching a thickness of 760 m in Cheshire). The later Jurassic was accompanied by deposition of alternating beds of argillaceous and limey sediments so that, for example, important economic deposits became emplaced such as Liassic clays used in brick making and cement stones utilizable in cement manufacture as well as iron ores and oolitic limestones important as building stones. At the top of the Jurassic is the famous Portland limestone and this is followed by the overlying Purbeck series. Without any break, i.e. conformable, this passes up into the Cretaceous capped by the Chalk, a fine-grained limestone mostly composed of algal fragments.

The Tertiary deposits are peculiarly significant in construction work because of the deposition of the Palaeogene, the older part, in two synclines in England, namely the London and Hampshire Basins. In the north, there was igneous activity and volcanoes erupted at Mull, Skye and Arran with great basalt flows in Northern Ireland at Antrim. Crustal stresses gave rise to the emission of Tertiary dyke swarms in northern Britain, these having been formed by magma of basic type ascending through fractures.

In the Quaternary the situation changed. Whereas in the Tertiary orogenic movements formed the Alps, Himalayas, etc., and a vast east–west trough, the Tethys sea existed in the area of the existing Mediterranean and became infilled with sediments, the later Quaternary was characterized in the Pleistocene by extensive glaciation and deposition of materials such as moraines, kames, eskers, erratics and boulder clay as noted earlier. Some of these comprise satisfactory foundation materials while others do not. As was seen above, boulder clay is unsuitable because of its heterogeneity.

Great changes in the surface drainage were produced by the glacial phases and so, for instance, the River Severn became diverted into its existing course below Shrewsbury. Formerly this river drained into the Irish Sea by means of the Dee Estuary. However, during the retreat of Irish Sea ice from the Cheshire plain the meltwater became impounded in front of the ice forming a glacial lake (called Lake Lapworth), the outlet of which was through a col at Ironbridge, in Shropshire where a gorge was cut by the overflow. Subsequently the Severn was prevented from following the pre-glacial course by drift accumulations, hence it became established in the existing channel through the Ironbridge gorge.

Old valleys of many rivers are filled with glacial deposits of various suitability for foundation work—to name a few, the Clyde, the Forth and the Mersey. Recent deposits were laid down subsequent to the retreat of the ice about 10 000 years ago in Great Britain. Alluvial deposits are included and these usually comprise sands, silts, clays and the odd gravel seam. Sometimes such alluvial beds attain thicknesses of 10 m or more. Such materials, e.g. the Thames terrace gravels, are often indifferent as foundation beds although occasionally they may be suitable.

6.8. GEOLOGICAL INVESTIGATIONS ON SITE

A preliminary literary search for borehole records and maps is mandatory for such an investigation although in the developing countries this is not always feasible. Where such information is available, the following procedure has been proposed by Dumbleton and West:[6]

(i) locate all relevant documentation, i.e. all the above-mentioned together with aerial photographs and so on, so deriving a geological picture of the environmental conditions above and below ground, expressing these in appropriate sections and perhaps also in block diagrams;
(ii) seek more information from such institutions as the Geological Survey or geological societies, local authorities, universities, etc., as well as from site engineers who may have worked in the relevant area;
(iii) revisit the site so as to collate all data and identify areas where engineering problems may arise as well as areas where further detailed investigations may become necessary;
(iv) compile a technical report incorporating geological and geophysical information with appropriate references to the literature, if any, and addresses of contacts who might be of use;
(v) evaluate construction requirements in relation to the proposed structure; these should be dovetailed into the environmental parameters emerging from the preliminary geological and geotechnical analyses.

The field study enables the geotechnical characteristics of the site to be understood and the existing geological conditions accurately to be described. There are three basic approaches:

6.8.1. Linear Approach

Samples should be taken along a line or column of ground, boreholes being included. Orientation may affect the latter significantly, e.g. in a region of vertical faulting, this will not be apparent from vertically directed boreholes. Consequently for every borehole, irrespective of its orientation, there will be a 'blind' region as regards data acquisition. Drilling and coring methods are well known and involve consideration of the type of ground, the instrumentation later to be installed, etc. Borehole support is necessary in incompetent materials and also the core recovery may be unsatisfactory at times. An important factor here is the diameter of the core barrel. EX, AX and BX core sizes give a core recovery of 50–70% in weak beds such as shales and with larger diameter barrels recoveries of almost 100% can be obtained. The use of the larger varieties necessitates drilling skill as well as a vibration-free rig and first class equipment including swivel type double tube core barrels. A diameter of 131 mm is suitable in many cases. The appropriate lubricant must be selected according to the nature of the ground, water being the commonest. Air can be utilized too and in some instances promotes better core recovery. This is so in shattered rocks or friable soils. Sample disturbance where sampling is effected can result from the relief of stresses following removal (often associated with desiccation or oxidation) or by mishandling (more obvious).

Shales are perhaps the most susceptible materials in terms of the effects of stress relief so that fracturing in the core may be much more intense than fracturing *in situ*. Integral sampling may be employed in such cases. This is done by drilling a hole to just above the sampling level and then driving a smaller diameter coaxial one into the sampling zone, the ratio of the larger to the smaller diameter being usually recommended as about 3:1. Then a perforated hollow reinforcing bar with a diameter about 5 mm below that of the smaller hole is inserted into the latter and bonded to the rock by grout injection through its hollow centre. The sample zone is over-cored using low drilling pressures and the sample retrieved. The method provides good samples, the grout strengthening the weak rock and infilling voids. In this way the separation between fissure surfaces can be examined. (Appropriate drilling data are supplied in Table 6.9.)

TABLE 6.9
DIAMOND DRILL CORE CASINGS

Size	*Outside diameter, mm*	*Approximate diameter of bit, mm*	*Approximate diameter of hole made for core barrel, mm*	*Approximate diameter of core, mm*
EX	46	47	38	22
AX	57	58·5	46·3	28·6
BX	72	74·5	60·3	41·2
NX	89	90·5	76·2	54

6.8.2. Areal approach

Whereas the linear approach is traversal, this approach derives from aerial photographs, remote sensing and geological mapping. In this way a two-dimensional picture of ground truth is obtained and such areas may be vertical as well as horizontal. Vertical surfaces are exposed in pits and trenches and show lateral variations also. Geophysical studies reveal vertical details also, especially the results of electrical resistivity and seismic surveys. Table 6.10 lists main geophysical methods.

As regards trenches, these may not be deep enough to reveal required geological detail and in this event, adits may be utilized. It must not be forgotten that in both cases (and also in pits) stress relief phenomena may occur as well as other disturbances, which tend to loosen the ground and constitute potential dangers. Horizontal areas are the object of investigations using aerial photography, side-looking airborne radar (SLAR), geological mapping, etc.

Geological maps attempt to represent geological conditions and structures of an area as apparent upon the planetary surface and some features on them may be visible on the ground whereas others are concealed by drift. A variety of scales is available, but the most detailed usually printed is about 1:10 000 (1:10 560 in the UK, i.e. 6 in to the mile). Such maps in England are referred to as '6 inch maps' and they constitute the base for detailed geotechnical investigations.

However, even more detailed maps may be used if considered necessary, e.g. in mapping a particular foundation area. The usual scale utilized is approximately 1:50 000, the nearest equivalent in the UK being 1:63 360 (1 in to the mile). This country is covered by 563 sheets on this scale. They are distributed as follows: 360 of England and Wales, 131 of Scotland and 72 of Northern Ireland. Each sheet is coloured and covers in Scotland an area of 24 × 18, i.e. 432 square miles, and elsewhere in the UK 18 × 12, i.e. 216 square miles. Solid and drift maps are available for most of these sheets, the former excluding secondary deposits such as drift which, together with the solid geology where exposed, are included on the latter. Maps on the 1:50 000 scale are very valuable in assessing the general geology of an area and assist in locating borrow pits and mineral deposits. Broadly speaking all scales of map are useful, even those on the 1:1 000 000 scale, which are synoptic maps demonstrating regional tectonic patterns. Maps on scales 1:2000 to 1:10 000 can be grouped as detailed while those in the range 1:20 000 to 1:50 000 constitute base maps. Scales such as 1:100 000 and 1:200 000 can be used also and in the UK there are the 1:253 000 and 1:625 000 scale maps (the former known as 4 miles to the inch or quarter-inch maps, the latter being termed 10 miles to the inch maps).

TABLE 6.10
GEOPHYSICAL METHODOLOGY

Method	*Operations*	*Parameters measured*	*Computed results*	*Applications*
Electrical and electro-magnetic	Ground self-potential Resistivity surveys Airborne and ground electromagnetic surveys Induced polarization surveys	Natural potentials Potential drop between electrodes Induced electro-magnetic fields	Anomaly maps and profiles Positions of ore bodies Depths to rock layers	Mineral exploration Site investigations
Seismic	Reflection and refraction surveys	Time for seismic waves to arrive at measuring point	Depths to relevant formations Velocity of seismic waves Seismic contour maps	Oil and gas exploration Site investigation Regional geology studies

Borehole logging	Insertion of equipment to make measurements of electrical, seismic and other parameters	Variations in the parameters studied	Velocities Resistivities Densities Gas Depth to water and its salinity K, U, Th contents, etc.	Oil, gas and water supply location Site investigations Regional geological studies
Radiometric	Ground and air surveys using gamma ray spectrometers and scintillation counters	Natural radioactivity levels in minerals and rocks Induced radioactivity	Radiometric anomalies Location of mineral deposits	Exploration for metals employed in atomic energy facilities
Magnetic	Airborne and marine magnetometers surveys Ground field	Variations in intensity of terrestrial magnetic field	Magnetic maps and profiles Depths to magnetic minerals	Oil and gas reconnaissance Location of some mineral deposits Site investigations
Gravity	Gravimetric land surveys Marine gravimetric surveys using submersible instruments	Variations in intensity of terrestrial gravity field	Bouger anomaly and residual gravity maps Depths to rocks of contrasting density	Oil and gas reconnaissance Detailed geological studies

Initially soil maps were employed in engineering work and these show types of soil present and also underlying the solid geology. Later on, engineering geology maps were produced which indicate the geotectonic situation and such geodynamic features as landslides and sinkholes in limestone areas. A separate hydrogeological sheet may be made so as to show surface and groundwater occurrences, rock permeabilities, areas of inundation, springs and swamps, river systems and lakes, etc. A third sheet, a reference map, can be used to record outcrops, test pits, wells and boreholes.

With all these data, a specialized map can be compiled dealing with specific problems of a site and, covering a larger area, to be zoned into regions possessing roughly the same set of engineering geology characteristics—invaluable in defining these in terms of particular types of construction work. Various scales are applicable—e.g. 1:5 000 is optimum for large civil engineering constructional work and 1:25 000 for regional planning (although urban planning usually requires a larger scale, say 1:5 000).

Needless to say, all engineering geology maps must be fully documented and associated with rock and soil samples, aerial photographs, results of laboratory testing and so on. The work involved is expensive and time-consuming, but the result will reduce the construction costs by much more than the cost of maps. The geologist in the field makes great contributions to such investigations and of course records four basic types of information, namely:

(a) rock types;
(b) boundaries between rock exposures;
(c) rock structures (dips read to 1° with directions in compass bearing form);
(d) hydrogeology.

All of these matters have been considered in Chapter 2 so that it is only necessary to add that the geological hazard map is also useful. This was first devised by the French Geological Survey under the Designation 'Cartes de Risques Geologiques' and shows such features as landslides, potential rockfall areas, erosion localities, etc.; usually plotted on a 1:20 000 scale, the hazardous locales are defined in colour, say red for scarps, green for places where debris might accumulate and so on. This type of map has been widely utilized in mountainous regions—see for example the work of Goguel and Humbert.[7]

6.8.3. Three-dimensional Approach

Attempts have been made to model the real geology of a region which involves the utilization of tests such as loading, shear strength, pumping and blasting. The amount of work involved in site investigations for a civil engineering construction need not normally be very great but if rock masses with very widely spaced joints and bedding planes occur, the scope may have to be broadened. This will entail the examination of relatively huge samples and increase the expense proportionately.

Once the appropriate scale of testing has been determined, it becomes necessary to select the optimum orientation in relation to the proposed structure—this can be difficult in view of the fact that almost all rock masses are heterogeneous in nature and thus their performance characteristics will vary according to the direction of testing. A basic rule is that such testing must be effected in undisturbed material and the most important parameters to be ascertained are deformability, shear strength and permeability.

In situ stress can be measured in order to derive either absolute or relative stresses, the former indirectly. This is done by measuring the strain present in a sample of rock after separation from original *in situ* stresses—a monitoring device is inserted into a borehole which is later over-cored as shown in Fig. 6.6. Relative stress technology involves counteracting strains with an applied pressure and this is carried out by cutting a slot into the rock which will close slightly thereafter because

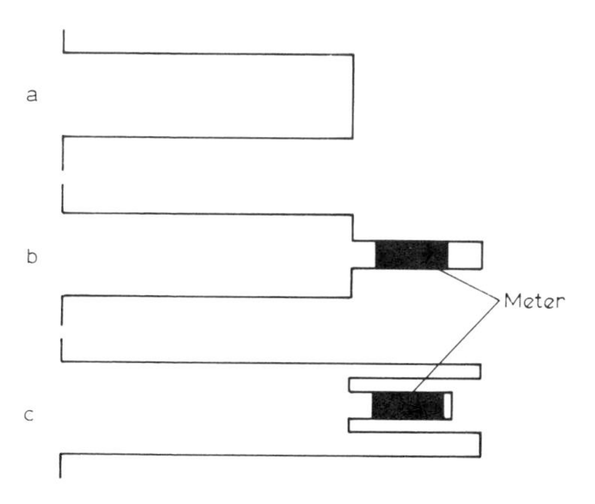

FIG. 6.6. Measurement of *in situ* stress by (a) drilling a hole in the rock, (b) using a coaxial hole as site for the deformation meter and (c) measuring the over-coring released around the meter.

of *in situ* stress. An hydraulic jack is inserted and pressure applied to return the sides of the slot to their former positions. The stress is calculated using values needed for completion of the test procedure.

To measure deformability a static load is applied to the ground and the resulting degree of deformation measured, results being interpreted on the basis of the theory of elasticity; values for Young's modulus and Poisson's ratio are thereafter given for the ground. The static load in question can be imposed over the area of a rigid plate or over the area of a borehole; an hydraulic jack may be used. In the plate bearing case, the subsequent deformations are recorded and an area as small as 1 m^2 may be sufficient. It is cheaper to load only a section of a borehole, again using a hydraulic jack—Rocha[8] has described a suitable approach.

The elastic modulus can be determined by measuring the response of the ground to dynamic loads, e.g. a hammer or an explosion. Such loads are of short duration and work within the elastic range of the material; also their propagation is not necessarily affected by water-filled voids and fractures. Seismic shock compressional wave velocities can be utilized with assumed values of 0·27–0·35 for Poisson's ratio to evaluate a dynamic Young's modulus for the purposes of the site engineer. The velocities of propagation of shock waves must be known when earthquake engineering or, for that matter, the control of blasting are involved. Some values are given in Table 6.11.

In applied seismic methods waves are generated at a shot point and their arrival at various stations on a line of traverse is detected using seismometers at the surface or in water using hydrophones. These

TABLE 6.11
SEISMIC VELOCITIES IN COMMON MATERIALS (AFTER REF. 9)

	Seismic velocities, $m\,s^{-1}$	
Material	*Compressional*	*Shear*
Sand	300–800	100–500
Shale	880–3 900	400–2 000
Sandstone	1 400–4 200	700–2 100
Limestone	3 500–6 500	1 800–3 800
Granite, massive	5 500–7 000	2 500–4 000
Granite, weathered	680–3 000	250–1 200
Air in voids	330	—
Water in voids	1 450	—

detectors convert ground or water motions into varying electrical currents transmitted along wires which link the detectors to a seismograph where outputs from the channels are amplified and recorded. The seismic record, the seismogram, embodies the ground or water motion within its trace. The seismic method is valuable because it enables the dynamic behaviour of the ground to be related to physical properties.

If compressional waves are generated and received, having radiated with a velocity V_1, the effects of fractures can be determined by obtaining the velocity V_2 which obtains in an unfractured specimen of the same rock mass under the same load (simulated by application of an axial stress) and moisture content. The ratio of these two velocities V_1/V_2 approaches unity if the ground is unfissured. The approach has been used to study the grout uptake of dam foundations by Knill.[10]

Seismic velocities may be correlated with the ease of excavability of ground.

Shear strength may be obtained using shear boxes, but there is no way of doing this with hard rocks. *In situ* testing may be effected by selection of a block of ground with an area of about $1\ m^2$ which is trimmed thereafter and then subjected to loading using pads of concrete in association with hydraulic jacks which apply normal and shear loads. The vertical and horizontal displacements are recorded during testing. Data may be obtained also by pressing a plate into the ground to failure. With soils such as clays the vane test may be utilized. This instrument has four thin rectangular blades about four times as long as their width and it is pressed into the ground and twisted in the soil at a rate of about $0{\cdot}1°$ per second. This results in the development of a cylindrical rupture surface at a certain torque, the value of which is measured and then employed to calculate the shear strength.

The determination of permeability is usually effected by the pumping test or packer test in engineering work. As indicated earlier the pumping test entails use of a well (sunk if necessary), surrounded by smaller diameter observation holes spaced along lines radiating out from it. The water level in the well is lowered by pumping and a cone of depression forms. Using values for the well discharge at given times and measuring the drawdown data in the observation holes at the same times combined with the known distances of these holes from the well centre, it is possible to calculate the permeability of the ground.

Alternatively, permeabilities can be calculated by the recovery method—this necessitates observing the rise in water levels occurring after the cessation of pumping.

The packer test needs only one hole into which is lowered a tube with one, two or four inflatable packers attached. Water is passed into the tube and using its rate of discharge into the ground, the area of the section of test and the head of water in the system, it is possible to determine the field permeability—the relevant calculations have been examined by Cedegren.[11] The great advantage of this packer approach is that it may be applied to existing boreholes and hence there is no need to drill new or additional ones. Often, engineers use a standard pumping test in which the flow will be the same as that anticipated during production. Aerial photography and other remote sensing techniques are becoming ever more important—see Chapter 10.

Geological sections are graphical representations of rock sequences which usually show the strata and tectonics on a vertical plane, frequently with an exaggerated vertical scale. To draw such a section it is first necessary to know the topography of the area and for this topographic maps may be used. If none are available, as is often the case in developing countries, then a survey of the ground along the proposed section must be made. After obtaining a true topological profile of the relevant area, it is possible to construct a geological section either from a geological map or directly in the field. The following observations are germane:

(i) The section prepared from a map will also incorporate data from the field such as borehole records. It is necessary to know the outcrops of solid geology, their boundaries and also the dips and strikes of the strata involved.

(ii) In the field it is possible to obtain valuable information from human excavations as well as from natural outcrops and exposures of rocks.

It is often very important to survey quarries, for example. This is done by choosing a base line at the bottom of the quarry face and setting stakes along it at say 2 m intervals. The line should be horizontal. The stakes compartmentalize the quarry wall. When this is inclined, there will be a distortion in field drawings and later this must be corrected. As well as drawings, photographs should be taken. Photogrammetry can provide precisely measured sections on inaccessible slopes. The direction of a section is significant because of considerations of dip and strike.

(iii) Cross- and longitudinal-sections must be distinguished, the former drawn at right angles to the strike and the latter parallel to it. Longitudinal sections can establish the lateral extent and dip of strata; they are especially important in studying hydrogeological conditions and also as part of a programme of mineral exploration. On the other hand, cross-sections enable the overall geological structure of a region to be recognized.

Of course it is also feasible to make oblique sections which do not fall into either of these two categories. In fact it is often unavoidable because a section may be drawn across strongly folded strata and hence a part must be oblique to the strike. Here the dip of strata towards the line of section must be determined correctly.

It is usual for sections: to be constructed with an exaggerated vertical scale, but only if the vertical and horizontal scales are the same can a true representation of the actual state of affairs be obtained. In some cases, e.g. with folded rocks, the use of exaggeration in the vertical scale will cause a distortion of the general geological structural picture and necessitate the recalculation of dips in order to compensate.

To determine the true thicknesses of strata on a geological map, it is essential to know the true dip and the angle of slope; from these and the width of the particular bed shown on the map, the relevant information can be obtained. One of three situations will apply:

(i) The bed dips into the slope (see Fig. 6.7(a)) and the thickness b is given by

$$b = \frac{d \sin(\alpha + \beta)}{\cos \beta}$$

(ii) The bed dips downslope with the dip less than the slope angle and this is shown in Fig. 6.7(b). The thickness b is then given by

$$b = \frac{d \sin(\beta - \alpha)}{\cos \beta}$$

(iii) The bed dips downslope, the dip is greater than the slope angle (see Fig. 6.7(c)), and the thickness b is given by

$$b = \frac{d \sin(\alpha - \beta)}{\cos \beta}$$

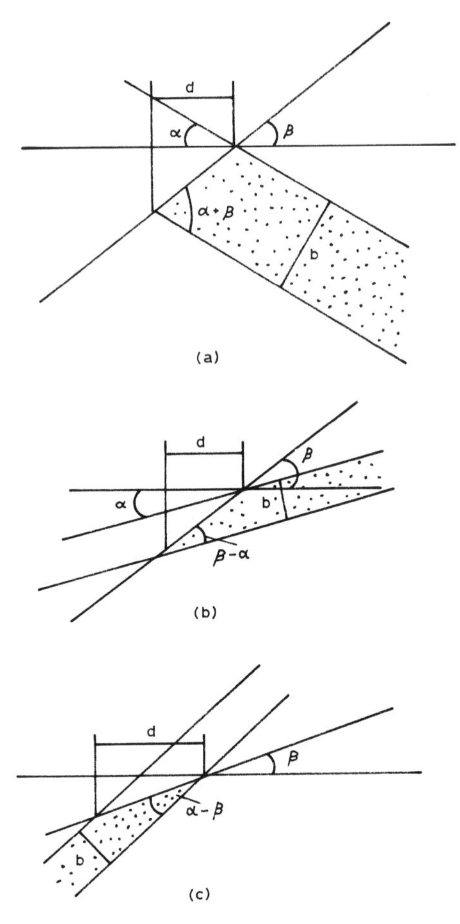

FIG. 6.7. Determination of the line thickness of beds with various dips: (a) into the slope; (b) downslope with smaller angle than the slope; (c) downslope with greater angle than the slope.

In all of the above, α is the dip of the bed and β is the slope angle, d being the surface width of the bed.

6.9. BOREHOLE INVESTIGATIONS

Test drilling activities will provide optimum data on subsurface conditions and a bore so constructed may be converted into a pumping well or used as an observation well either for pumping tests, as mentioned above, or for measuring water levels. Boreholes may be effected by the methods alluded to earlier, i.e. by percussion (referred to as the cable tool technique) or rotary operation (termed the hydraulic rotary technique). The information derived will follow up initial investigations using augering. Augering is effected in soft ground and at shallow depths is limited by the ground water table. Hand-operated augers are effective to 6 m or so, but machine-driven ones can work at up to 24 m or more.

Careful logging of all boreholes must be made and records kept of the various strata and rock types encountered together with details of their thickness and state of weathering. Details of the water present should be included as should details of core recovery, results of water pressure tests and standard penetration tests. Logging is also possible using resistivity measurement, potentials, calipers, temperatures and radioactivity.

Resistivity logging involves measuring the electrical resistivities of the surrounding media in an uncased well and it is effected by lowering the current and potential electrodes into this and observing variations in the readings with depth. Relevant factors are the presence of a fluid (water), the diameter of the well, the nature of the rock strata encountered. Resistivity can be expressed as follows:

$$\rho = \frac{RA}{L}$$

where ρ is the resistivity, R the resistance, A the cross-sectional area and L the length. In the CGS system, the units of resistivity are ohm-$m^2\,m^{-1}$. Determination of the parameter in this way involves measuring the potential difference between the two electrodes resulting from the application of a current through two *other* electrodes outside, but in line with, the potential electrodes. An orthogonal network of circular arcs is created by the current and equipotential lines if the resistivity is uniform in the subsurface region below the electrodes and the measured potential difference is a weighted value over the particular subsurface region controlled by the shape of this network.Thus the measured current and potential difference give apparent resistivity over an unspecified depth, an increase in this occurring if the space between the electrodes is increased (a different apparent resistivity is also thereby attained). Since it is usually the case that the actual subsurface resistivities vary with depth, the apparent resistivities also change as the electrode spacings increase, although not in the same way. The effect of resistivity changes at great depths affects apparent resistivity to only a very small degree compared to that at shallow depths; thus the method is not really effective for assessing actual resistivities when depths exceed hundreds of metres. The electrodes comprise metal rods inserted into the ground and they handle the actual current. The potential electrodes are porous cups with saturated solutions of copper sulphate and they inhibit electrical fields from arising around them. Either an a.c. low frequency current or a reversible d.c. current may be used to minimize polarization effects.

Standard electrode spacing arrangements are applied and include the recommendations of both Wenner and Schlumberger. These are employed together on occasions; for example on the Bumbuna damsite in Sierra Leone experience showed that while the Wenner data are easier to calculate in the field, are simpler to obtain (the distances between the potential electrodes and between the current and potential electrodes being equal) and are ideal for rapid reconnaissance work, the method is more susceptible to lateral subsurface variations than is the Schlumberger approach (although here the field calculations are more difficult). In the Schlumberger array, the potential electrodes are close together and the apparent resistivity is given by:

$$\rho_a = \pi \frac{(L/2)^2 - (b/2)^2}{b} \frac{V}{I}$$

where L and b are the current and potential spacings respectively, a is the distance between adjacent electrodes, V is the voltage difference between the potential electrodes and I is the applied current. Theoretically $L \gg b$, but in practice it is only possible to obtain satisfactory results when $L \gg 5b$.

As regards the Wenner array, the potential electrodes are located at the third points between the current electrodes and the apparent resistivity is given by the ratio voltage:current multiplied by a spacing factor. The apparent resistivity can be expressed:

$$\rho_a = 2\pi a \frac{V}{I}$$

Resistivity logging is discussed in greater detail in *Ground Water.*[12] Resistivity curves can provide information of value to the engineer regarding the lithologies of various beds of rock as well as facilitate the detection of the fresh water/saline water boundary. The actual resistivity of groundwater depends upon the ionic concentration and mobility of contained salts, the latter relating to molecular weight and electrical charge. There are, of course, substantial differences between various chemical compounds, e.g. a solution of common salt is much more mobile than one of calcium bicarbonate. Temperature inversely affects resistivity because as it increases so does the ionic mobility of groundwater and this is taken into account in standardizing resistivities at 25°C.

Resistivity measurements may be utilized to determine porosity through the relationship:

$$n^m = \frac{\rho_w}{\rho}$$

where n is the porosity, ρ is the formation resistivity in situ, ρ_w is the water resistivity and m is a void-distribution coefficient (cementation factor) (see ref. 13).

Electrical resistivity logs can also be used to determine the optimum placing of well screens and also in locating aquifers.

Potential logging measures natural electrical potentials occurring within the Earth and these are termed self- or spontaneous-potentials (*SP*). Appropriate measurements can be made using a recording potentiometer connected to two like electrodes, one usually being lowered into a borehole and a second being connected to the surface of the ground. Interpretation of the resulting logs is complex, reflecting uncertainty regarding the actual cause(s) of the phenomenon. Actual recorded values can range up to hundreds of millivolts and positive values occur when flow from a formation into a well is examined, negative ones occurring for the reverse flow. Logs are read in terms of positive and negative deviations from an arbitrary norm, perhaps a thick and impermeable formation; hence they indicate permeable zones best defined where the differences between them are most marked. Care must be exercised when near to urban areas where for example artificial earth currents may be induced by electrical instrumentation. Spontaneous potentials derived from electrochemical potentials (arising from fluid concentration differences) can be expressed thus:

$$SP = M\frac{\rho_f}{\rho_w}$$

in which ρ_f is the drilling fluid resistivity in ohm-m, ρ_w is the groundwater resistivity and M is a factor depending upon the chemical compositions of the two fluids and also upon the nature of the formations adjacent to any particular aquifer. A value of $M = 70$ has been found to be satisfactory in a number of cases. Using this value and measuring SP and ρ_f enables the groundwater resistivity to be assessed.

Caliper logging involves using a hole caliper in order to measure well diameters down a well. A suitable instrument has four extensible arms with an electrical resistor which is motivated by these. The technique is to insert the device with the arms folded into a well until its bottom is reached. Then the arms are released by detonating a small charge and the average hole diameter logged as a continuous graph by the recording of resistance changes while the caliper is being raised.

Radioactive logging (RL) is an important tool of subsurface exploration using boreholes and wells. It requires equipment composed of three components, namely a surface unit, a set of logging sondes controlled by it and a cable on a winch for transmission of data to the surface and for powering the sonde and sources. RL methods applicable to hydrologic investigations sometimes significant in engineering work include gamma logging (GL), gamma–gamma logging (GGL), neutron–neutron logging (NNL), and neutron–gamma logging (NGL). In all, as the logging sonde is (continuously) moved along the borehole traversing the relevant soil and rock section it provides a log of the parameter under measurement at the surface.

(*i*) *Gamma logging*: This records the natural radioactivity of the rocks penetrated by a boring. Radiation from radioactive elements (these include the uranium and thorium series and potassium-40) and their decay products is detected. The relevant gamma energies are between 0·24 and 2·62 MeV. Quantitative estimates of the natural radioactivity levels in rocks are referred to the equivalent radium, RaEq (i.e. a quantity of radium which would emit the equivalent radiation dosage). Baronov[14] has provided useful data; it is recorded in Table 6.12. Data for radioelement contents of certain sedimentary rocks are given in Table 6.13. These tables show that homogeneous clay-free organogenic and quartzose sediments, such as anhydrites, coals, limestones, dolomites, quartz sands and sandstones, are the least radioactive. On the other hand, claystones and shales possess the greatest and almost constant radioactivity. These facts are useful in delineating permeabilities of strata.

(*ii*) *Gamma-gamma logging*: This involves using a sonde containing a gamma-emitting source and an appropriate gamma detector. As the sonde moves along a borehole, the detector measures scattered gamma

TABLE 6.12
MEAN VALUES AND VARIATION RANGES OF RADIOACTIVITY IN SELECTED SEDIMENTARY ROCKS

Rock	*RaEq* (10^{-12} g g^{-1} *of rock*)
Anhydrite	0·5
Brown coal	1·0
Rock salt	2·0
Dolomite	0·5–10·0
Limestone	0·5–12·0
Sandstone	1·0–15·0
Clayey sandstone	2·0–20·0
Clayey limestone	2·0–20·0
Carbonaceous claystone and shale	3·0–25·0
Claystone and shale	4·0–30·0
Potassium salt	10·0–45·0
Deep sea clay	10·0–60·0

radiation and, since this is a function of the density of the environmental rocks and fluids, the GGL method provides a log of the bulk density for the unsaturated and saturated zones in a borehole. The extent of lateral investigation is around 15 cm and factors exist which greatly influence the results, particularly the diameter of the borehole, the presence or absence of casing and the quantity of drilling mud encountered. Some of the undesirable factors can be countered by, for example, using a borehole without a casing and mechanically pressing the sonde against the wellbore or collimation of the radiation source and detector.

TABLE 6.13
RADIOELEMENT CONTENTS OF SEDIMENTARY ROCKS

Rock	*Radium*, 10^{-12} g g^{-1} *of rock*	*Uranium*, 10^{-6} g g^{-1} *of rock*	*Thorium*, 10^{-6} g g^{-1} *of rock*
Sandstone	Up to 1·5	Up to 4·0	—
Quartzite	0·54	1·6	—
Clay	1·3	4·3	13·0
Claystone and shale	1·09	3·0	—
Limestone	0·5	1·5	0·5
Dolomite	0·11	0·3	—

(*iii*) *Neutron-neutron logging*: This records scattered neutron radiation emitted by a fast neutron source in the sonde as this is moved along the borehole. NNL is based upon the functional relationship of the large cross-section of hydrogen atoms for slowing (thermalizing) fast neutrons; therefore an NNL log is a record of the hydrogen content of the medium along the profile of the borehole and this can be interpreted as indicative of moisture.

(*iv*) *Neutron-gamma logging*: This records the gamma radiation emitted by nuclei in a formation when these capture thermal neutrons emitted by a fast neutron source in a logging sonde. The neutron-capture/gamma yield in an inhomogeneous medium depends upon the moderating characteristics of the medium, its density and the number of gammas emitted per neutron capture. The presence of chlorine in a medium for instance causes a high neutron-capture/gamma yield because it emits about 2·4 gammas per neutron capture. In hydrogeological work, NGL can be utilized in order to identify the boundaries between fresh and saline waters; it is also used in the stratification of chloridic waters in an aquifer. The method is also applicable to the profiling of the moisture content of water-bearing strata and to the estimation of the content of some rock-forming elements (in the latter case neutron activation principles are usable).

All in all, RL methods can be very useful and indeed correlating GGL with the caliper log can permit the determination of quantitative values and qualitative variations in the density curve for both unsaturated and saturated zones. It has been found that the gamma background ratio for sands, sandy loams and loams and clay is 1:2:3 and the clay particle content follows the empirical relationship:

$$\text{Relative clay content} = \frac{I_\gamma - I_{\gamma\,\min}}{I_{\gamma\,\max} - I_{\gamma\,\min}}$$

where I_γ is the measured radiation of the rock, $I_{\gamma\min}$ is the gamma radiation of pure sands or sections with the highest sand content of the profile being examined and $I_{\gamma\max}$ is the gamma radiation of pure clays or sections with the highest clay content of the profile being studied.[15]

Temperature logging is, like RL, important in subsurface exploration using boreholes and wells. A recording resistance thermometer may be used in order to obtain a vertical traverse measurement of the groundwater temperature. Ordinary temperatures increase with depth following the normal geothermal gradient of about 1°C per 30 m depth. Departures from this provide information regarding unusual conditions: abnormally low temperatures may demonstrate the presence of gas; very high temperatures may indicate the presence of local heat sources which are found in volcanic and seismic areas, regions of hot springs, etc., i.e. the hyperthermal regions of Armstead.[16]

High enthalpy areas can be divided into semi-thermal regions with geothermal gradients up to about 70°C km^{-1} depth and hyperthermal regions, i.e. those with geothermal gradients many times greater than those found in non-thermal areas. High and outwardly-directed heat flow in hyperthermal fields may be ascribed to:

(a) hot spots in the Earth's mantle;
(b) localized thinning of the planetary crust;
(c) local high concentrations of radioactive minerals in the crust;
(d) exothermic chemical reactions;
(e) frictional heat associated with the differential motion of rock masses sliding over each other at geological faults;
(f) latent heat released on crystallization or on solidification of molten rocks;
(g) direct ingress of intensely hot magmatic gases pressuring their way through bedrock faults into an aquifer.

Magmatic water (i.e. juvenile water) includes water derived from volcanoes and differs from precipitation (meteoric water) in its isotopic composition. The ratio of hydrogen to deuterium in precipitation is about 6 800 : 1 compared with a ratio of 6 400 : 1 in magmatic (juvenile) steam. As a result of this difference it is possible to ascertain the relative proportions in a mixture of meteoric and magmatic waters. However, the magmatic water content of known hyperthermal fields is small, probably never exceeding 10%.

As regards heat sources, the famous hyperthermal field in California, 'The Geysers', is now described as an example. It is reported that a body of magma over 5 km deep is the likely heat source here, this lying directly under Mount Hannah, it was detected by studies of recorded seismic waves travelling through the magma chamber from distant earthquakes. This agrees with gravity work which has demonstrated a mass deficiency under the mountain. Seismic waves traversing the zone underlying the steam-producing region of The Geysers and the volcanic area found to the north and east of it undergo considerable slow-down (about 15%) at a depth of 15–20 km; this is in accordance with the fact that molten rock can cause the same effect. The zone underlying the volcanic area is no doubt partially molten and under the geothermal energy production zone there probably occurs a highly fractured steam reservoir overlying the magma. This latter fractured portion could slow down seismic waves also as noted.

REFERENCES

1. TSCHEBOTARIOFF, G. P., 1951. *Soil Mechanics, Foundations and Earth Structures*, McGraw-Hill, New York.
2. BOWEN, R., 1981. *Grouting in Engineering Practice*, 2nd Edn. Applied Science Publishers, London; Halsted Press, New York.
3. LIZZI, F., 1982. *The Static Restoration of Monuments, Basic Criteria—Case Histories*, Sagep Publishers, Genova, Italy.
4. ZARUBA, Q. and MENCL, V., 1976. *Engineering Geology: Developments in Geotechnical Engineering, 10*. Elsevier Scientific Publishing Co., Amsterdam.
5. MCGREGOR, K., 1967. *The Drilling of Rock*, C. R. Books Ltd, A. Maclaren & Co., London.
6. DUMBLETON, M. and WEST, G., 1971. Preliminary sources of information for site investigations in Britain, *Road Research Laboratory, Report LR 403*.
7. GOGUEL, J. and HUMBERT, M., 1972. *Carte des Risques Geologiques Pour La Commune de Peisey-Nancroix (Haute-Savoie)*, BRGM, Orleans.
8. ROCHA, M., 1966. Determination of the deformability of rock masses along boreholes. *1st Int. Cong. Int. Soc. Rock Mechs*, Lisbon, Vol. 1, Paper 3, p. 77.
9. BLYTH, F. G. H. and DE FREITAS, M. H., 1974. *A Geology for Engineers*, Edward Arnold Ltd, London.
10. KNILL, J., 1969. The application of seismic methods in the prediction of grout take in rock. *Proc. Conf. on* In Situ *Investigations in Soils and Rocks*, Brit. Geotech. Soc., London.
11. CEDEGREN, H., 1967. *Seepage, Drainage and Flow Nets*, John Wiley, New York.
12. BOWEN, R., 1980. *Ground Water*, Applied Science Publishers, London; Halsted Press, New York.
13. JONES, P. H. and BURFORD, T. B., 1951. Electric logging applied to groundwater exploration. *Geophysics*, **16,** 115–39.
14. BARONOV, V. I., 1957. *Spravočnik po Radiometrii*, Gosgeoltehizdat, Moskva.
15. FERONSKY, V. I., 1968. Stratification of aquifers. In: *Guidebook on Nuclear Techniques in Hydrology*, Tech. Repts Ser. No. 91, IAEA, Vienna.
16. ARMSTEAD, H. C. H., 1978. *Geothermal Energy*, E & F. N. Spon, London.

CHAPTER 7

Earth Movements and Non-diastrophic Structures

7.1. EARTH MOVEMENTS

These may be natural or manmade; there are many varieties, which can be shallow or deep-seated in origin, and they show a wide range of magnitude. All may be stated to occur where stresses within a slope exceed the strength of its component materials; in common with the movement of weathering products, mass movements along definite planes take place under the influence of gravity and constitute with the former a gravity transport phenomenon. Movements occurring in surficial layers are triggered by disturbances controlled by such surface parameters as precipitation and temperature change; those movements which take place at depth on the other hand may involve the displacement of enormous masses of rock and will be independent of these parameters, being instead influenced by adverse, deep-seated stresses. Thus movements may be microscopic, as in creep, or catastrophic, as in landslides, and the damage produced is similarly variable. Entire towns have been destroyed by falls of rock and of course slope movements can, and often do, cause difficulties in the construction of dams and tunnels.

Landslides may be very dangerous in quarrying. Frequently quarries are not correctly excavated and this alone can cause many problems. Underground quarries are comparatively rare but they are necessary when the overburden is very thick; the Catacombs in Rome are the result of such excavations in ancient times. In the UK and France (Paris) similar quarries are found and indeed some served as bomb shelters in the Second World War.

Shelf quarries are the easiest to work in—the base is not placed at the floor of the valley since accumulated sediments there obscure the actual deposit and have to be removed periodically thus adding to the expense. Obviously the higher the floor of the quarry the better because the thickness of overburden usually thins as one ascends. The dip and strike of the strata are important because if the former is into the slope, more work is necessary to remove blocks of stone. If the dip is towards the quarry floor, it is easy to extract blocks and slide them down bedding planes. In the situation where the strike is at 90° or so to the face of the quarry, a cutting may be made parallel to this so that working may proceed up the dip. A dangerous situation may arise if the dip is sharply into the floor because loosened blocks of stone may crash down. Alteration of the working face so that it is oblique to the strike of the strata will help to reduce the dip at right angles to the working face, one side remaining braced on the intact part of the slope. Slope failure is not then so likely. In the case of crushed stone quarries the dip of the strata is not so significant. The workface is usually high and blasted often using a system of vertical holes. Jointing may cause problems however—columns may incline towards the quarry floor and overhangs may then form.

Various geological factors produce movements and these are listed and discussed below.

7.1.1. The Properties of the Materials: the Parameters of Cohesion and Friction

Cohesion and friction can be combined with the effective pressure normal to a plane and utilized in the empirical law of Coulomb in order to predict maximum resistance to shear on that plane. The usual expression is:

$$\tau_f = c' + (\sigma_n - u) \tan \phi$$

where τ_f = shear, c' = apparent cohesion in terms of effective stress, ϕ = angle of shearing resistance in terms of effective stress, σ_n = total pressure normal to the plane under consideration and u = pore pressure at the relevant place. In the field, materials of any kind will possess discontinuities on a large spectrum of sizes from microscopic (mineral boundaries) to macroscopic (joints). The smallest sizes are designated fabrics and the largest structures, but these terms are misleading in this context in that the basic characteristics of both are the same, namely both are discontinuous. Mineral boundaries constitute the minutest fabric of importance in site engineering and they are particularly well developed in rocks such as schists and slates although their significance is perhaps greatest in quick clays which are highly sensitive to strain and rapidly collapse and reduce the strength of the material if subjected to it. As well as mineral orientation, it is also necessary to observe the strata and fracture state. As regards the features alluded to, of course cohesion is absent in many field situations, e.g. between open joint or fault surfaces as well as in non-cohesive soils. Friction is also variable depending, as it does, on surface roughness and mineralogy as well as the applied load.

All geological structures represent the end products of earth processes and resulting rocks relate to gravity so that the more closely the direction of a slope concords with this the more stable it will be and vice versa.

Folding is a phenomenon that occurs frequently and it has a great effect on the shear strength of stratified rocks. Morgenstern and Tchalenko[1] have showed that shearing a soft clay will produce a set of shear surfaces which ultimately become arranged parallel to the direction of shear, sliding thereafter occurring along such slickensided surfaces with the shear strength of the clay reduced to its residual value. Folding also produces such low shear strength surfaces in appropriate rocks and some may be observed in the Siwalik Series in India, as reported by Henkel.[2] Additionally, zones of this type cause problems for the stability of excavations as at the Mangla Dam site in (West) Pakistan, described by the author.[3] Actually, such zones are also found in the Carboniferous coal measures and can constitute a hazard to open pit stability.

7.1.2. Alteration in the Gradient of a Slope

This may arise from a variety of causes such as undermining by stream erosion or excavation. Very rarely the slope angle may be increased (steepened) by tectonic forces such as uplift, but a change in the internal stress of the rock mass will occur then and this can disturb formerly existing conditions of equilibrium.

7.1.3. Vibrations due to Seismic Tremors or Explosions

These can affect the slope equilibrium which results from a change in stress due to oscillations of different frequencies. In loose materials, they may disturb intergranular bonds and in this way diminish cohesion. If soils are saturated and are sensitive clays or fine-grained sands, grain rotation or even displacement is sufficient to promote sudden liquefaction.

7.1.4. Water Content Changes

These can lead to such a saturated state by penetration of precipitation into joints that hydrostatic pressures are produced and in soils pore pressures are increased. In fact, rainfall measurements definitely show that recurrent movements of slopes have taken place in years of unusually high rainfall. If clays are involved in such climatic conditions, the effects may be highly undesirable if the rainfall takes place after a drought or prolonged dry spell because such soils will have been desiccated and thus have shrunk. Consequently the introduction of precipitation will be facilitated by percolation through cracks and fissures. As well as such effects due to rainfall, groundwater may play an important part in regard to soil properties. Flowing groundwater will exert a pressure on the particles of soils and thus make slopes less stable; in fine-grained materials such as silts, the water may flush away particles altogether and thus promote cavity formation which leads to a decrease in the strength of the soil concerned and thereafter a reduction in the stability of the slope of which it is a part. Confined (artesian) groundwater also exerts an uplift pressure on overlying beds.

Any water, be it surficial or groundwater, so long as it is in the liquid state will tend to reduce the strength of geological materials in a number of ways as follows:

(i) it may alter the mineral composition by for example hydrating anhydrite to gypsum;
(ii) it may infill pores and reduce capillary tension, binding mineral grains together;
(iii) it may increase the bulk density and thus change stresses within the mass;
(iv) it may produce seepage forces tending in some cases to cause instability.

Considering pore pressures in a little more detail, it may be noted that water below the water table exerts an hydrostatic pressure on the material and this acts as a force equivalent to an overall tensile stress at the grain contacts in granular beds or point contacts in joints and similar fissures. Reference to the Coulomb expression above shows that an increase in pore pressures results in a decrease in shear strength of the material if this is non-cohesive. If the soil is cohesive, then an increase in pore pressures to such an extent that they become equal to the normal stress, such that $\sigma_n - u = 0$, will make the strength of the material entirely dependent on the cohesive strength. Holm[4] has described an interesting case of the utilization of a drainage system to increase the stability of a slope on the west side of the Oslo Fjord in Norway. This slope was a natural one of 20° incline about 10 m high and 300 m long, excavated in horizontal sand and gravel layers overlying 10–18 m of Norwegian quick clay. This rested on a dense sand and gravel blanket with bedrock below it. Water in the bedrock supported a piezometric surface 5 m above ground level at the shoreline and so water leaked steadily out of the lower half of the slope. A retaining wall, etc., were to be constructed at the top of the slope and the stability had to be increased by decreasing the pore pressures in the basal sand and gravel layers. A system of vertical sand drains was put in to function as relief wells and the pore pressures decreased considerably while they were under construction.

Turning to the subject of seepage forces, these can be great enough to dislodge mineral grains for poorly cemented materials such as silts and afterwards wash them away. This is a process termed internal erosion and it entails loss of underlying support so that overlying layers will normally collapse by sliding, as happened at Castle Hill in Newhaven, England, as described by Ward.[5] Here there was seepage of surface water draining from road and topsoil into fine yellow sand lying between clay beds. This sand became saturated and groundwater flowing through it was washing away its finer parts at the seaward face about 2·7 m below the level of the drain. The overlying strata failed and remedial work had to be effected. This included sealing off the drain and protecting the seaward exposure of the sands with a suitably graded filter. Such filters are made up of granular material with a grading such as to produce a network of pores inhibiting the transport of fine material above a certain size. Requisite grading size is obtainable from Terzaghi's filter rule for filters which states that a material satisfies the essential requirements for a filter if its 15% size (D_{15}) is at least four times as large as that of the coarsest material layer in contact with the filter and not more than four times as large as the 85% size (D_{85}) of the finest adjoining layer of material.

7.1.5. Frost

This can have considerable effects in fissures in rock because of the expansion of water on freezing and the resultant pressure tending to open these features and lower the cohesion of rocks even further. This subject is discussed in detail in Chapter 1. However it can be stated here that in clays and sands with clays in them, laminae of ice form and when these melt there is an increase in the water freezing at the surface and the result is that the water table rises and the overall equilibrium is disturbed.

7.1.6. Weathering Due to Physical (Mechanical) and Chemical Processes

Both processes disturb the cohesion of rocks and it has been found that a number of landslides have for example involved chemical processes such as ion exchange in clays, hydration, etc. Manmade weathering by the removal of trees (deforestation) is of course very adverse to slope stability because tree roots retain the soil *in situ* or at least permit only soil creep. Such deforestation also interferes with the water regime in subsurface layers. Devastating consequences can result as has been observed by the author, for instance in the Kandi Tract running northwest–southeast along the Siwalik Hills in Punjab, India, and also in Central Anatolia near the Soganglu Valley south of Ürgüp.

Of course weathering acts over many years so that both the short and long term stability of slopes can be affected. Generally speaking the materials most liable to attack are soft sediments, i.e. soils and the infillings of fractures in rocks. The overall effects of weathering on such soils is exemplified in the behaviour of the Keuper marl as described by Chandler.[6] Both the liquid limit and the natural moisture content of this material increases with effective weathering whereas the bulk density, permeability, apparent cohesion in terms of effective stress, angle of shearing resistance in the same terms all decrease. The rate of such changes will depend almost equally upon climate and the original mineral composition.

Many examples of slope failure due to effects of chemical changes could be mentioned. For instance, Matsuo[7] has referred to a railway cutting at Kashio in Japan which failed after being stable for a decade. Clays, clayey sands and gravels were involved in the sliding and groundwater emerged from the toe of the slide at variable rates ($1{\cdot}3$–$50\ \text{cm}^3\ \text{s}^{-1}$) with a much higher free carbonate content than the precipitation of the region. The relative quantities were in fact $2{\cdot}44\ \text{mg litre}^{-1}$ compared with $0{\cdot}039\ \text{mg litre}^{-1}$. The removal of carbonate was shown by laboratory tests to have lowered the strength of the slope and so promoted failure. It was proposed that calcium should be recharged into the ground by spreading an aqueous solution of calcium salts on the surface.

Two of the most significant and rapid chemical changes which can occur in geological materials are the expansion of anhydrite on hydration to gypsum and the decay of pyrite or marcasite through oxidation to sulphuric acid. These changes can take place in a few weeks and greatly reduce the stability of slopes.

Addition of water may cause swelling and its removal shrinkage, with an accompanying tensile stress. Usually this occurs between the ground surface and the capillary fringe just above the water table, i.e. in the unsaturated zone (the zone of aeration). There are coefficients of swelling and shrinkage which can be applied to the assessments of geological materials. Both utilize a ratio dividing the change in volume on wetting or drying by the original volume, either dry or wet. The maximum swelling or shrinkage in sediments has been observed to occur normal to the stratification in sediments as was demonstrated by Murayama.[8]

Freeze–thaw causes periodic expansion of ground, but affects only about 20% of the total land area of the Earth (the circum-polar areas). The phenomenon was much more widespread during the Pleistocene and produced changes then which still influence the stability of sloping ground—hence the importance of understanding periglacial factors in construction work involving slope stability.

7.2. CLASSIFICATION OF SLOPE MOVEMENTS

Appropriate data relating to various rock types are given in Table 7.1. Slide movements are greatly influenced by factors listed in the table and they may be classified according to mode and rate of movement, shape of the plane of sliding, type of material which slips, etc. Various schemes are given below.

1. In 1925, Terzaghi proposed a classification based upon the physical properties of the relevant rocks and this proved to be useful in site engineering.
2. In the USSR, Savarenski's classification is utilized—this is based upon the shape of the sliding plane and has the categories asequent, consequent and insequent.

   The categories may be defined as follows:

   (a) asequent landslides develop in cohesive and homogeneous soils along curved and roughly cylindrical surfaces;
   (b) consequent landslides occur along bedding planes, joints or planes of schistosity and dip downslope;

TABLE 7.1

ϕ VALUES AND UNCONFINED COMPRESSIVE STRENGTHS (DERIVED FROM REFS 9, 10 AND 11)

Material	ϕ_d *(assumes some cohesion)*	ϕ_{ult} *(cohesionless surface)*	*Strength, MN m^{-2}*
Igneous rocks			
Basalt	47°	<45°	150–300
Dolerite	47°	45°	100–350
Gabbro	47°	45°	250–300
Andesite	31–35°	28–30°	
Porphyry	40°	30–34°	
Diorite	35°	31–33°	150–300
Granite	>35°	31–33°	100–250
Sedimentary rocks			
Sandstone/Greywacke	27–38°	25–34°	20–170
Siltstone	43°	43°	10–120
Shale/Mudstone	37°	27–32°	5–100
Limestone	>40°	33–37°	30–250
Metamorphic rocks			
Schist	Variable and probably low		100–200
Gneiss	Variable, lower than quartzite		50–200
Quartzite	44°	26–34°	150–300
Infilling materials	*Average ϕ value*		
Calcite	20–27°		
Breccia	22–30°		
Rock aggregate	40°		
Shaley material	14–22°		
Clay	10–20°		

(c) insequent landslides cross the bedding and usually are large scale with sliding planes extending deeply into the slope.

3. In the USA the classification proposed by Warnes[12] has been used and its essential features were accepted by the Swiss Association for Rock Mechanics in 1964.
4. In Czechoslovakia Nemčok[13] and others have proposed a schema dividing landslides into categories based upon the mechanism of their occurrence and also the relevant rate of movement. They

referred to creep, flow, sliding and fall. There is a number of drawbacks, e.g. very slow movement may take place along a slide plane.

5. Zaruba and Mencl[14] have proposed a schema based upon age; i.e. in Czechoslovakia Quaternary cover deposits are separated from pre-Quaternary bedrock; landslides in the latter are divided according to their type of movement and also the nature of the relevant rocks. A summary is given below:
   (a) Slope movement of superficial deposits, including both slope detritus and weathered material, produced mainly by subaerial agents.
      (i) talus creep;
      (ii) sheet slides;
      (iii) earth flows;
      (iv) debris flows, muren and liquefaction of sands.
   (b) Slides in pelitic rocks (clays, marlstones, claystones, clayey shales, etc.).
      (i) along cylindrical surfaces;
      (ii) along composite sliding surfaces;
      (iii) caused by the squeezing out of soil which underlies rock.
   (c) Slope movements of solid rocks.
      (i) on predisposed planes such as bearing planes, joints or faults;
      (ii) long term deformation of the slopes of mountains;
      (iii) rockfalls.
   (d) Special types of movement.
      (i) solifluction, a slow downhill movement of soils or talus cover consequent on the alternate freezing and thawing of contained water;
      (ii) slides in sensitive ('quick') clays;
      (iii) subaqueous slides.

Slope movements develop with time and various factors may alter chronologically so that the movement concerned may undergo several stages of the development, passing from say an initial fissuring of ground through downslope travel to final deposition. Landslides may be ancient or modern and if the former, may be regarded as fossil if they cannot be revived today. Such ancient and fossil landslides may become buried and in this event are so termed. Clearly, from an engineering standpoint, the most important divisions will be active and dormant:

(i) Active landslides provide many surficial recognitional characteristics, the surface features being fresh and such aspects as tree displacement and damage to structures are obvious.
(ii) Dormant landslides may be regarded as potential and they may be overgrown or eroded so that evidence of previous sliding is not apparent. However, the underlying parameters causing sliding are still there, hence the sliding can be resumed at any time. If the original sliding originated under conditions not now operative and unlikely to return, then such a landslide may be regarded as being both dormant and stabilized.

A perfect instance of a revived ancient landslide causing tremendous damage and destruction is afforded by that which took place at the Vaiont Dam two decades ago (the reader is referred to Chapter 9 of this book for brief details). The Vaiont River flows through a steep gorge sculpted in the northern Italian folds of the Alps and the valley involved is a syncline believed to be Tertiary in age. The relevant rocks are Jurassic and Cretaceous sediments mostly of the calcareous type. There is a thick stratified limestone series with occasional shaley beds comprising the Jurassic, the Cretaceous being composed of well bedded limestones grading into limestones and marls towards the top. The Pleistocene glaciation eroded a large mass of rock, scouring the glacial valley along the synclinal axis. Stress relief occurred parallel to the valley as this superincumbent was later removed with the accompaniment of new jointing and opening of some bedding planes. Landslides took place and it is quite possible that they dammed the valley initially to be eroded away later. One slide crossed the valley leaving its leading edge on glacial sands and gravels which cap the bedrock in this locality. After the glacial retreat the valley was elevated and so the river cut a gorge 200–300 m deep in its floor, an action again causing stress relief, a phenomenon persisting until today.

Groundwater had produced many solutional features in the limestones as well.

Following an hydroelectric scheme for the valley, a concrete arch dam was constructed and completed in September 1961. A number of geological investigations was effected, starting as early as 1928, but in October 1960 an intensification of effort occurred because more rapid movement into the reservoir area was noted on the south slope of the valley within 400 m of the damsite. With this occurred the formation of a large tension gash extending along this south slope, the 1960 scar. The slide was estimated to comprise about 237×10^6 m^3 of rock and to be moving on a zone of sliding surfaces located some 200 m below ground. The rate of movement of the front was 8–10 cm daily and the western portion was moving more rapidly than the eastern—probably due to different subterranean circumstances. Of course such a vast moving mass could not be remedied except by reducing the water pressure in the slide. The slope was to be drained by adits and the water level in the reservoir lowered, this having risen during impoundment to 85% of its full height. However the water level in the reservoir actually rose an additional 10 m, filling it to 89% of its final projected height, before the projected drainage commenced. The stability of the slope thus related to the level of the reservoir. On 4 November, 7×10^5 m^3 of material slid from the toe of the slide into the reservoir in 10 min, the 1960 slip. By January 1963, lowering of the reservoir level by 50 m had been effected and until September of that year its level had varied between this position and one 15 m lower. Through all of this time, the groundwater level remained almost constant. Thereafter the slide appeared to be stable and filling of the reservoir began again so that by February 1962 the water had reached the 1961 position, i.e. it was 50 m higher. Movement then increased, but was minor (about 12 mm daily). Lowering was recommenced and the reservoir dropped back to its level of January 1962. Subsequently the toe of the slide travelled 125 cm further into the reservoir. By March 1963 it was observed that maximum movement took place when rock was flooded for the first time. It was believed that if the levels of water were raised in stages, equilibrium could be attained in the sliding mass or at least it would move too slowly for any problem to arise. Hence the water level was raised again in April 1963 and then a mass movement took place to the extent of approximately 2·5 mm daily—this went on until the former maximum level had been attained. The rate of movement increased, although to only about one-third of the rate recorded at the time of the original flooding of the valley. The filling process was halted for a time in mid-July 1963 and then begun once again in August 1963 (rock movements having been seen to be less than the ones observed a year earlier). Heavy rains then fell and the reservoir water level got to an all-time high (but was still below the approved critical maximum level). The slow movement of the slide increased so that lowering was once again undertaken at the end of September 1963. As before, the idea was to arrest the movement. The rate of lowering was slow, however, no more than 15 m weekly, so that by October 1963 the water level had diminished to the point at which it had been in November 1962 and June 1963. On the night of 9 October, actually at 23.38 GMT, there occurred a failure of catastrophic proportions lasting a full minute during which the entire mountainside descended with sufficient momentum to cross the 99 m wide gorge and ride 135 m up the further side of the valley. Moving at about 24 $m\,s^{-1}$, hundreds of millions of cubic metres of rock slid for 20 s with a shock registering on seismographs all over Europe. The rock mass, travelling at 84 $km\,h^{-1}$ sent out an enormous wave which flowed over the crest of the arch dam and reached the Piave River with the accompanying inundation of a number of villages, the largest being Longarone.

Thousands of people were killed but the dam itself survived. Several inquiries have been held subsequently, but it is still not established why the disaster occurred in legal terms, although Jaeger[15] in 1969 proposed a technical solution. According to him, failure occurred initially in the 1960 scar, a zone of tension, the value of friction diminishing to a low level. The zone nearer to the gorge experienced rotation in its upper levels and was anchored at depth. The safety factor for such a system diminishes with a decrease in the ratio of the angles of friction and one must have been low in order to permit observed deformations in the upper part to take place. Consequently the value of the other became critical to overall stability. This would be decreased by an increase in the uplift forces and the factor of safety is postulated as being between unity and some value above unity on the basis of periodic movements which are related to rises in the water level in the reservoir. The process must have continued until cohesive forces on the lower slide section were reduced sufficiently to allow failure by fracture. Prehistoric and historic sliding in the Vaiont area indicates that the contemporary disaster may represent one phase in progressive failure which has been proceeding for millenia. No doubt such a slide would have occurred naturally but the construction of the dam and reservoir must have accelerated and perhaps magnified it (see Fig. 7.1).

As regards slope movements in surface deposits generally, their examination involves a description of the latter and the first approach to the problem was made as long ago as 1944 by Hollingworth *et al.*[16] They took most of their examples from the Jurassic rocks of the Northamptonshire area in England and listed such features as cambers, dip and fault structures, valley bulges and slips. Nowadays geomor-

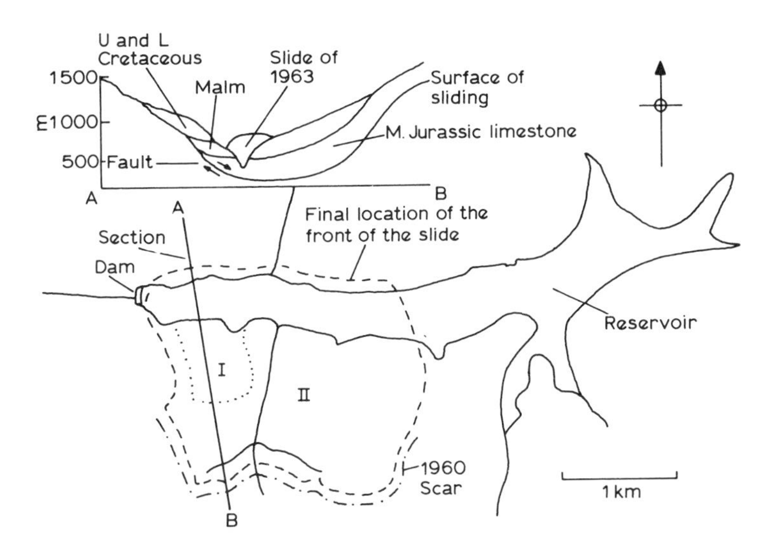

FIG. 7.1. Landslide at Vaiont near Longarone, northern Italy (9 October 1963). (I, 1960 slide; II, 1963 slide.)

phological mapping is being widely utilized in engineering projects and the technique can be of assistance in noting superficial features, notably landslips. It is a rapid, highly cost-effective technique achieved through a field mapping programme dependent upon aerial photographs being available. Landforms are interpreted in terms of their origins, material composition and associated contemporary geomorphological processes and problem or site-orientated maps are produced from the primary geomorphological survey as Doornkamp and others[17] have indicated. However geomorphological mapping cannot be regarded as an adequate substitute for geological mapping or proper site investigations although it can add useful information to that obtained by such means. The various types of geomorphological map are listed:

(a) Morphographic maps showing landforms identified by their technical names e.g. drumlins and their shapes.
(b) Morphogenetic maps showing the origin and development of landforms with appropriate genetic descriptions, e.g. a sandy alluvial plain.

(c) Morphometric maps showing the dimensions of landforms with details of slope steepness, magnitude of relief, density of the valley network, etc.
(d) Morphochronologic maps showing the distinctions between landforms according to their times of initiation.

This follows the work of Verstappen.[18]

7.2.1. Shallow Slope Movements

These include the creeping of talus which results from minute processes.

Freezing in the winter season promotes the upheaval of surficial layers as well as the loosening of fragments of rocks and when the spring thaw arrives the moved material does not return to its original position but rather tends to move downslope under the influence of gravity. Normally talus lies at an angle of repose between 25° and 35° and as noted, the debris comprising it may move as a result of temperature change.

Creep is a very slow movement which in the case of clays entails plastic deformation without any development of what normally constitutes a slide surface, but instead a zone of many partial movements exists. It is confined to the surface, in fact to a layer the thickness of which relates to the maximum depth attained by temperature variation and variation in moisture content throughout the seasons. When detrital layers creep, strata become bent terminally reflecting their disruption. This is particularly dangerous where shales, sandstones, limestones, gneiss and granite are concerned and must be watched out for by engineers on site because such bending may promote a tendency to slide on the surface of the affected beds as well as constitute a hazard for the geologist who may take false dips of such beds as the true value.

As well as such fragmental movements sheets of rock may move which may be as much as several metres in thickness. Of course they are usually stable in the dry season, moving only when precipitation acts, perhaps during the spring thaw. In clays or clayey deposits such sheet sliding movements are particularly common, e.g. in the flysch regions of Europe involving Cretaceous and Tertiary sediments.

Additionally, earthflows and mudflows may take place and the former sometimes occur when a sheet slide along cylindrical surfaces takes place. Both categories involve masses of saturated or partly soaked material in suitable morphological sites such as inclines, valley sides and cliffs. Naturally, earth slides tend to adapt themselves to the relief of the area in which they take place and hence follow gorges or channels. As precipitation soaks into the material, this becomes heavier and its shear strength diminishes. Earth slides move more rapidly than sheet slides and are very abundant in some regions such as the Carpathian flysch belt. Some may be enormous, for instance that near Handlova in Slovakia in 1960 which attained a total length of nearly 2 km over several weeks and involved a volume of material exceeding 20×10^6 m^3. This particular earthflow destroyed 150 houses, disrupted amenities such as roads and water lines and dammed a river valley as reported by Zaruba and Mencl.[14]

Talus flows are rapid and in mountainous areas may be termed muren (debris avalanche). The solidus to water ratio is about 1 : 1 and their movement is rapid, sometimes fast enough to engulf moving trains. In the temperate zones they are confined to mountain areas. Mudflows may take place in temperate and arid zones. In the latter they originate by intermittent torrential rainfall in valleys or wadis in which there is insufficient vegetative cover for protection against such erosive activity. Volcanic mudflows may arise in appropriate regions and in fact one buried Herculaneum after the explosion of Mount Vesuvius in AD 79, when Pompei was destroyed. Other similar disturbance phenomena in slopes are caused by the liquefaction of sands as a result of internal water movement (sometimes caused by the lowering of the level of a reservoir). Also, vibration may liquefy sands as a result of the following sequence of events: firstly there is a rearrangement of the grain distribution in the sands which reduces porosity and expels excess water; this being unable to escape, there is an increase in pore water pressure as a secondary effect; consequently the friction between the grains is reduced and so the sands become liquid for a time.

7.2.2. More Deep-Seated Slope Movements

Landslides may take place in clay-bearing rocks along cylindrical slide surfaces when the shear strength of these is exceeded. Such slumps possess quite characteristic forms. The curvature of the slide surface entails a rotational movement and so the surface of the mass which has slumped is inclined towards the slope with the basal area showing a concave shape. In the tongue of such a slump, transverse cracking may occur and become waterlogged so that the equilibrium of the slope is even more disturbed. Such a slump can be large or small and in any event can grow. Rainfall will play an important part as regards this slump; the shear strength of such argillaceous rocks is also involved.

Radley Squier and Versteeg[19] have given an excellent example of the interrelationship between precipitation and renewed slide movement in a clayey silt of stiff consistency derived from the decomposition of basalt. The landslide in question is deep and located at Portland, Oregon, USA. In the upper part of the slope a curved detachment surface developed which, lower down, followed the surface of weathered basalt. The length of the landslide is approximately 200 m and it is about 300 m in width. The clay layer in the slide is roughly 25 m thick. This is actually an old slide area which had its stability disturbed by building work involving excavation for a new road and carried out in 1957. Heavier precipitation during 1958–59 promoted new movement and in this a delimitation of the head scarp occurred by cracking. In the period 1963–70, a study was effected and it became apparent that the movement was triggered whenever the winter rainfall was in excess of 200 mm per month. Stabilization of the landslide was achieved by the use of a system of relief wells and drainage borings together with a buttress of riprap placed on the front of the landslide.

Fossil landslides of the same type as that in Portland exist in Egypt in the upper part of the Nile valley where a massive limestone and marly shale complex of Eocene age slid along curved surfaces. Near the floor of the valley, there is a set of subhorizontal clayey shales, the Esna shales, containing a shear surface within it. Landslides are believed to have taken place during some of the Pleistocene pluvial periods which were a climatic regimen far removed from the present arid conditions.

Soft rocks may be squeezed to produce slope movements—Hollingworth and others[16] have described the phenomenon under the term 'bulging' as occurring in open mines in iron ore deposits near Northampton in England. The phenomenon has been noted in Rumania and elsewhere.

7.2.3. Slope Movements Confined to Specific Environments

7.2.3.1. Solifluction

This combines both flow and slip and involves the surface layer on a frozen substratum; hence it occurs mostly in the Arctic and in high mountainous areas. Thaw in such ground occurs only to a depth of 50 cm or so during the brief summer and meltwater accumulates because of the impenetrability of the ground which remains frozen and effectively impermeable. This overlying waterlogged bed thereafter flows downhill under gravity as a dense sludge. Solifluction was much commoner during the Pleistocene ice stages than it is now—the resultant slope detrital material remains today and is often a mixture of fragments of rock in a sand or clay matrix lying within what was the periglacial zone.

7.2.3.2. Sensitive Clay Sliding

This is rapid; such clays are marine in origin, having formed flat areas after regression of a sea which now lie at altitudes of up to several hundred metres above sea level. The strength of these clays decreases as the salt content is reduced in the pore water. This salt is removed by osmosis of percolating water which may be derived from precipitation. The decrease in salt content is accompanied by a decrease in the strength of the clays because the bonding between the particles and water is loosened. Additionally the liquid limit decreases and the sensitivity, the tendency to strength loss by moulding, increases. The water content of the clay does not change, but the strength loss results in a sliding movement in which the material behaves as a viscous fluid, almost like a grout. The phenomenon is particularly dangerous because it can take place in nearly flat areas (slope angle less than 5°) at great velocity.

Landslides of this kind are common in eastern Canada, Norway, etc., as well as in Alaska where they caused the great damage during the March 1964 earthquake in the town of Anchorage. This is located on a littoral plain composed of a 20 m thick sand and gravel bed lying on clay. The seismic shock liquefied a 7–10 m thick bed of sensitive clay extending to the area of the town at sea level. The result was that the overlying stiff clay and *its* overlying sand and gravel bed floated seawards on the liquid clay. At that time of year the ground was frozen and hence the rocks moved in large blocks without splitting up. Thus several houses were transported metres away from their former sites without too much damage (except of course for the destruction of underground service lines). The blocks moved horizontally and the sensitive clay was squeezed before the slide into ridges. At the head of the slide a set of deep grabens originated (see Fig. 7.2). These are limited by antithetic features.

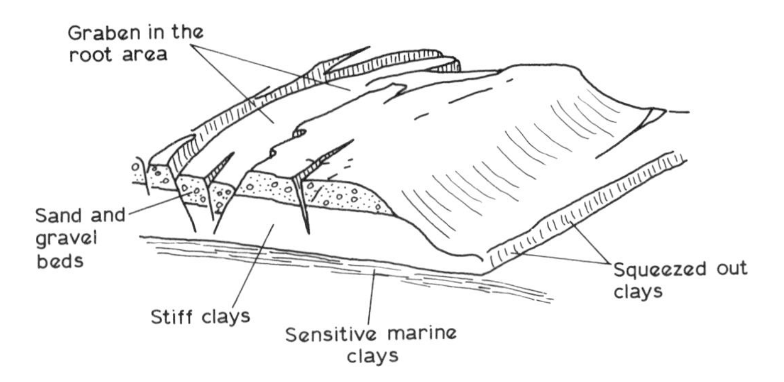

FIG. 7.2. The translational landslide triggered by the liquefaction of sensitive clays during the Alaska earthquake of March 1964. (After ref. 23.)

7.2.3.3. Subaqueous Slides

These are created by the slippage of unconsolidated sediments on a sloping sea bottom, which mostly comprise of clayey or silty and calcareous muds or fine sands. Such slides can be of practically any size and contain structures ranging from bending of beds to complex slumps. Possibly the optimum conditions for the origin of subaqueous slides are to be found in deltas where deposition is rather fast and the slopes are steep. Movement may be stimulated by seismic tremors. In unconsolidated clayey sediments, there will arise an excessive pore pressure which triggers sliding even on gentle inclines. In this case no external stimulus is required.

Subaqueous slides may initiate turbidity currents carrying disturbed sediments mixed with water far out into sedimentary basins. If later consolidated, these constitute a rock category, namely turbidites. The rock greywacke is typical of this category; it is a badly sorted material with many coarse angular grains of quartz and feldspar with mica. Greywackes form in unstable environments, for instance during the infilling of a geosyncline and they show graded bedding, the sediments passing from coarse to fine from the bottom up (a result of differential settlement). On the undersides of such greywackes may occur sole markings showing the direction of the currents which were operating at the time of formation.

Turbidity currents may rupture submarine cables and this occurred as an after-effect of an earthquake near Newfoundland in 1929. They reached a velocity of $4{\cdot}4\ \mathrm{m\,s^{-1}}$ (just under $16\ \mathrm{km\,h^{-1}}$) as calculated on the basis of the time lapses between successive cable failures. Subaqueous slides have been recognized in fossil form in many geological formations, especially in the flysch series.

7.2.4. Other Movements

In valley bottoms block slides may arise and these will occur where a plastic layer underlies a block of heavy rock. Such a layer, if thin, can promote movement and this may be accentuated by tilting of the overlying block. An appropriate situation could be, for instance, a set of sandstones overlying a soft marl. Block slides have been recorded in the area flooded by the Bratsk Reservoir near Lake Baikal in Siberia. The velocity of movement is usually very slow so that it may be described as a type of creep and of course there is no definite sliding surface but it is a situation in which the block sinks into a soft underlying layer which then undergoes plastic deformation with accompanying slow downslope movement.

In contrast, solid rock slides along pre-determined surfaces such as bedding planes or fault planes or joint planes and the phenomenon occurs where these dip downslope with a disturbance of their continuity at the foot of such a slope.

In the case of stratified rocks, bedding planes exist and the dip cannot exceed the maximum angle at which stability is possible. Undercutting of such beds by a river or through the work of Man such as excavations will result in them holding their position by friction alone; this can be diminished in a number of ways, such as through freeze–thaw or by the hydrostatic pressure of water in joints if there is no possibility of free outflow for such water.

Occasionally tectonic uplift can take place in a region and this can cause an increase in slope angles with concomitant failures where this is too great. In high mountain areas, rock sliding along bedding planes can be disastrous because the differences in elevation promote accelerations practically equal to those encountered in rockfalls.

A vast slope failure of the above type occurred in the Mantaro Valley of the River Mantaro in Peru on 25 April 1974 as recorded by Kojan and Hutchinson.[20] From the summit of a very steep slope, a mass of Permian sandstone with interbedded marls slid along bedding planes and a subsequent flow of debris extended for over 8 km, blocking the valley and creating a temporary lake which retained approximately 670×10^6 m^3 of water. The actual volume of the rock mass is estimated at about 1×10^9 m^3 and the elevational difference of 1 500 m induced a velocity of 120–140 km h^{-1} so that the entire event occurred in 3 or 4 min during which time a number of villages were destroyed and 450 people killed. About six weeks later, the dam formed by the debris became overtopped with water and a couple of days later it was washed away completely. The flood wave was 35 m in elevation, but nobody was killed because of prior evacuation. Several factors operated in this event. One was that landslides are fairly common in this particular valley, e.g. one took place in 1945 at which time the valley became obstructed by a large slide in jointed granodiorite (see the report of Snow[21]). The already-cited Vaiont is another instance, a catastrophe which rendered useless one of the largest arch dams in the world (an Italian report of this has been made by Selli and Trevisan[22] (among many others)).

Rock slides of the above type are rather common in many parts of Europe, for instance in the Carpathian flysch areas where the basic cause is the nature of the rocks and the general topography. The deposits are soft and the slope angles lag behind the erosion rate in terms of adjustment to this. In consequence the slopes become steep, sometimes steeper than the stratal dip. Thereafter, if the dip is towards the valley, excellent conditions for the promotion of a rock slide are set up. Of course, the activity of constructors may start slippage as well—excavations for railways and roads, etc., and quarry work may be affected.

There is another type of slippage which is rather long term and this is gravitational sliding which takes place only in rocks which can undergo plastic deformation. These include phyllites, mica schists, chlorite schists and the like, which can be deformed by differential movement on slopes, movement which may take place along planes of bedding, schistosity or jointing. The phenomenon is sometimes termed gravitational folding and can be observed in young mountains such as the Alps and Carpathians. If it is recognized in time, difficulties in the construction of say tunnels or dams can be foreseen and appropriate measures taken to deal with it.

Rapid fall of blocks is rather different and comprises rockfalls which are characterized by high velocity. They can involve loose individual blocks or large rock masses and can eventually coalesce to form talus (scree) fans at the foot of mountain slopes. The inclination of such fans may vary from 30 to 40° according to the slope and dimensions of the relevant rock fragments. Sometimes velocities attained during descents are as much as 200 km h^{-1}. As might be expected, rockfalls of this type are common on steep slopes of mountain ranges and especially in glaciated regions with deepened valleys. Thus it is not surprising that they have been recorded in the Alps, the Carpathians, the Himalayas, the Andes and Rockies. Probably the biggest is that recorded from the Bartanga River Valley in 1911. The total rock mass involved was $4{\cdot}8 \times 10^9$ m^3 and this blocked the valley and created a lake 75 km long and 262 m deep. This part of the Pamir Mountains is notable for rockfalls, although not usually ones quite as large as this. The origin of it and others may be seismic shock. It has been stated that the 1957 earthquake in the Baikal graben caused rockfalls in the Mujiskii khrebet, some 220–230 km away from the epicentre, these covering an area of 150 000 km^2. Many other instances from such mountainous areas could be cited and so too could cases from fjords in Norway where the phenomenon can trigger swash waves up to tens of metres in height. These may cause coastal flooding. Appropriate remedial action can reduce the danger and may include the utilization of cables or rock bolts.

7.3. ANALYSING SLIDES

7.3.1. Slab Slides

These occur because upper beds are bound loosely to lower ones or else are different from them structurally. If the component of gravity pushing the sliding part downslope is represented by T, and f is the coefficient of static friction between the upper and lower beds, then since T together with N, a force normal to the plane of sliding, comprise the weight of the slab tending to slide, the expression

$$T = fN + C$$

is applicable, where C is the bond resistance. As well as such successive, translational slides developing downslope, the opposite effect may occur and a retrogressive slide may tend to develop upslope. This is composed of a set of simple rotational slides which follow each other rapidly.

7.3.2. Rotational Slides

These are best manifested in homogeneous materials and the smallest value of the ratio of the resisting forces to the driving forces constitutes the relevant factor of safety for the slope concerned. If this is less than unity, then the slope is definitely unsafe. In checking slope

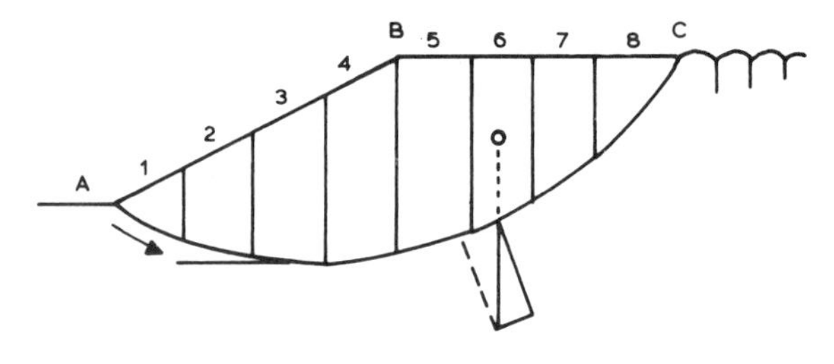

FIG. 7.3.

stability, it is necessary to know both ϕ, the angle of shearing resistance, and the unit cohesion c.

The method of slices is applicable: The slope is subdivided into several slices and the stability of each one is investigated. The principle is indicated in Fig. 7.3.

Rotational slides may also occur in semi-homogeneous materials such as varved clays or gravelly clays or in fissured, jointed materials.

7.3.3. Pore Pressure Sliding

This involves investigation of the piezometric surface of groundwater. It occurs where groundwater is constricted (artesian) and borings are made into the aquifer.

7.4. STABILIZATION MEASURES

It may not be possible to completely stabilize a situation, but an attempt must be made and this often entails restriction of human interference. In all cases a rigorous investigation is necessary. This will follow the same principles as any engineering site investigation although the following emphases are advisable:

(a) development of an idea of the type of slope movement to be expected and determination of the probable position of the slide surface if failure takes place;
(b) analysis of all surfaces of weakness which could promote sliding such as planes of bedding or faulting or jointing;
(c) where sliding has occurred, the position of the slide surface (determined from boreholes) should be ascertained; lines of stakes may be inserted to record vertical and lateral motions

over periods of time and there should be a thorough examination of the root of the slide;

(d) in dangerous slopes, determination of the supply channels feeding them with precipitation must be made and also the subterranean avenues of supply of groundwater must be discovered;

(e) stress measurements should be effected on the ground.

Subsequent to all the above work, a statical analysis may be made, although this will not always furnish reliable results.

As regards slope stabilization this can be effected by using one or several of the following approaches:

(a) A change in shape of the slope—This involves unloading of the potential slide area with removal of the load to the toe.

(b) Drainage of the slide area—As water exerts hydrostatic pressure which can function in a lateral sense in pores and fissures in rocks or as an uplift pressure on the impermeable caps of permeable beds, it is necessary to remove it from areas of potential sliding. A permanent drainage system is proposed. Subsurface drainage may be resorted to and this will require the installation of trenches infilled with permeable material (if the downslipped beds are not too thick). Water from depths exceeding a few metres may be drained by borings sometimes constructed to hundreds of metres of depth. Horizontal borings may be used.

(c) Anchorage is sometimes desirable—If there is no possibility of reducing the angle of inclination or of installing adequate drainage, walls may be constructed so as to retain the slope. Behind such retaining walls groundwater may accumulate, if it is not drained away, so that waterlogging may follow during the spring thaw. Subsequent slippage which may overflow the wall can occur.

Clearly, therefore, the stabilization of a slope capable of sliding by using a retaining wall is not useful unless adequate drainage is provided. There is another means of handling this problem, however:

(d) Such a slope may be covered with a thin layer of permeable material so that water in the near surface layer will not freeze in the winter.

An excellent instance of the stabilization of an ancient landslide occurs on a motorway near Zürich in Switzerland. Here the road runs on the concave bank of the Limmatfluss and has been constructed over pre-existing routes. Lower Molasse rocks occur and comprise marls, silts and sandstones. (In regard to the word 'molasse' it is important to note that it comes originally from Switzerland and refers to sediments derived from the erosion of mountains after the last stages of an orogeny (mountain-building movement).) In Switzerland itself, molasse includes conglomerates, breccias, arkoses and shales formed in inter-montane basins (non-marine) and similar deposits have been recorded from the Triassic of the eastern USA (the Newark Sandstone) which is derived from the Appalachians and from the Old Red Sandstone (Devonian) of England which is derived from the Caledonians. Sediments originating from the erosion of developing fold structures which later deform constitute flysch and in Switzerland such sediments are marine and comprise conglomerates, breccias and argillaceous rocks. As regards the above-mentioned stabilization problem, the landslide, which is a sizeable one, overlies the molasse and dates from the last Würm interstadial. Researchers demonstrated that any interference with it would provoke new movement so that precautionary measures had to be taken. Road cuttings were supported by cast *in situ* piles which were anchored into the bedrock. Anticipated new movement would have been triggered by groundwater outflow from the sandstones (also from a fault zone) so that permeable strata in the bedrock were drained horizontally using wells and borings. The motorway was carried on a bridge spanning the area considered to be hazardous and the piles of the foundation of this bridge were driven well into the bedrock. The cutting for a lower level road was treated with anchored piles also and the upper portion of the incline relieved by a box-beam structure of reinforced concrete.

REFERENCES

1. MORGENSTERN, N. and TCHALENKO, J., 1967. Microscopic structures in kaolin subjected to direct shear. *Geotechnique*, **17,** 309–28.
2. HENKEL, D., 1966. The stability of slopes in the Siwalik rocks in India. *Proc. 1st Int. Cong. Int. Soc. Rock Mechs*, Lisbon, Vol. 2, pp. 161–5.
3. BOWEN, R., 1981. *Grouting in Engineering Practice*, 2nd Edn, Applied Science Publishers, London; Halsted Press, New York.
4. HOLM, O., 1961. Stabilization of a quick clay slope by vertical sand drains. *Proc. 5th Int. Conf. Soil Mechs and Fdn Engrg*, Paris, Vol. 2, pp. 625–7.

5. WARD, W., 1948. A coastal landslip. *Proc. 2nd Int. Conf. Soil Mechs and Fdn Engrg*, Rotterdam, Vol. 2, pp. 33–7.
6. CHANDLER, R., 1969. The effect of weathering on the shear strength properties of Keuper marl. *Geotechnique*, **19,** 321–34.
7. MATSUO, S., 1957. A study of the effect of cation exchange on the stability of slopes. *Proc. 4th Int. Conf. Soil Mechs and Fdn Engrg*, London, Vol. 2, pp. 230–3.
8. MURAYAMA, S., 1966. Swelling of mudstone due to sucking of water. *Proc. 1st Int. Cong. Int. Soc. Rock Mechs*, Lisbon, Vol. 1, pp. 495–8.
9. HOEK, E., 1970. Estimating the stability of excavated slopes in opencast mines. *Trans. Inst. Min. Metall.*, **79A,** 109–32.
10. HOEK, E. and BRAY, J. W., 1974. *Rock Slope Engineering*, Inst. Min. Metall., London.
11. ATTEWELL, P. B. and FARMER, I. W., 1976. *Principles of Engineering Geology*, Chapman and Hall, London.
12. WARNES, D. J., 1950. Relation of landslides to sedimentary features. In: *Applied Sedimentation*, John Wiley, New York.
13. NEMČOK, A., PAŠEK, J. and RYBÁŘ, J., 1972. Classification of landslides and other mass movements. *Rock Mechanics*, **4,** 71–8.
14. ZARUBA, Q. and MENCL, V., 1969. *Landslides and their control*, Elsevier-Academia, Amsterdam, Prague.
15. JAEGER, C., 1969. The stability of partly immersed fissured rock masses and the Vaiont rock slide. *Civ. Eng. and Pub. Works Review*, pp. 1204–7.
16. HOLLINGWORTH, S. E., TAYLOR, J. H. and KELLAWAY, G. A., 1944. Large scale superficial structures in the Northampton Ironstone field. *Q. J. Geol. Soc., London*, **100,** 1–44.
17. DOORNKAMP, J. C., BRUNSDEN, D., JONES, D. K. C., COOKE, R. U. and BUSH, P. R., 1979. Rapid geomorphological assessments for engineering. *Q. J. Eng. Geol., London*, **12**(3), 189–204.
18. VERSTAPPEN, H. TH., 1970. Introduction to the system of geomorphological survey. *Koninklijk Nederlands Aardrijkskundig Genootschap Geografisch Tijdschrift Nieuwe Reeks*, **4**(1), 85–91.
19. RADLEY SQUIER, L. and VERSTEEG, J. H., 1971. The history and correction of the Omsi-Zoo landslide. *Proc. 9th. Ann. Eng. Geol and Soils Eng. Symp.*, Boise, Idaho, pp. 237–56.
20. KOJAN, E. and HUTCHINSON, J. N., 1974. *The Rock Slide Debris Flow on the Mantaro River, Peru*, Geol. Soc. Am., Bull.
21. SNOW, D. T., 1964. Landslide of Cerro Condir-Senecca, Department of Ayacucho, Peru. *Eng. Geol. Case Histories*, **5,** 1–6.
22. SELLI, R. and TREVISAN, L., 1964. La frana del Vaiont. *Annali del Museo geologico di Bologna*, **32,** 68 pp.
23. HANSEN, W. R., 1964. Effects of the earthquake of 27 March 1964 at Anchorage, Alaska. *US Geol. Surv. Prof. Paper 542-A*, Washington.

CHAPTER 8

Tunnelling

8.1. TUNNELS

In the construction of tunnels, subterranean power plants, underground railways and similar projects, geological considerations assume a great significance in terms of the selection of appropriate routes and optimum methods of execution of the necessary work. They involve in-depth studies because of course the near-surficial geology may change appreciably further down and operations of this type involve not only extrapolation from solid rock outcrops at the surface, but also the installation of appropriately sited boreholes, galleries, test pits, adits, etc., all aimed at facilitating the making of a detailed geological map of the project area with an associated set of trustworthy geological sections on an optimum scale.

Possibly the most interesting geological investigations regarding such projects have been made in the Jura mountains and the Alps where explorations have been carried out for the building of many famous tunnels. One such construction is the Grenchenberg tunnel in the Juras which on its route passes through two peaks, Graitery (1240 m) to the north and Grenchenberg (1405 m) to the south. Work entailed driving through Middle Triassic to Upper Jurassic rocks as well as Tertiary molasse and a preliminary profile was prepared by Rollier in 1902. He indicated that the tunnel would cut through two major anticlines of almost symmetrical structure which formed the above-mentioned two peaks. He assumed that the intermediate Chaluet ridge was a double syncline disrupted by a thrust fault along which the Jurassic overrides Tertiary molasse of the southern limb of the Graitery anticline. The profile made by him differs from later ones, partly because the actual tunnel follows a slightly different route. Between 1912 and 1915,

Buxtorf and others made detailed investigations before, during and subsequent to the construction and so the intricacy of the structure became apparent and was slowly unravelled. These studies demonstrated that the structure of the molasse zone at the northern tunnel opening was interpreted correctly, but the Graitery anticline turned out to be more complicated than Rollier had believed, its apical part being overturned slightly to the north and faulted, the synclinal bend disturbing its northern limb being discovered during the driving of the advance heading. As regards the Chaluet zone between the two peaks, Buxtorf supposed that the Tertiary molasse formed a simple syncline on to which a block of Jurassic rock slid from the southern slope of Graitery. A set of boreholes driven before the commencement of tunnelling demonstrated that the syncline was divided by a thrust fault along which the Jurassic outcrops. Interestingly, therefore, the original idea of Rollier proved to be closer to the truth. The tectonics and geology of the Grenchenberg anticline were even more complex. Buxtorf's 1912 profile is not in accordance with the actual situation. When the advance heading had been completed, it was found that Jurassic rocks of the southern limb had been thrust southwards on the Tertiary molasse and also that the core was disrupted by the overthrust of the Chaluet zone. Pre-execution geological profiles demonstrate clearly how hard it is to forecast tectonic details in advance because many structural details only became apparent after the central portion of the tunnel had been constructed. The cooperation between engineering geologists and engineers is mandatory for the solving of such tectonic problem situations. In the Alps, there is a number of deep tunnels worthy of examination from this standpoint:

(*a*) *The Simplon Tunnel* is 19·8 km long and connects Brig in Switzerland with Iselle in the Val di Vedro in Italy, conveying the railway between the towns of Brig and Domodossola. The associated mountain has an elevation of 3567 m while the portals are at 676 and 624 m, the crown being at 694 m. The rocks involved are slate, gneiss and calcareous micas and their maximum temperature is 55·5°C. The construction took place during the years 1898–1906.

(*b*) *The St Gotthard Tunnel* is 19·9 km long and was constructed from 1872 to 1882 through gneissic, mica schist, serpentine and hornblende rocks. Portal elevations are 1127 and 1091 m, the crown being at 1136 m. The maximum rock temperature is 30·5°C.

(*c*) *The Mt Cenis Tunnel* is 12·8 km long and the construction period was from 1852 to 1872. The rocks involved are limestone, calcareous schist, gneiss and sandstone. The portal elevations are 1248 m and 1129 m, the crown being at 1274 m.

(*d*) *The Aarlberg Tunnel* is 10·25 km long and was excavated from 1880 to 1883. The portal elevations are 1281 m and 1198 m, the crown being at 1289 m. The maximum rock temperature is 18·5°C.

(*e*) *The Lötschberg Tunnel* is 14·6 km long with portals at 1200 m and 1181 m and the crown at 1220 m. It was driven through granite during the period 1906 to 1911 and has a maximum rock temperature of 33·9°C. This tunnel gives an excellent additional instance of unforeseen circumstances. It was thought that only granite would be encountered during the driving, but about 2 km in from one portal an ancient glacial gorge was penetrated. This was infilled with debris and followed by the Kander River. Approximately 6000 m^3 of material rushed in and 25 workmen were killed. The course of the tunnel had to be changed, increasing the final length by 0·5 km.

Turning to other parts of the world, another excellent example of such a catastrophic occurrence is now cited—the Las Raices Tunnel in Chile. This is almost 4·5 km long and it was cut through the Andes Mountains in order to extend the trans–Andean railway. All the initial surface indications suggested that competent porphyrite would be encountered throughout and so no water difficulties were expected. Drilling disclosed that the porphyrite, although generally firm, was in places fractured and fissured with clay bands and some infiltration of water. Suddenly, at 525 m from one face of a section junction with a smaller section, a break-in occurred. Inflowing mud trapped 42 men who had to be rescued by means of a specially constructed tunnel. About 1900 m^3 of material blocked the tunnel and created an overlying hole. It was ascertained later that the trouble arose from the presence of a granodiorite boulder, the tunnel having tapped and drained an ancient glacial bed formerly filled with river deposits.

Clearly a number of aspects have to be studied carefully with regard to tunnelling work and even then problems may arise. Tunnels have been constructed by Man for thousands of years; in Malta there are flint-cut tunnels in sandstone which are believed to be at least 5000 years old. The earliest known submarine tunnel is one under the Euphrates River which is 3·6 m wide and 4·5 m high. It is thought to have been dry-cut following diversion of this river. The Romans constructed tunnels and they utilized fire to heat the rock until cracking took place when it was cooled. This facilitated excavation. In limestones vinegar was employed as a coolant and promoted chemical disintegration. One Roman tunnel constructed to drain Lake Feccino was 3·5 km long and $1{\cdot}8 \times 3$ m in cross-section and it is recorded that 30 000 men took over a decade to complete it.

8.2. MAJOR PROBLEMS IN TUNNELLING

Problems can be minimized if a reconnaissance study is effected prior to construction and detailed investigations are made during the work. Of special importance are the following:

(i) Rock characteristics and the general areal geology, factors influencing the route of the tunnel and its alignment.

(ii) Hydrogeological studies of the tunnel route so as to estimate where in the tunnel and in what quantity water inflows may occur. If corrosive waters are likely to be present, the tunnel may have to be abandoned.

(iii) With regard to the alignment of a tunnel, once the general direction has been decided, it is necessary to analyse parameters which control the exact positioning. For instance, a tunnel should not be sited in fractured rock or in heavily weathered rock or in regions of faulting and folding. If, as is often the case, such situations must be accepted, then certain steps may be taken to minimize deleterious effects. Thus, for example, in a faulted area tunnels should be aligned at right angles to such faults.

(iv) In locating the entrances to tunnels, great care must be exercized in order to avoid disturbing stable slopes by approach cutting work. Any slides could delay the project.

(v) Tunnelling methods must be decided, i.e. the approach to drilling. This will have a bearing on costs and depends upon several factors such as the drillability of rocks and the ease or difficulty with which they can be excavated. The actual site equipment necessary is thereby determined.

(vi) Lining of the tunnel may be required and aggregate material for this must be located and should, if possible, be close to the site. The pressure anticipated on the lining must be estimated.

(vii) The effects of the tunnel as an hydraulic structure on the adjacent area while it is being drilled and thereafter must be examined. This is especially important because tunnels are effectively drains and drawdown of a water table in an area may cause water to vanish from wells and springs in the vicinity of the tunnel.

Tunnelling procedures commence with consultation of geological maps and appropriate documents together with field studies. Normally test pits and boreholes have to be constructed so as to expose the bedrock and reveal jointing, fracture zones, faults, etc. The data so obtained must be added to the maps so as to convert them from purely geological to engineering-geological guides—these are of great importance in improving the knowledge of the structure of an area that has been obtained from preliminary work.

One of the most valuable tools for getting information is the advance heading which is not used invariably in modern tunnelling practice. Olivier-Martin and Kobilinski[1] have provided an instance. When the 13·5 km long diversion tunnel of the River Isere in the French Alps was excavated in a crystalline schist complex, the work was interrupted when the talc entered talc schists and anhydrite. The tunnelling method had to be changed and at first horizontal test bores were used for investigating. These were not successful however because of small core recovery, hence an exploratory drift was drilled ahead to a length of 1 km.

Obviously the importance of adequate knowledge of geological structural conditions along a tunnel route can hardly be overstressed—in order to gain this, where a shallow tunnel is involved, advance headings can be replaced by vertical boreholes. As for deep tunnels, horizontal test holes driven ahead of the face may substitute. This presupposes that, as is usually the case, tunnels are driven through the crust of the Earth. There is however the odd occasion when a tunnel is driven through artificially made ground and instances of this are afforded by the 14th and 16th Street tunnels in New York City. However, as a general rule the geology and structure must be determined and so must the characteristics of the rock to be encountered. This enables the site engineer to estimate fairly accurately what cover of rock must be allowed between the tunnel line and the ground surface in for instance a water tunnel wherein water is conducted under a pressure head (i.e. a pressure tunnel). The necessity for and design of an artificial lining for a tunnel, if this is necessary, can be determined and also the likelihood of having to grout the rock adjacent to the tunnel. This subject has been dealt with in detail in *Grouting in Engineering Practice.*[2]

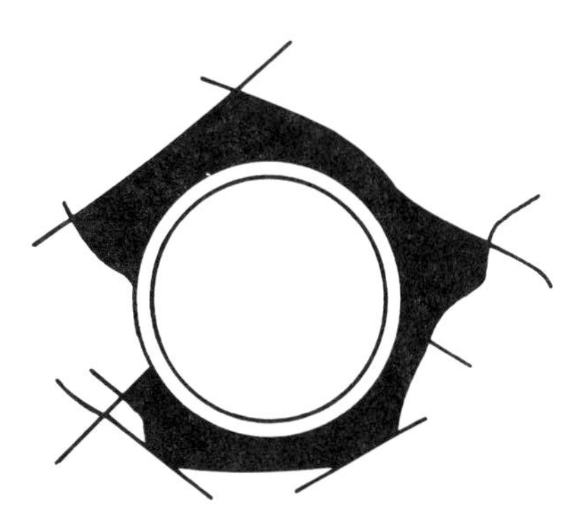

FIG. 8.1. Overbreak in a gallery in a jointed and stressed rock.

The engineer may also be able to calculate the percentage of overbreak which he may encounter during construction (see Fig. 8.1). However, geological information is also vital to the contractor. Entire construction programmes and construction methods depend upon the material to be encountered and the main hazards, e.g. groundwater, are determined solely by geological conditions. Usually Precambrian rocks are very difficult to excavate because they are crystalline and compact, but Palaeozoic rocks are easier and so cheaper to excavate. On the other hand, recent formations increase construction difficulties because generally tunnelling becomes more difficult as the rocks become younger than Palaeozoic. To penetrate such materials, interesting techniques, such as the freezing of water in water-bearing strata (in Siberia and as early as 1884–86 in Stockholm, for example), have been adopted.

As regards geological investigations and the alignment of the proposed tunnel, if a geological map on a scale of at least 1:25 000 is unavailable, then any initial reconnaissance investigation must involve a detailed mapping programme of quite a wide area along the line of the future tunnel. Of course mapping of such a narrow strip of the area in question will not give a sufficiently complete picture of the geological conditions if the tunnel is to be, for example, a deep mountain one. Here the entire drainage area of the designed tunnel must be mapped. Mapping entails:

(a) the plotting of all outcrops together with relevant dips and strikes—this is particularly relevant in the valleys and gorges transecting the mountain range, emphasizing those running parallel to the route of the proposed tunnel.

(b) in cases where surface outcrops of the solid geology are unavailable, extensive subsurface exploration will be necessary; these conditions arise for instance where surface boundaries of rock strata are covered by thick deposits of drift—here test pits may be dug.

After effecting the above activities, it is feasible to make a geological map of the relevant region and also include with it a number of longitudinal profiles, the actual number of these depending on the available time and money, the nature of the tunnel and the number of alternative routes proposed for it. At any rate, several cross-sections of the tunnel area are advisable.

With a deep and long tunnel, it will be extremely difficult to predict the geological conditions which will arise; this is partly due to the fact that geological approaches do not always suffice for solving the problem through geological profiles. It is also partly due to the many small inaccuracies which may be found in most topographical maps. Curiously, mountainous areas with complex tectonics are often easier to unravel than simpler looking regions where there is a certain degree of repetitiveness about the stratigraphical and tectonic conditions.

Sometimes, however, the geologists get it exactly right; an excellent example of this has been cited by Zaruba and Mencl.[3] Two geological profiles of the Harmanec Tunnel in Slovakia are considered, one prepared in 1937 prior to the construction of the tunnel, the other compiled in 1939 after the excavation of the advanced heading. The first was made using detailed mapping and a number of deep boreholes. The geology of the area was found to correspond almost exactly with the geological predictions of Zaruba and Andrusov.[4] The tunnel is mostly driven through Middle Triassic limestones and dolomites of the Choc nappe which was overthrust on to the Lower Cretaceous marly limestones of the Lower Subtatric nappe. Between the two nappes in question there is a layer composed of shattered dolomite and dolomitic breccia which probably belongs to the lower one. During excavation, thrust surfaces were often exposed and these confirmed the assumption of the overthrusting of older Triassic rocks on to younger Cretaceous complexes. The two profiles alluded to above differed most markedly between 30 and 31 km where the surface of the thrust runs higher than was anticipated. In consequence

TABLE 8.1
PRESUMED AND ACTUAL LENGTHS OF VARIOUS ROCK FORMATIONS IN THE HARMANEC TUNNEL

Rock type	*Presumed*	*Actual*
Dolomites	50%	47%
Triassic limestones	16%	19%
Cretaceous marly limestones	34%	34%

the limestone sequence recurred four times along the line of the tunnel and the thrust surface was met about 100 m nearer to the Harmanec portal. In addition the downwarp of the thrust between 31·5 and 32·5 km was larger than anticipated. However, the minor inconsistencies balanced each other out and so the portions of the single rock formations encountered in the tunnel did not differ much from their predicted lengths as shown in Table 8.1.

One of the most significant findings arising from geological investigations into tunnel alignment and construction is that the actual lengths of proposed tunnels are less important in terms of economic considerations than the characteristics of the rocks to be encountered. This reflects the fact that the expense of excavating in rock of an unsuitable nature can be greater than excavating in suitable rock by a factor of at least two. Unfortunately, non-geological considerations often force the tunnel builder to construct in highly unfavourable areas.

The hydrogeology of the region in which a tunnel is to be constructed is a key matter because of the fact that groundwater is a serious potential hazard in such work. An influx of groundwater into a tunnel can increase the costs of construction by at least 20% because it will necessitate installing water-proof lining and treating the drainage behind this lining. This water may be hot enough to scald and kill. A case in point is provided by the Tecolote Tunnel where in January 1951 drillers were almost drowned and suffered burns after a methane explosion at the work face. These events have been discussed in *Grouting in Engineering Practice.*[2] Water present at the level of the tunnel can be drained temporarily, but the usual approach is to install barriers and increase the air pressure in the working area, which holds back groundwater during tunnelling. However such an operation involves risk especially because of the possibility of 'the bends'. Another problem arose during the construction of the Victoria underground line in London. At one particular locality compressed air was in use in

Woolwich and Reading sands which contain water and underlie the London Clay. This diverted 'fossil' air entrapped in the voids of these sands into a nearby tunnel which was being constructed simultaneously and, since this tunnel was oxygen-deficient, fatal injuries could have resulted (although fortunately they did not). Similar phenomena have been noticed elsewhere, for instance in Seattle, Washington and in Melbourne, Australia.

Methane is always a possible hazard and in this connection the Furnas hydroelectric project, Brazil, of 1963 is noteworthy. Here, after an explosion in an outlet tunnel which caused very great damage with accompanying difficulties in remedial construction, two men died through asphyxiation. Methane has been recorded in several sites in the USSR, e.g. the Kama Dam, as well as in Rumania and elsewhere. Thus, constant attention to the potential hazards is necessary at all sites where it may occur; fills composed of refuse and occurrences of buried natural organic matter, for example, may contribute the gas to groundwater.

In the case of large scale subterranean structures such as power stations located in large caverns, drainage of the lining is essential because it cannot resist groundwater pressure. On the other hand, sometimes external water pressure is appropriate in water pressure tunnels. In sites below river valleys, drainage is not feasible because of very high inflows.

It is usually necessary to estimate the volume of inflows of water at specified points along the proposed axis of the projected tunnel. The actual nature of these must be determined, i.e. whether dispersed influxes or concentrated flows are to be anticipated. Budgeting in advance must be undertaken so as to be able to defray the costs of appropriate and remedial measures in the course of construction; these may include water proofing and the employment of temporary or permanent collectors together with their dimensions. Determinations of the sections of proposed tunnels in which increased water inflows may be anticipated is vital in the case of downwardly driven tunnels. Some figures for water influxes are cited in Table 8.2.

On this matter of water influx, the railway tunnel between Bologna and Florence in Italy may be mentioned. This is driven through the Apennines and was effected from both portals as well as from an inclined adit midway along the route. There was such a high influx of water into the downwardly driven section that the pumping station had to be enlarged more than once. In total there were 37 pumps with a

TABLE 8.2
TYPICAL WATER INFLOWS INTO TUNNELS

Tunnel	*Inflow rate, litres* s^{-1}
Simplon, Alps	up to 1 200
St Gotthard, Alps	up to 230
Weissenstein, Alps	up to 450
Grenchenberg, Jura	up to 800
Harmanec, Slovakia	up to 800
San Jacinto, California	up to 3 000
Tecolote, California	up to 680

discharge of 1200 litres s^{-1} before the tunnel was completed. Of course the inflow of groundwater into tunnels can take place through several inlets, the most important being as follows:

(a) Fractures resulting from the tectonic disruption of rocks (including joints and faults).
(b) Porous and permeable rocks such as sandstones and conglomerates. It must be borne in mind that crystalline rocks such as granite may become permeable from a practical point of view if they are heavily weathered. Also, some volcanic rocks may be highly permeable, e.g. vesicular basalts in which water occurs in the actual vesicles which were originally filled with gases.
(c) Karst openings in karstified limestones, which may be very wide. Some of the limestone was karstic in the Harmanec tunnel and in January 1938 the face of the advanced head became flooded by a concentrated influx of about 400 litres s^{-1}.

Of course it is extremely difficult to predict how much water will flow into a tunnel, but a guesstimate can be made if the area of the catchment is known accurately and also the annual precipitation. The former is not always or even often coincident with the periphery of the associated geographical basin. If, as is frequently the case, permeable and impermeable strata alternate, then the limit of the catchment area is defined by the outcrops of the lowest impermeable stratum to be encountered when the tunnel is driven. The infiltration quotient must also be known and this relates to evaporation and loss by outflow. Obviously there is a great deal of room for variability in view of the wide range of conditions of climate which may occur in an area. The relevant factors include rainfall intensity and distribution as well as duration and, in appropriate regions, the ratio between rainfall and snowfall. Air temperature and relative humidity are considered also as are geomorphology and geology. Clearly, if steep inclines of impermeable rock strata are present, runoff will be promoted and there will be no infiltration. By contrast, if there are gentle inclines of permeable, weathered rock debris and rock, then infiltration becomes dominant. Vegetation is significant as well.

If permeable rocks overlie impermeable ones forming a basin, investigation of the level of the groundwater is required. If a proposed tunnel is to be sited above this level, there would be practically no groundwater flow into it. However, if the tunnel level is to be below the groundwater table, then very concentrated inflows of water are to be expected and they are very hard to foresee, i.e. as regards quantities. The only remedy is to lower the groundwater table to the level of the tunnel.

In the construction of tunnels intended to convey water, i.e. water pressure tunnels, it is necessary to estimate the pressure of the groundwater because knowledge of this factor is essential for determining the appropriate thickness of the lining.

Once the siting of the proposed tunnel has been decided, the next step will be the selection of the alignment. Sometimes catastrophic events compel changes in this as was the case with the Lötschberg tunnel, alluded to earlier, where a glaciofluvial infilling of the valley of the River Kander was encountered 170 m under the valley. In the case of the St Gotthard Tunnel a similar disaster was confronted during its construction; the reasons only came to light when in the 1940s boreholes, drilled from the tunnel at between 2·7 and 8 km (where high pressure had been encountered during tunnel driving) revealed that the thickness of the rock overburden was only 40 m, the remaining 280 m to the surface comprising saturated glacial drift of the Andermatt Basin. Inaccurate estimation of the thickness of alluvial deposits of the Rausbach caused difficulties in the construction of the Weissenstein Tunnel near Gänsbrunnen in Switzerland in 1907, cave-ins occurring in some places which caused sinkholes to develop.

Deeply weathered rocks are also dangerous and tunnels sited in the slopes of valleys may run into danger when natural horizontal stresses are released as a result of the work. Fault zones are deleterious, constituting zones of weakness, but they can be detected at the surface if they produce depressions in the relief of the landscape. In regions of major faults, tunnels should be aligned so as to cross these at right angles if construction cannot be avoided. If this is done, it will also be necessary to change the direction of the tunnel axis as required.

As regards stratification, there is a preferred orientation of the tunnel—this is to have its axis at right angles to the strike of strata dipping in the driving direction. If the strata dip into the excavation, the situation is not so good because if blasting is carried out, blocks may slide from the face although the walls of the tunnel may be quite stable. (Gelatin dynamite was once used for blasting purposes, but blasting gelatin replaced this shortly after Alfred Nobel discovered in 1875 that nitroglycerin dissolved in collodion cotton produces a gelatinous mass more powerful than dynamite—blasting will be discussed later in greater detail.)

With reference to Fig. 8.2, if tunnelling is carried on through an anticline perpendicular to the strike of the strata, a wedged rock is encountered and this blocks the excavation work. The collapse and falling in of rock blocks may occur if the rock is densely fractured. On the other hand, if tunnelling is effected in a syncline, inflows of water are to be anticipated, particularly in the case where there is an alternation of permeable and impermeable rocks.

If the situation of driving a tunnel not perpendicular to the strike of strata but rather parallel to this be considered, a different set of circumstances arises. The oblique transgressing of strata can involve the initiation of asymmetrical pressures upon the supports and consequent large overbreaks. If the tunnel axis should happen to coincide with an anticlinal axis, the actual amount of the dip of the strata becomes of paramount importance. If the dip is gentle then rock pressure will be prevented as the roof is arched, but if the dip is steep then the strata will not obtain support in the walls of the tunnel and the load of loosened strata will apply its total weight to the roof. In the case where the tunnel axis coincides with the axis of a syncline, rock

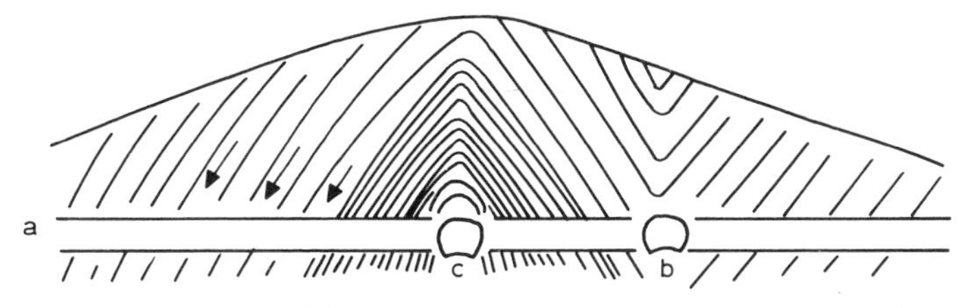

FIG. 8.2. Tunnels driven (a) perpendicular to the strike of the beds, (b) along a synclinal axis and (c) coincident with the axis of an anticline.

blocks have a tendency to slip into the tunnel from its walls and of course permeable beds will conduct water.

One of the worst possible geological conditions is encountered where vertical strata occur parallel to the axis of the tunnel—this necessitates considerable support for the roof and in addition there is even the chance that a rockfall may take place. It is apparent that structural geological considerations in tunnels driven in hard rocks are of the first importance and the parameters of significance are dip, strike, fractures, etc., all of which must be investigated thoroughly.

Of course not all tunnels are excavated in hard rock. Some are soft ground tunnels. All comprise conveyance systems, however, open at both ends and all must be driven using appropriate excavating methods. In the case of hard rock tunnels, probably the most essential matter to look into is that of the pressure which will be exerted on the supports of the finished tunnel. Static analyses can establish the feasibility of a proposed structure in engineering and the engineering geologist is in an excellent position to assist site engineers in tunnelling using his knowledge of the behaviour of rocks in general and their anticipated behaviour in the case of a particular project.

Although sometimes the engineering geologist may be expected to produce some sort of quantified assessment of such rock behaviour patterns, this request reveals a lack of understanding of the use of the earth sciences in engineering. In such a situation the geologist can produce only an opinion based upon what Soviet geologists term a 'method of analogy', i.e. another way of saying 'guesstimate'. Of course engineering methods may be applied to rocks as well as to manmade materials such as concrete, but in any event they are simplifications of the circumstances actually encountered in the field because the intricacy of nature is far too complex even for Man to understand completely even using his most advanced technologies. It must always be remembered therefore that formulae or equations applied to field situations have only a very restricted applicability or validity (this is true not only as regards tunnelling, but also in respect of many other engineering activities, including grouting). It is also important to realize that no type of statical analytical solution can assist in certain conditions in a tunnel. Thus, for instance, if planes of weakness parallel to the axis of a tunnel meet above its crown, forming a structurally weak apical region, collapse may well occur. Such a state of affairs arises in rocks which possess extensive fracturation, e.g. joint systems or faults. Such rocks are often igneous or metamorphic in nature. It is impossible to as it were x-ray the rocks in advance so consequently in dealing with them it is possibly a good idea to assume that such a set of circumstances may exist and, moreover, exist all along the length of the proposed tunnel. In practice this is often precluded through financial restraints, to some degree at least, because if it were done *in toto* the expense of driving a 'safe' tunnel would be prohibitive. The best that can be done in this imperfect world is to make the assumption anyhow, utilize driving methods and the follow-up supporting measures which will give prior warning of collapse and also possible obviate this danger. Obviously rock bolts, cables and steel supports can play a valuable part in these endeavours. Consultation with an experienced engineering geologist is highly recommended in such cases and for that matter in all tunnelling work. Rock and roof bolts are utilized in hard rock tunnels.

In rock bolting the tunnel roof may be bolted to a firm bed situated at a certain height above it in the suspension roof support (the 'sky-hook' method—see Fig. 8.3). Holes are drilled and special bolts with a wedge at the end are inserted and tapped with a hammer. A small plate of steel is placed over the outer end of the bolt and a hexagonal nut is screwed on and lodged against the plate, the nut then being tightened up and the bolt driven into place. The wedge ensures that the bolt finishes up spread apart so that friction is developed between the bolt and the rock, suspending the roof and preventing

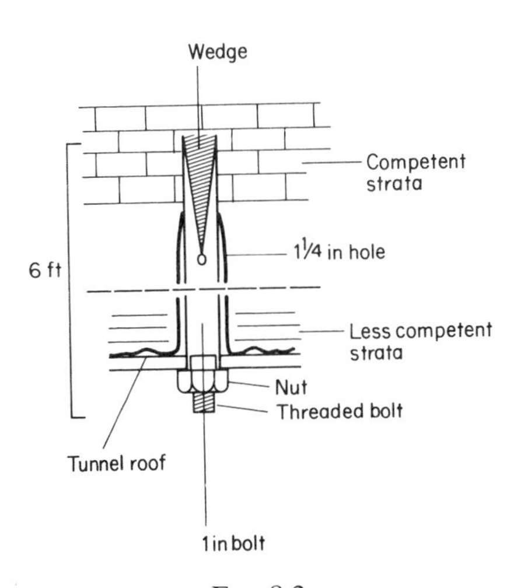

FIG. 8.3.

downward caving. The diameter of the bolts may be 3·5 cm or more (smaller ones do not have enough resistance to driving). The advantages of rock bolts are that they are not easily dislodged, improve clearance in the cross-section of the tunnel, promote better ventilation and support the roof immediately adjacent to a working face.

Resins are employed nowadays in the use of rock bolts and in fact resin anchors are commonplace. Prior to utilization of a resin anchor system, calculations of the loads ultimately upon it have to be made, using a representative factor of safety. If the loads are likely to be less than 25 tons, rebar bolts may be used. If they are to exceed this figure, EN16T or stronger bolts will be required. If stainless steel bolts are used and full bonding is attained, resistance to corrosion is at its maximum. The rock bolts may be grouted into a convoluted steel or plastic tube and sometimes heat shrink sleeving, slightly embedded in the resin used for anchorage, is a protection against corrosion. There are two ways of installing resin fixings, namely:

(i) using resin capsules;
(ii) using resin injection and pourable resins.

Bearing plates may be bedded with a polyester mortar in order to attain optimum contact with the rock surface.

Figure 8.4 shows how a second arch effect is created in constructing a tunnel using rock bolting for reinforcement purposes.

Figure 8.5 shows construction details of rock tunnels, (1) being unlined with a rock bolt, (2) being concrete lined with pressure grouting and (3) being brick-block lined (not much used nowadays). In type (2) the pressure grouting adds strength to the lining and the rock mass. Hardly any free response of the lining with respect to the surrounding rock is possible during seismic or other shock situations.

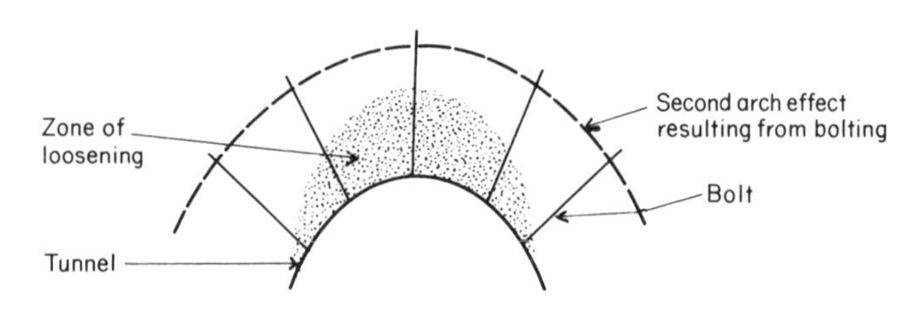

FIG. 8.4.

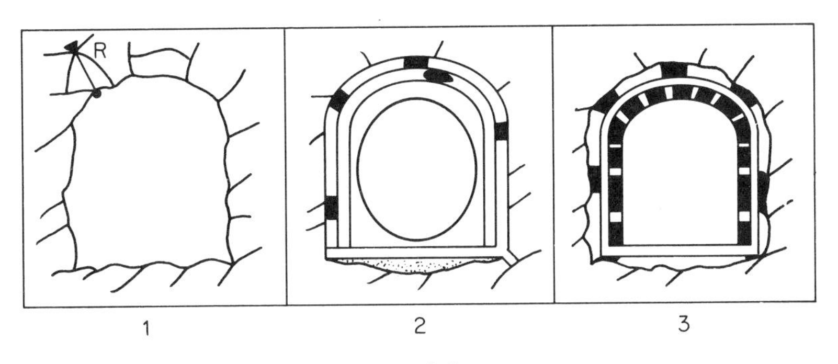

Fig. 8.5.

The question of damage to rock tunnels through earthquakes is very interesting and Dowding and Rozen[5] have compared rock tunnel response to such motions with calculated peak ground motions for 71 cases in order to determine damage modes. 42 of these recorded damages, ranging from cracking to closure; the tunnels, being located in California, Alaska and Japan, are railway and water links and are 3–6 m in diameter.

It is not only seismicity which can cause damage, but also fault movements. Either may cause ground failure in the form of liquefaction of materials or landsliding at the tunnel portals. As noted above, fault movements may be obviated by siting tunnels well away from intersections with active faults, but in the case of earthquakes it is much more difficult, especially as the effects can affect long stretches of tunnel.

The actual construction technology is important in understanding response to earthquake shaking. There are four common types of tunnel lining configurations and these are:

(i) unlined or fitted with rock bolts;
(ii) temporary steel set support with wooden blocking;
(iii) permanent concrete lining engulfing the temporary support, pressure-grouted;
(iv) final masonry lining, not in use now.

While soil masses may be regarded as continuous, rock masses are characterized by intact blocks bounded by discontinuities and these constitute weaknesses governing rock mass response. However, when rock blocks are prevented from moving by a lining, failure occurs through the intact blocks themselves. In earthquake engineering, the peak ground acceleration is the most widely accepted index of the ground shaking intensity and damage. Of course it is not the actual acceleration which causes the damage, but its use as an index affords an efficient method of determination of the imminence of large scale damage levels.

The tunnels examined by Dowding and Rozen[5] were constructed between the late 1800s and today and while two were as small as 2 m diameter, the vast majority came into the 3–6 m range. Detailed geological data were available but were used only in 23 of the 71 tunnels. 12 tunnels were in competent rock, 11 in sheared, weathered or broken rock and 3 in soil-like materials. The remaining 45 were located in non-soil materials. The 71 cases involved 13 different earthquakes with intensities varying from 5·8 to 8·3 on the Richter scale. Six took place in California, six in Japan and one in Alaska, focal depths varying from 13 to 40 km. Reports of damage were divided into three groups, namely:

(i) shaking;
(ii) fault movements;
(iii) ground or portal failure.

The three response levels are stratified as regards calculated peak surface motions. There were no reports of falling stones in unlined tunnels or cracking in lined tunnels up to 20 cm s^{-1} and 0·19 g. Up to 0·25 g and 40 cm s^{-1}, there are only a few instances of minor cracking in concrete lined tunnels. Between 0·25 g and 0·52 g or 80 cm s^{-1}, there was one partial collapse associated with landsliding and involving a masonry lining. Based on the various case histories, Dowding and Rozen drew the following conclusions:

(i) The collapse of tunnels from shaking occurred only under extreme conditions and there is no damage in either lined or unlined ones at surface accelerations up to 0·19 g with only very few instances of minor damage up to 0·25 g. Up to surface accelerations of 0·5 g, there are no examples of collapse due solely to shaking.
(ii) Tunnels are safer than overground structures for given intensities of shaking.
(iii) More severe localized damage is to be anticipated when a tunnel is crossed by a fault which is displaced in the course of an earthquake.

(iv) Tunnels in poor soil or incompetent rock which have stability problems during excavation are more susceptible to damage during earthquakes, especially where wooden lagging is not grouted after construction of the final liner.

(v) Lined and fully grouted tunnels will crack only when subjected to maximum peak ground motions linked with rock drops in unlined tunnels.

(vi) Tunnels deep in rock are safer than shallow tunnels.

(vii) Total collapse of a tunnel is found to take place only where there is association with movement of an intersecting fault.

8.3. BEHAVIOUR OF ROCKS AROUND A TUNNEL

There are some statements which should be made regarding the behaviour of rocks around a tunnel. If it is assumed that only vertical forces exist in rock masses, after a tunnel has been driven and a consequent sub-circular opening created, then these vertical forces must exert their influence through the abutments of the structure. Above the crown of the tunnel in its roof, a rock arch will develop. Rock is compressible and this results in this arch undergoing displacement with the arising of a zone of horizontal tensional stress in the roof. After the tunnel has been excavated, the vertical sections of the walls acquire a stress concentration approximately equal to three times the intrinsic vertical stress. Elastic analysis gives

$$\sigma_z = p_z(1 + B^2/2r^2 + 3B^4/2r^4)$$

where $\sigma_z = 1{\cdot}26p_z$ at a distance $r = 2B$ where p_z is the intrinsic vertical stress. The pressure normal to the vertical cross-section is given by the expression

$$\sigma_x = p_z(1 - 3B^2/r^2)B^2/2r^2$$

which gives a tensile stress $\sigma_x = -p_z$ in the roof ($r = B$) and a maximum compression stress $\sigma_x = 0{\cdot}07p_z$ at a distance of $1{\cdot}5B$ above the roof ($r = 2{\cdot}5B$). If the situation in the tunnel includes proximity of the rock arch to the tunnel wall with extensive development in a high cover of rock and high horizontal tension in the roof, then the settlement therein is given by the expression of Borecki and Kidybinski.[6] Settlements are usually quite small, say a few centimetres, but normally exceed calculated values because of the fact that at tension the deformation modulus is small. In this discussion it is assumed that there is a high cover of rock over the tunnel. Where this is not so and there is only a thin rock cover, there is a change of the state of stress accompanied by considerable shear strain.

Obviously the roof of a tunnel is a very important feature of it. Rockfalls may take place from the roof and this is particularly likely when tunnel crossings are involved. The implication is that support should be given at such points even if apparently competent, unweathered, crystalline rock is involved. The wider the tunnel, the higher is the region of tensorial stress which usually extends to about one-third of the width above the opening. Even in small tunnels there is still a tensional region. In general stress is divisible into normal and tangential or shear components and the former is regarded as negative if it is tensional in character—the positive counterpart, of course, is compressional tension in the roof which diminishes with horizontal compression. If the magnitude of the horizontal inherent stress is about a third of the vertical stress, then no tensional stress at all will be present in the roof and hence no tension zone will develop. Although favourable, this does not mean that no collapse of the roof can occur.

Horizontal stresses are always to be found in rocks in nature but in areas of slope, such as the sides of valleys or in bends along rivers, they become rather small—as a result of this it becomes more likely that a roof will cave in and such a cave may extend up as far as the surface (this may mean for tens of metres) because of vertical jointing originally arising from stress relief that occurred when the valley was eroded.

In tunnels it may be noted that several different regions may develop in the enveloping rocks and these include the following:

(i) loosened and somewhat de-stressed rock near to the walls;
(ii) rock arch with increased stressing;
(iii) rock mass in a state of unchanged stress away from the tunnel.

When a lining is installed, further plastic deformation of the rock is to be avoided, so it must possess a resistance and also be thin enough to be able to adjust itself to variations in the pressure of the enveloping rock. Both the lining and the loosened region of the tunnel rock surround constitute a unit able to resist the pressure of the rock arch.

Rock pressure will decrease with increasing rock deformation, but there are factors modifying this rather sweeping statement. For instance, in some loose rocks there may be clay contents and these decrease the strength so that rock pressure will increase and not decrease. Deformation of the walls will set up high pressures in the roof.

Now, from these considerations, it is clear that measurements ought to be made in order to ascertain the thickness of the region of loosened rock and also the depth in the rock mass at which the rock arch arises under conditions of tunnelling. For this purpose a number of techniques are available, among them the use of the extensometer which can determine the displacement of measuring points fixed in boreholes. These are fastened in place and have wires running to the mouth of the borehole which may be drilled from the tunnel. In the case of shallow tunnels, this may be done from the surface after which the displacement of fixed points measured during the driving of the tunnel is effected. Microseismic measurement may also be employed and this depends upon the effects of increasing pressure in a rock mass. Such a pressure increase may be caused by tunnelling and it will result in an increase in both the modulus of deformation and also in the velocity of seismic waves propagated around the tunnel; these waves can be initiated by blasting and the measured velocities are plotted so as to indicate the area of increased stress, namely the region of the rock arch. The results may not be very precise, but they do give an indication regarding this feature. Obviously the rock arch will develop as blasting goes on to convert advance galleries into complete sections of tunnels. Thus the prediction of the size of the rock arch is difficult and what is needed ideally is continuous monitoring through appropriate testing as the tunnel is under construction.

With reference to blasting, this subject has been considered by a number of authors, among them Langefors and Kihlström.[7] It involves drilling appropriate holes and this process is intimately connected with the type of rock involved. Holes may be of variable size and in quarries are commonly 7·6–12·7 cm diameter and up to 18 m deep. Gelignite with a suitable detonator is utilized.

Numerous attempts have been made to estimate the volume of rock in the loosened zone of a tunnel in hard rock which is liable to fall. Some entail classifying such rock on the basis of inherent strength characteristics so that, for example, quartzite is taken to be about 1·33 times stronger than granite and about 2·5–3 times stronger than hardened sandstone. Clay is taken as having about 5% of the strength of quartzite or basalt. However this is not a particularly good scheme because of the considerable variability in strengths shown by natural materials *even of the same type.* Terzaghi[8] has proposed an empirical approach stating, e.g. that dense stratified rocks without clay contents show a loosened zone with a height of approximately 25% of the width of the tunnel.

8.4. EMPIRICAL METHODS IN TUNNELLING

In practice, these often appear to offer the only solution in view of the many imponderables involved. The experience of the engineering geologist and the tunnel engineers on site is of course an invaluable aspect of the project because such personnel will know from previous work quite a lot about the behaviour of hard rocks during such operations and for instance will be aware that limestone with its frequent slabby occurrence is comparatively safe as regards cave-ins (slabs, as it were reinforcing one another) whereas granite is susceptible to such events as a result of the fact that blocks of this igneous material fracture in a brittle manner and can cave in without warning with very unpleasant consequences.

Figure 8.6 shows the stress state of a rock arch above a tunnel. This is also subject to stressing in compression and shearing with accompanying volume change. If the rock is dilatant then the compression in the rock arch will increase. On the other hand, displacement of rock towards the opening of the tunnel may take place. It is a fact that rock

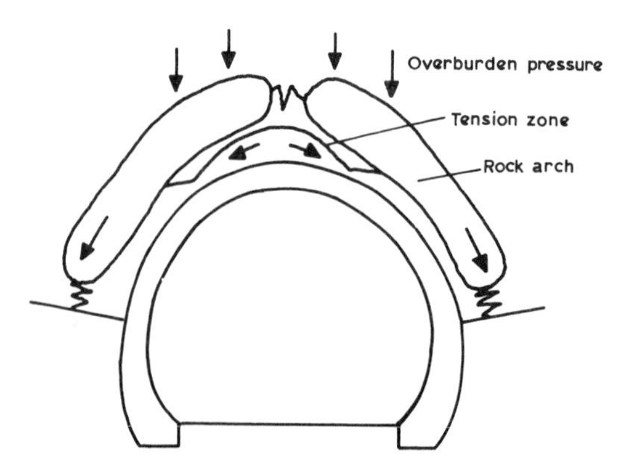

FIG. 8.6. Zone of tension developed below the induced rock arch above a tunnel.

cannot span an opening so that individual support points do not suffice and some sort of flat support is required.

Ductile rock deformation causes displacements also and these may extend quite a long way from the actual tunnel. This effect may be encountered at great depths, in levels occurring at critical depths in fact. In strong rocks such as Carboniferous slates, this critical depth may be as much as thousands of metres below ground.

In flysch-type rocks an association phenomenon occurs, namely pseudo-contractant behaviour. Concentrated stress forces slates out from the walls of the tunnel so the rock volume diminishes to promote ductile behaviour. This can be obviated by installing a flat supportive system which can be done by guniting or by putting in reinforced concrete back-grouted immediately after placement. Very hard and compact rocks resist increases in stress such as those originating from stress relief in the walls when a tunnel is constructed. Of course if the stress concentrations attain abnormal levels, e.g. when a tunnel is very deep, even the strongest rock will give and buckling will result. The opening is of concave shape so that spalling plates must be so shaped as to be elongated in the direction of the tunnel axis. Another, very unpleasant, phenomenon of a similar kind is that of rock bursts which are usually sudden and often accompanied by a bang—in the Alpine tunnels such events are termed ‘bergschlag’. In the Simplon Tunnel, even after it was opened to railway traffic, rock bursts occurred and caused bending of rails and the upheaval of ballast (hence causing several derailments). If a rock burst is large, it may be accompanied by cave-in of the tunnel opening and shock waves may be transmitted through the surrounding region.

Remedial measures against some of the above-mentioned hazards range from guniting (a reinforcement measure) to rock bolting and cable anchorages. The former is valuable because gunite layering develops high strength rapidly. The latter are important in many situations although there is no really satisfactory theory regarding their action. There are of course formulae for the designing of rock bolt systems in tunnels and these take into account the dip of the strata, the coefficient of friction along the bedding planes and the shear strength of the rock itself. However in order for these formulae to be applied correctly, many factors need to be known which are not normally known to the tunnel designer! The pragmatic approach again appears to be the answer and has been widely employed in the USA in order to determine the fields of tension and compression around tunnels. These depend upon the reaction of the rocks to tunnelling. Whenever an underground opening is constructed in a rock mass, the original stresses become redistributed around the opening causing the development of intense pressures near the advancing face. After the completion of stress redistribution, there are still remnant low pressures in the tunnel arch and in the tunnel walls, all due to the loosened zone of broken rock. Rock is usually not supported here and must be supported by a lining or rock bolts. If the latter are employed, they must be anchored beyond the zone of tension. Field trials were effected by John Knill and his associates on a basalt boulder at Edinburgh to try and elucidate the stresses developed in rock by rock bolts. Grouted and ungrouted bolts were utilized. The stress plugs which were employed gave only spasmodic results, but they nevertheless indicated that generally the stress in the rock falls off quickly with distance of the bolt from the anchor. In one case a bolt was loaded to 44·5 tons but plugs within 1 m of the anchor and only 0·3 m from the bolt showed no change in the state of stress. Laboratory tests which were carried out in conjunction with the field tests demonstrated that radial stresses are induced in the surrounding medium on loading the bolt, but there is also an important axial stress due to wedging action when there are irregularities in the borehole. The exact manner in which rock bolts function is not absolutely clear, although Knill and others think that small radial stresses applied by rock bolts modify the overall stability of an excavation as a consequence of the fact that they reduce the possibility of new cracks forming; alternatively, they may alter the directions in which such cracks orientate themselves in the event that they do form.

Rock bolts also act to prevent rock already fractured from falling away from the excavation face. If on the side of an excavation a rock slab acts as a strut, bolting it to the rock behind, it will reduce the effective length of the strut and lessen the chance of buckling taking place. It is interesting to note that a rock bolt can be useful even in loosened rock because, where the anchor point holds together two or more rock pieces, these can bond together and initiate arch action more easily than if they were separated. Grouting may increase the life span of a bolt system; it is also very valuable for treating the rock surrounding the bolts as well as for disallowing compression of the rock immediately under the bearing plate.

In Europe the empirical rule is that rock bolt lengths should be at least half the blasting round length or one third of the tunnel width, whichever is the greater. The assumption is made that the rock bolts will be inserted as quickly as feasible after firing of the round. If the rock is much fissured, it is desirable to use wire mesh suspended from the rock bolts in order to contain minor falls of rock. A good example of underground rock bolting is given by the Alpine Mont Blanc Tunnel, a road conveyance tunnel completed in 1962 and linking Entreves in Italy (elevation 1 358 m) with Hameau-des-Pélerins near Chamonix in France (elevation 1 183 m). The tunnel traverses the mountain for 11·71 km and contains a two-lane illuminated highway for cars. During construction both headings ran into difficulties and occasionally small pilot headings and extensive timbering had to be utilized to further the tunnel. Even in the granite of the main mountain massif, extensive rock bolting and netting were required in order to support both the roof and the sides. On the Italian side, the cross-section of the tunnel (approximately 72 m^2) was being advanced at 3·9 m per round in this type of rock. However the face itself also had to be rock bolted to keep it stable while the actual drilling proceeded. From 15 to 25 rock bolts were used for every round and the Italians placed more than 54 000 rock bolts and used 500 tons of protective netting in order to keep the rock in place. On the French side of the tunnel matters were not so adverse and it proved possible to drill full face headings throughout. However, despite this, a large number of rock bolts had to be emplaced in the granite and from time to time appreciable delays of up to 10 h occurred between rounds while the face was secured. In parts of the tunnel, bolts had to be inserted every 0·2 m^2. In such cases as many as 50 rock bolts were needed for the actual face. After two years the French workers had installed 72 000 rock bolts of lengths varying from 0·9 to 1·95 m (excluding those needed in the face). At completion, more than 900 000 m of rock bolts had been installed in the tunnel, a record for such activities.

Of course rock bolting may also be used regarding underground structures other than tunnels and some of these may be mentioned here. For instance the method was applied to the Poatina Power House in Tasmania where a power station 90 m in length, 25·5 m high and 13·5 m wide is located 150 m underground not far from the foot of a 900 m high escarpment. It is mostly excavated in mudstone but the roof extends into an overlying marl formation, the bedding planes being practically horizontal. The compressive strength of the marl *in situ* is approximately 34 500–41 400 kN m^{-2}. Exploratory excavations demonstrated that a semi-circular or semi-elliptical roof would be liable to failure by shear along horizontal bedding planes near the crown. Taking into account cohesion and friction, failure was to be anticipated when the shear strength was exceeded by the shearing stress on the horizontal plane. Underground testing established that the cohesion was 10 350 kN m^{-2} with an angle of friction of 22° and vertical and horizontal loads of 8280 kN m^{-2} and 16 560 kN m^{-2}, respectively. It was possible therefore to show that a semi-circular roof would be unstable except at the crown and so a trapezoidal shape was adopted. Additional work demonstrated that after completion of the work, compressive stresses in excess of 69 000 kN m^{-2} would develop in the corners between the sloping and horizontal sections of the roof with low tension zones in the vertical sides. High compressive stresses would develop as well in the bottom corners of the excavation. To relieve the stresses at the corners, slots were drilled in the rock through adjoining holes 5·1 cm in diameter with a 0·625 cm pillar between them to a depth of 0·9 m. Alternate holes were plugged with wood dowels just after drilling. When the required quantity of stress release had taken place as shown by stressometers and by measurements of the internal diameter of the open holes, the latter were grouted with rapid-setting mortar. The maximum tensile stresses on the centre line of the roof were approximately 897 kN m^{-2} and so both the roof and the walls were rock bolted using bolts of lengths 3·6 and 4·2 m set at 0·9 m centres. These rock bolts were grouted into place and pressure grouting was effected through the rock bolt holes. Three tubes were inserted in the hole with each rock bolt, two being grout tubes and the third being there in order to remove displaced air. The walls and the roof were gunited so as to prevent deterioration of the rock surfaces by drying out. After construction, movements of roof and walls were monitored using long, untensioned, ungrouted rock bolts. Those installed in the roof were 4·2 m long and the ones in the side walls 15 and 30 m in length. Measurements made across the excavation demonstrated that the walls had moved towards each other by almost 2·5 cm by the time the construction job had ended. Within the walls a dilation of about 0·6 cm was measured over the 30 m distance. This is less than the wall closure hence the dilation must have extended further into the wall.

Rock bolting is also utilized on slopes.

Great care must be exercised whenever tunnelling is undertaken near to other underground works. A case in point is the Breakneck Tunnel in New York (part of the then NY Central Railroad). Here an existing double track tunnel which was used extensively and only partly lined had to be enlarged and a new double track tunnel constructed parallel to the existing one with a 9 m wall between them. The rock material involved was very hard granite gneiss, the north portals being close to one of the siphon shafts conveying the Catskill water supply from the deep water transportation tunnel under the Hudson River. Excavation had to be performed so as not to affect this siphon shaft. This required a delicate approach because the compact gneiss could transmit explosion-induced vibrations over large distances without damping. The job was done by drilling 6·1 cm holes 1·8 m deep and setting them around the final outline of the tunnel section spaced at about 8 cm centres. The 1·9 cm gaps between each pair of holes were then broached after which the isolated core of rock was broken up and removed by standard methods. The operation was completed entirely satisfactorily.

8.5. TYPES OF TUNNEL

The various methods of tunnelling depend upon the nature of the materials with which the site engineers and engineering geologists have to deal. Most tunnels in hard rocks are built by blasting. In the full face method the entire area of the tunnel cross-section is blasted out at each round whereas the heading and bench method is different, the heading being carried on ahead of the bench which acts as a working platform. Both are shot at one round, the bench charges being detonated. A method applicable to poor rock is the side drift one in which drifts are driven ahead of the principal excavation and their supports are removed just before the central part of the tunnel is shot. A multiple drift variant involves use of more than two drifts. In every method the space between the rock surface and the supports produced by the blasting should be packed with stone or concrete and wedged. In many cases, concrete linings are installed and after their emplacement all possible back packing voids ought to be cement mortar grouted. Sometimes different components are utilized, e.g. in the Pennsylvania Turnpike Tunnels voids were filled by blowing in powdered slag, chemicals being subsequently pumped in so as to form a hard mass after reaction—in this case however the excavations were made in solid rock. In bad rock, face or excavation benches are provided with a gunite layer soon after excavation—this enables savings to be made in the thickness of a concrete lining. In the full face driving method, steel rib supports may be used in conjunction with the construction of a concrete lining far behind the excavation. This facilitates greater driving velocities.

Turning to tunnels excavated in soft rocks and soils, the situation is rather different and indeed it is possible to differentiate between these and hard rocks from an engineering standpoint by allusion to the difficulty or ease of operation in them in the following form of categorization:

(i) less easily excavated requiring no temporary supports with unrestricted lengths of excavated and unlined sections; full face driving is possible and a light lining sufficient in such competent hard rocks;
(ii) less competent rocks needing heavy support and thick lining with short sections only opened and multiple stage face driving;
(iii) saturated and incompetent rocks needing elaborate methods such as shield tunnelling or complex multistage advance with fore-poling.

Terzaghi[8] has given the following classification, which has been used widely:

(i) hard and compact rock;
(ii) hard stratified or schistose rock;
(iii) massive, moderately jointed and well-interlocked rock;
(iv) hard, jointed, well-interlocked but less competent rock;
(v) rock which is heavily fractured or densely jointed, incompetent;
(vi) crushed rock and non-cohesive sand;
(vii) ductile rock at moderate depth;
(viii) ductile rock at considerable depth;
(ix) swelling ground.

The best help which the engineering geologist can give to the tunnel engineer is to indicate to the latter the outcrops of the rocks which are to be encountered in the course of construction and effect *in situ* and laboratory testing on samples from these or examine the behaviour of similar rocks in a proximate tunnel, if one exists already. The engineering geologist may have to indicate how much explosive is to be used in a tunnelling project and this depends partly upon the nature of the rock. However the actual driving method to be utilized is also relevant. Earlier, when tunnels were driven using air hammers, holes were located so as to derive the maximum effect from the smallest possible charge. The rate of consumption of explosive was then about $0{\cdot}5\ \mathrm{kg\,m^{-3}}$ rock and up to $0{\cdot}8\ \mathrm{kg\,m^{-3}}$ where very hard rock was involved. Zaruba and Mencl[3] have stated that modern full face driving using long drill holes and a rigid pattern of drilling requires some $2\ \mathrm{kg\,m^{-3}}$ excavation.

The method of driving a tunnel relates closely to the overbreak situation. Overbreak refers to the quantity of rock actually excavated beyond the perimeter previously fixed by site engineers as the finished excavated tunnel outline. It must be kept to a minimum because payment is usually made per unit volume of excavation. The amount of overbreak depends mainly on the nature of the rock mass undergoing tunnelling so that, given equally good driving methods, the importance of preliminary geological investigations becomes apparent because they facilitate estimation of that amount. Overbreak is of course normally confined to hard rock tunnels. It rarely occurs in soft ground tunnels because such tunnels can be taken out almost precisely to the neat line of excavation and construction equipment can be designed to accurately fit this line. Close grained igneous rocks usually break closer to the theoretical section than sedimentary ones or metamorphic rocks. An overbreak will relate also to the detailed nature and structural arrangement of rock strata where encountered.

Generally speaking it is hard to determine overbreak in advance. With such stratified rocks however, the tendency towards overbreak is often encountered if they are hard enough to need extra explosive. In practice, blasting techniques used nowadays tend to reduce overbreak by applying a dense pattern of holes, a set of uncharged rim holes and a dense, precise sequence of blasts (known sometimes as millisecond blasting). The result is frequently a perfect cross-section (smooth blasting) which is also conducive to the stability of the rock walls.

The most adverse consequences of excessive overbreak are found in water pressure tunnels because here the overbreak must be filled with dense concrete. In fact concrete linings are used routinely in pressure tunnels. Only the very best quality concrete should be employed because mass concrete has almost negligible tensile strength so that, when tension is applied, it tends to form minute cracks. This is less likely with high quality concrete which is well placed and compacted to maintain the imperviousness of the tunnel. In manufacturing the concrete concerned, rock excavated from the walls of the tunnel may be utilized as aggregate if possible so that a geological opinion of this should be solicited prior to excavation—during construction, therefore, the geologist can map the roof and walls with data usually plotted on a scale of 1 : 200. Detailed mapping of this kind is valuable because if the record is retained, it can be used in repair work years later. Recently the work has been speeded up by utilizing both geologist and continuous photographic recording.

Ventilation is important in tunnels which are to be used by human beings and also as regards their involvement, determinations of temperature are significant. Rock temperatures increase at the rate of about 1·8–3°C per 100 m depth (the geothermal gradient which increases under mountains and in geothermal energy areas). Surfaces of equal temperature (geoisotherms) are almost parallel to the surface of the ground in flat areas, the geothermal gradient decreasing where valleys occur and increasing under mountains.

Dividing tunnels into hard and soft rock varieties, instances of the former have been cited earlier, but there are many others such as the Great St Bernard in the Alps which was completed in 1962, lies 1830 m above sea level and covers 5·6 km—it shortens the distance across those mountains by about 9·7 km. The previously mentioned Mont Blanc Tunnel does a better job of shortening distances—it reduces the journey from Paris to Milano by no less than 314 km. The longest railway tunnel in the world is the Cascade Tunnel of the Great Northern Railroad in Washington, USA, which is 9·8 km long and was built in 1931. The longest water supply tunnel driven from only two headings is probably the Adams Tunnel, an integral part of the Colorado–Big Thompson Transmountain Diversion Project. It is 21·1 km long and, unlike for instance the Simplon and Mt Cenis Tunnels which have horseshoe-shaped cross-sections, it has a circular cross-section.

As regards soft rock tunnels, they are driven in materials with much lower tensile and shearing strengths and the water table is an important factor since, unlike the sudden, sporadic and violent influxes of water sometimes encountered in hard rock tunnels, the combatting of water may be a more or less continuous process. One remedial approach is to lower the water table which can be done by sinking well points. Supports for the tunnel may be unnecessary if excavation is being effected in firm soft ground such as a cemented sand. In flowing ground, however, poling boards are employed to support roof and sides, the work going on by shifting these as required. Breasting is utilized to facilitate work in cohesive running ground.

The commonest method of excavation of soft rock tunnels is that involving the use of the shield. This is a circular steel box or ring usually possessing a transverse diaphragm. The front edge is a cutting edge and the rear end extends just back over the finished lining, which often comprises cast iron rings. The shield is advanced by means of hydraulic jacks which react against the completed lining to the rear. In firm beds, the work progresses in a series of steps. In soft materials, the shield is pushed through the facing soil and some of this flows through into the tunnel through openings in the diaphragm, the balance of the soil materials being displaced upwards. The shield is larger than the tunnel lining so that in working, an annular space is created and this must be grouted. This is usually done with a dense suspension of cement and sand in equal proportions.

Many shield-made tunnels exist, for instance the Holland Tunnel in New York which was the first specially planned for heavy motor traffic, parts of the Paris Metro and the Tower subway under the River Thames at Tower Hill in London, this latter constructed using the Greathead shield and opened in 1870.

Allusion may be made to an appropriate section of the Paris Metro, namely that between Charles de Gaulle–Etoile and the River Seine along which the tunnel traversed Palaeogene sand underlying limestone. This section extends from Etoile to La Défense with one double track 10 m in diameter. It was driven utilizing a mechanized shield under compressed air. With a single track tunnel of course, there are less problems with rock pressure and stability of the excavation face and if two single track tunnels had been used, the one driven first could have been employed as exploratory for the second one. In the section between Etoile and the river, high pressure and gas emanations occurred, the gas arising from lignite layers intermediate in the sand. Unfortunately small amounts of such inflammable material ignite under compressed air as happened here when large scale ignition of the lignite took place and blocked the driving. Extensive grouting was necessary in order to separate the area of danger by means of a grout curtain. So as not to deepen the level more than required, floating caissons were utilized to cross the bed of the Seine. To the west of La Défense station, the local urbanized conditions facilitated excavations of shallow covered cuttings and diaphragm walls were used as protective devices.

A recent soft ground tunnel of which the author has written[2] is the second Dartford Tunnel where the ground treatment involved waterlogged sands and gravels as therein described. Some others may be mentioned. One is the long Penge Tunnel constructed by driving through the London Clay and serving the Southern Railway en route to the Kent coast. Interestingly all the excavated material from this was removed to a brickmaking plant located especially at one of the portals and there, after moulding, fired to make building bricks subsequently applied to constructing local railway buildings, etc. Another is the Post Office tube in London which is 10·4 km long and 2·7 m diameter, contains a 60 cm gauge track automatic railway with regular trains transporting mail widely in the City.

Tunnels may also be considered in terms of their function, e.g. for people, vehicles, gas or water. The transit of sewage is an important usage. The sewers of Paris have been famous for much of the city's history.

Boston, Massachusetts has tunnels for sewage as well as for drainage and water supply and there is a complex bedrock geology locally with the Boston Basin comprising Roxburgh conglomerate and Cambridge argillite, the latter being quite competent as evidenced during the construction of the main drainage tunnel. This is 11·4 km long and much of it is under water in the harbour. It was constructed between 1954 and 1959 with two main headings connecting three shafts of which two were used as main construction shafts. On the mainland the tunnel is 3 m diameter and in the harbour 3·45 m diameter with a concrete lining often up to 60 cm thick. The city tunnel extension which was constructed between 1951 and 1956 is the same length, also penetrating solid bedrock and having a diameter of 3 m. This conveys water from the Chestnut Hill Reservoir (western part of Boston) to the southwestern district of Malden and only about 6% of it needed steel supports. It transgresses one of the main folds in the Boston Basin and much valuable geological data resulted. 106 faults were encountered, mostly with small apparent displacements. The Malden Tunnel was constructed during 1957 and 1958 and is under 1·6 km in length.

Some tunnels have been constructed for highways and normally prove to be both difficult and expensive. Occasionally they demand development of new equipment as was the case in the UK with the Monmouth to Mitchell Troy section of the twin A40 road tunnels running under Gibraltar Hill composed of Old Red Sandstone which dips from 30° to almost vertical across their line. Encountering friable claylike material and marls, calcite veins and extensive jointing, the contractors designed a new type of shield, actually a structural steel frame with an extended hood cover plate and steel fingers which proved capable of excavating as much as 2 300 m^3 of rock in a week. These South Wales tunnels were successfully completed despite groundwater in many of the joints and a maximum ground cover of approximately 30 m.

Tunnels that take highways may be for trains or cars—the Paris Metro has been mentioned above. It may be added that a great deal of information regarding the effects of geology on driving subterranean lines was gained in constructing an east–west express line in Paris. Earlier Metro lines were sited at shallow depth in Quaternary deposits but the new east–west one is located in these only at the Seine and at both ends, most of the line lying in the Palaeogene limestone and marl. Only rarely was underlying clayey sand and lignite encountered. The line was sited between 25 and 35 m below ground in order to avoid existing amenities and the usual running velocity of 100 km h^{-1} entailed a curvature radius of 500 m. The line was in fact sited without regard to the surface street system layout. The section Charles de Gaulle–Etoile–La Défense has been described and it is only necessary to mention that the method of excavation at the Auber Station necessitated the excavation of an underground opening 228 m long, 39 m wide and 19 m below the existing Metro line. The lower part of the station was located in limestone and marlstone and the arch in Tertiary sand. Intensive grouting was utilized to seal rocks and the sand and gravel above the arch were grouted first, using a fan of holes drilled from the top heading. Grouting of the limestone and marlstone followed so as to seal the areas of the future abutments. Subsequently side galleries were driven so that grouting of sand above the roof and limestone below the sole could be achieved. A cement and clay mixture was utilized for limestone and marlstone, the Tertiary sand being resin-grouted and the Quaternary sandy gravel hardened using silicic acid gel.

An excellent example of complicated tunnelling through difficult geology is provided by the Washington (DC) Metro, USA, built after a century of dreaming and a decade of design and construction as well as a daunting set of problems, both geological and non-geological. The latter included the geotechnical hazards of building under an established city containing old Federal erections some of which are two city blocks long. However James Caywood, former Project Manager for the Metro contract, stated that the most difficult engineering problem was constructing large underground stations in rock. There are 11 of these, each 210 m long, 13·5 m high and 19·5 m wide. Taking 3 years to finish, each cost about US$30 million. Leaking of water into the stations was hard to eliminate and bentonite panels were used to seal coffers. Geologically, Washington DC lies on the boundary between two very different major physiographic provinces (see Fig. 8.7). This necessitated the employment of almost every kind of construction technique including cut-and-cover, earth and rock tunnelling, retained cut and surface and elevated structures. The northwest part of the Metro is in Piedmont Province, a rather thin cover of residual overburden on crystalline bedrock. The bedrock inclines southeast at 11–23 m km^{-1}. The southeast part of the Metro is in the Coastal Plain Province, a broad and wedge-shaped mass of sediment deposited on the sloping surface of the bedrock. A northeast–southwest 'fall line' separates these dissimilar geological formations, the term 'fall line' referring to a line connecting points where rivers descend by rapids (falls) from resistant Piedmont rocks to softer coastal plain sediments. North and west of the fall line almost all construction is in solid bedrock and although this is inherently competent, extensive weathering (sometimes to considerable depth) required placement of some of the large rock tunnel stations also at considerable depth so as to ensure their location in sound bedrock. That section, including Zoological Park, Cleveland Park and Van Ness stations, was constructed wholly in solid bedrock. On the other side of the fall line, a wide spectrum of sediments occurs so that a number of soft ground construction forms had to be used. Often cut-and-cover could be employed whilst some of the soft ground tunnels were shield-driven. As the shield advanced, support was provided by steel ribs and timber lagging, a cast-in-place reinforced concrete liner later affording permanent support.

As the subterranean Washington Metro lines approach the fall line, they reach a point where mixed-face tunnelling had to be used—this is a particularly difficult type of construction. Mixed face tunnel headings were partly in soft ground and partly in rock. An instance is a section south of Farragut North station downtown. Here surface disruption by cut-and-cover work was impermissible. A mixed face tunnel was driven under a major park within a block and a half of the White House and there were no adverse effects. To facilitate commuter movement, the transit system was laid out radially with individual routes crossing the entire metropolitan area and only a few feeder lines and the locations of the stations were selected after consideration of existing and projected land use, historic patterns of bus and car travel

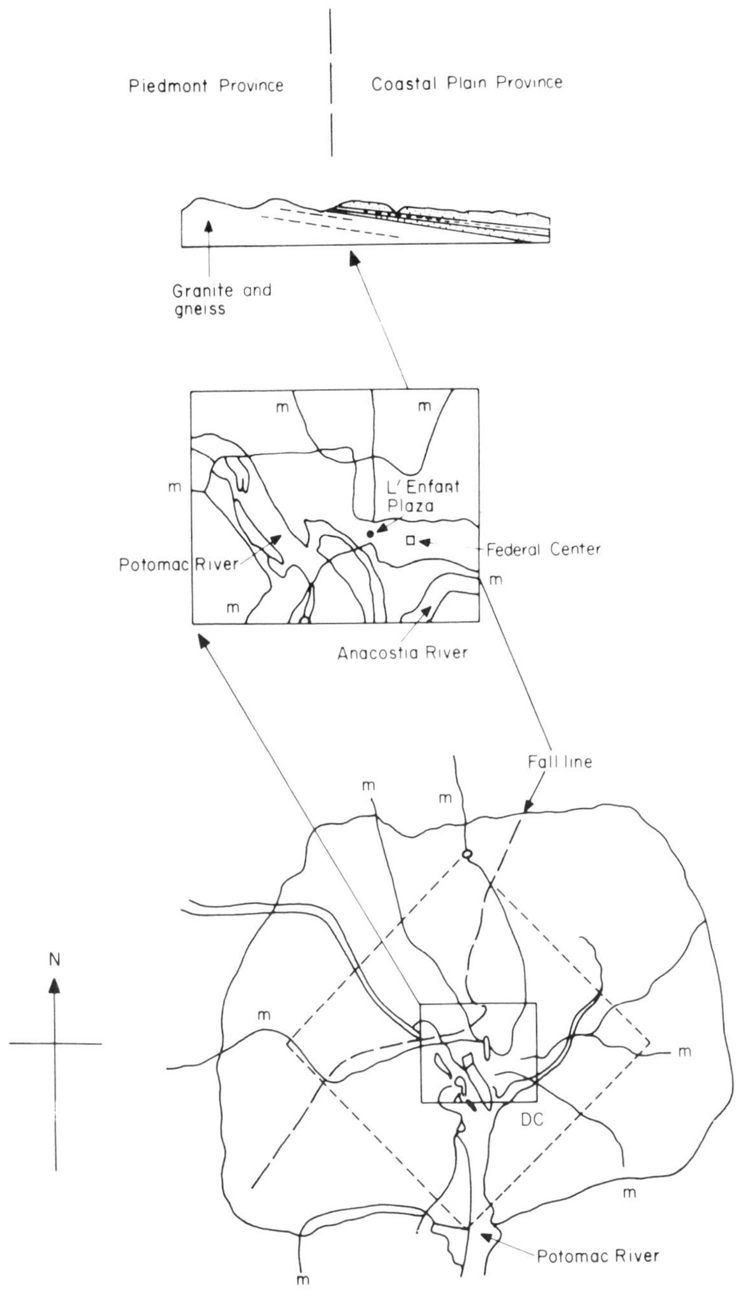

FIG. 8.7. Washington Metro. (Reprinted from ref. 17, by permission of Blackwell Scientific Publications Ltd.)

and the focal points of the arterial street systems. As the Metro is near to large monumental buildings in many places, underpinning was one of the most critical aspects of the job. For instance at Judiciary Square, the cut-and-cover construction ran diagonally under the District of Columbia Court of Appeals building and this three-storied structure was underpinned utilizing a corner pick-up system which has post-tensioned cast-in-place concrete beams. During beam construction, exterior columns were supported temporarily by steel underpinning beams which pierced the basement walls on each side of the columns. Construction proceeded in stages. After a group of beams had attained 27·6 N mm^{-2} (4000 psi) design strength, they were post-tensioned so as to transfer the foundation loads from the temporary supports to the beams. Blasting and vibration restrictions near such monumental buildings made it necessary to use a tunnel boring machine to tunnel the 5·73 m diameter, 3000 m long twin tunnels into rock parallel to and under Connecticut Avenue, one of Washington's main streets. Subsequently, by drill and blast methods, about 30 000 m^3 of rock were removed for station excavation. After this the rock face was lined with steel sets, rock bolts and shotcrete to give almost immediate support for the blocky and fractured rock. Finally the pre-cast concrete-finish liner was installed.

Underpinning was used extensively throughout the project. For instance, again at Judiciary Square, the entire north and east face of the National Portrait Gallery of two blocks length had to be so treated. The Gallery is built on a rubble foundation, not upon caissons. To do the underpinning job, the contractors were forced to go under the footings—a very difficult, but in this case successful, operation. Most Federal buildings like the Gallery are long and wide, but not high—there is a statutory restriction to a maximum of thirteen storeys on buildings in the capital, hence the necessity for underpinning. Within the Metro every attempt was made to reduce noise and vibration. This was done by using continuous welded rails, resilient track fasteners, floating track slabs, careful wheel maintenance and strict vehicle specifications.

The Washington Metro is the first transit system to utilize floating concrete slabs on a widespread basis in order to reduce wayside noise and vibration. These floating slabs are employed near hospitals, dwellings, courts and other structures sensitive to vibration. The slabs have been installed on neoprene pads and their effect is equivalent to increasing the distance between the Metro and adjacent buildings by

55 m. Sound-absorbent panels are suspended under all platform overhangs and they reduce ambient noise levels in train compartments considerably. Additionally, train cars are suspended on specially developed air springs which cushion the vibration transmitted to the track, inflating and deflating so as to adjust the door sill exactly to the level of the edge of the platform. All these measures combine to ensure that trains enter and leave the stations with very little of the noise normally associated with rail transit systems. The whole Metro, yet to be completed, will be very expensive—to date, costs have exceeded US$7 billion—but this was inevitable since President Johnson had called for it to be the best transit system in the world.[10]

Fernando Lizzi[11] patented the reticulated root pile structure (palo radice) and this is based upon the idea that reinforcing elements can be introduced to strengthen soils in a manner analogous to that in which steel bars strengthen concrete in reinforced concrete. In a reticulated palo radice structure the soil supplied the gravity and the pile is appropriately arranged on a three-dimensional plan and supplied lines of force such that the whole can support compressive, tensile and shear stresses. The approach is based upon a pile–soil interaction and vice versa and if the piles are not too far apart, a type of knot effect can arise, the network of piles encompassing and supporting the soil. After consideration of the job requirements and the strengthening of the soil to be anticipated, an appropriate design may be set up and this will take into account a density factor attributed to the actual structure. The following must be determined:

(a) number and diameter of the piles necessary for soil retention enabling the soil to behave as a unit;
(b) number and diameter of the piles required to ‘stitch’ the various layers among them and to provide rooting in the soil beneath.

Generally speaking, it is better to utilize a denser distribution of small diameter piles instead of a more scattered one with larger diameter piles. Of course there will be exceptions to this, e.g. when reticulated structures necessary to protect landslides comprising rocky and fractured materials are involved. Such materials possess low equilibrium and if inserted into a reticulated palo radice structure, they may become cooperative. In this event, the general encompassing of the soil is not so important. The shear resistance of the actual piles is best utilized if diameters greater than normal are employed together with substantial reinforcement. The use of reticulated root piles is now common and in fact they were utilized in Paris as a retaining structure for the protection of buildings during tunnel excavation work. They were also used in Washington and Barcelona as well as for the Naples rapid transit system.

An important car tunnel is Queens Midtown Tunnel in New York which provides a convenient access to the East Side Airlines Terminal. The underwater section is almost $1\frac{1}{4}$ km in length and the twin tubes are each 9·75 m in diameter. Drilling went on in complex geology including Manhattan schist, Inwood limestone, Fordham gneiss, Hell Gate dolomite and Brooklyn injection gneiss as well as disintegrated rock and glacial drift. These materials are highly variable in competency and entailed the use of compressed air at pressures up to 2·6 kg cm^{-2}, uncomfortably close to the safety limit. At one point the heading showed 2·44 m of coal overlying $2\frac{3}{4}$ m of mixed sand, clay, gravel and boulders and 2·44 m of bedrock of high competency. Despite such peculiar, indeed almost unique, admixtures the project was completed satisfactorily.

The author has described the Blackwall Tunnel and the Mersey Tunnel in *Grouting in Engineering Practice.*[2] These are vehicular, but another British tunnel underlying the River Thames is used for cables. This is the first Thames Cable Tunnel which starts from the Tilbury power station and runs across the river to Gravesend. It is about 1·6 km in length with a diameter usually of 3 m and is enlarged at its end sections to 4·6 m diameter (adjacent to two access shafts). The tunnel was constructed about $24\frac{1}{2}$ m below the lowest elevation of the river bed which did away with the need to use compressed air, the whole tunnel being excavated through the Chalk. Of course, sinking the two shafts was expensive and difficult and there had to be special arrangements for cooling the cables in the tunnel. The shafts possessed diameters of 6·1 m each and were sunk to depths of 38 and 40 m respectively. Water was rather close to the surface and the north shaft was constructed by pumping water out of deep wells accompanied by recharge through reverse wells which maintained adequate groundwater levels. The south shaft was made by treating the ground chemically prior to excavation. The uppermost surface of the Chalk was found to be badly disintegrated down to 5 or 6 m and of course this delayed project progress. The disintegration was considered to have arisen as a consequence of permafrost conditions in the UK during the last glacial period.

Another category of tunnels worthy of mention are access tunnels, which were common in the Middle Ages especially with regard to escape routes from castles. A well known modern instance is given by the tunnel constructed as part of a restoration project at the Tennessee State Capital building in Nashville, Tennessee, in the late 1950s. Before it was built, there were 103 steps to ascend in reaching the legislature floor and the access tunnel was driven from the level of the street in order to reach a new lift service (two lifts were installed). The length of this tunnel was a mere 73·6 m, but the work entailed great care because the work was effected through limestone and right under the building. Thus necessary drilling and blasting were carried out in association with monitoring seismographs which were observed continually and in fact demonstrated no appreciable vibration. A safety door was placed across the entrance to the access tunnel in order to prevent the noise of blasting being heard outside in the street and a concrete lining which was faced with marble was installed. The finished section was 3·35 m high and 3 m wide. In this case there was no grouting of the limestone, but this is often necessary with this rock, for instance in the Mississippian limestone foundation for the Gateway Arch and adjacent Mansion House Center in St Louis, Missouri. Also, grouting is sometimes effected in other materials; this was the case in the Blackwall Tunnel where the gravel crown was treated with semi-hard sodium silicate-based single fluid grout as recorded by Caron *et al.*[12]

Another access tunnel now cited was constructed in association with the building of a very long railway tunnel, the $18\frac{1}{2}$ km long Apennine Tunnel located between Bologna and Florence. For this work, construction shafts were inserted from a point near the village of Ca di Landino which is situated in the Apennines practically above the centre of the tunnel in question. Subsequently 1863 steps were made in one of the diagonal construction shafts in order to provide access to the overlying village.

Subaqueous tunnels include the famous East River Tunnel for conveying gas which was constructed before the First World War and had to undergo grouting as a drastic remedial measure. This has been described by the author previously.[2] The first to be constructed was under the River Thames at Wapping. This was a long project extending from 1825 to 1843, partly because great difficulties had to be overcome with rather poor equipment; not the least of these referred to a section of the river bed lowered by earlier dredging of gravel. The engineer involved was the renowned Sir Marc Isambard Brunel.

Water tunnels include the Tecolote in California and of course there are many examples in London, New York and other cities which could be cited. In connection with sewage disposal, Detroit now has a 45 km long tunnel.

One of the important functions of water tunnels is to transfer reservoir water from source to an hydroelectric power plant at a higher head. Such a construction constitutes a water pressure tunnel and it must be located below the minimum operational water level due to the fluctuation of the reservoir water. When the reservoir is full, the water head may be as much as scores of metres above the tunnel so that the lining of the tunnel supports the rock and is also stressed by the internal water pressure which is partly transmitted to the rock. The design of the lining must therefore be adequate not only to resist the rock pressure, but also to interrelate with the capacity of the rock to interact with it so as to resist the internal water pressure. If groundwater occurs in the rock, the effect of the lining may be reduced and consequently the hydrogeological conditions must be well known during the planning stage of such a tunnel project. If a subterranean power station is combined with a lengthy tailrace tunnel, this and the cavern are costly. The cavern can only be utilized when geological conditions are favourable and of course the rock should be free of groundwater if possible. The lining of the tailrace tunnel can be thin and even confined to the roof, which saves money. With such a set-up, water is taken in from the reservoir and there is no danger of leakage, but when the tunnel is emptied the lining must be capable of resisting the pressure of rock as well as the full head of water. The middle part of a water pressure tunnel is susceptible to water loss only in the situation where the groundwater level is below the water head in the tunnel—in such a case water can escape towards the tunnel outlet. The end of the tunnel should be provided with a circle of drilled grout holes, therefore, where it meets the steel lining of the outlet section. Longitudinal drains which conduct water during the construction of the concrete lining should be poured by concrete. Obviously the groundwater level in the outlet area stands below the water head in the tunnel and water escaping from the latter might destabilize the slope and thus the outlet structures. The rock mass round this outlet section ought to be handled like an earth dam therefore and a large section of the outlet should be steel-lined. In an empty tunnel this steel lining would be loaded by water pressure from the rock and this water must be drained by boreholes and galleries. However this drainage system should not be installed proximate to the actual tunnel because cavities induced by erosion might be hazardous to the stability of the steel lining. As regards the rock itself, the greater its modulus of deformation the more effective is its participation in the resistance of the lining against internal water pressure. A coefficient of rock resistance has been introduced in the USSR related to the designing of water pressure tunnels and it is equal to the ratio of the internal water pressure to the increase in tunnel radius. Elastic analysis has shown that the key parameter is E, the modulus of rock deformation and the greater this is the better. Displacement of the walls of tunnels by water pressure can be investigated in order to establish values for E. This is usually done by means of a specially driven lateral adit (because the measurement cannot be made directly in the tunnel through considerations of its operation). In fact the pressure is being exerted on the surface of the rock by means of flat jacks rather than water, these being set around the circumference and bearing against a stiff steel structure (the TIWAG method described by Lauffer and Seeber[13]). It has been found that values of E so obtained are frequently greater by about 33·5–34% than values obtained by plate-loading tests. The gap between the rock surface and lining and also the zone of loose rock should be grouted, a procedure which may increase the modulus of deformation. As regards the siting of the openings in the area of the outlet, several problems arise. This is true of the surge tank wherein dynamic effects can result from the halting of water flowage when valves are closed rapidly. To minimize these, this structure is located near to the outlet. The inclined pressure from penstock to power station via the tunnel is a costly and significant feature which must be resisted by foundation in stable rock.

Benson[14] has cited a case of displacement of a conduit because of slope instability. Here, four pipes, each of diameter 1·5 m, were located on a chloritic gneiss slope covered with detrital material. The head was 200 m. Soon after coming into operation, a disturbance was noticed and displacements reached 10 cm in 3 months. The conduit had to be relaid in the tunnel but, as sound rock was 60 m deep concomitant with deep weathering on the slope, the longitudinal profile of it was not straight.

The cavern for the power station must be excavated often and this is very expensive; hence the engineering geologist and site engineer must try to seek economies. These can be achieved in various ways, e.g. by substituting several small chambers for one big cavern. Obviously an optimum orientation with respect to joint systems must be attained. Borings alone may not supply sufficient information regarding joint sets and in fact it is sometimes necessary to drill exploration galleries for the purpose of obtaining more data. Other hazards may be present and may be avoided by locating the cavern so as to avoid rock with very high stresses. Consequently it is necessary to measure inherent stresses prior to the design of subterranean power stations. Usually such structures are situated below the level of the valley floor.

Carati[15] has given an excellent instance of the effects of geology on the design of an underground power station in the Premadio scheme in the Italian Alps. Two possible designs were investigated, one comprising a steel-lined conduit situated on an 850 m long inclined tunnel together with a cavern located near the slope toe in phyllite and a tailrace tunnel of 730 m length. The second was a vertical shaft located in Triassic dolomite and a 1 400 m long tailrace tunnel. The first was selected because, among other advantages, it was concluded that this approach offered the minimum possible effect on the hydrogeology of the area near the Bormio spa.

Often the excavation of caverns is facilitated by competent rocks, but sometimes the rock is unstable so that both roof and walls need supporting. Wall anchoring may be required. Anchors may increase frictional forces in bedding surfaces or joints as well as introduce new forces into the rock mass. In such a manner is the stress state in the rock around the cavern improved. Müller[16] has used this method in Liassic jointed limestone of Lower Jurassic age at the Braz Power Station protected by nearby Kössen slates from water percolation from overlying dolomite.

Pumped-storage power plants in mountainous areas usually involve construction in flat plateaux. The difficulties arise because construction of all the various elements in a single valley profile is not feasible. Different approaches have to be made. The Cruachan pumped storage power plant in Scotland had the situation of its cavern determined by geological considerations. Tailrace and adit tunnels traverse phyllite and hornfels enveloping a granite massif which supports the storage reservoir. The adit tunnel is 1 200 m long and was driven as an exploratory gallery in order to investigate the cavern thoroughly. This latter is sited in granodiorite and the rock is of excellent quality throughout so that no lining was required. Consequently the walls were protected by the piers of the crane runway alone. The cavern roof spans 26 m and it is protected by concrete vaulted strips 2·3 m wide and the free rock surface intervening between them and the 20 cm width is provided with polyvinyl chloride slabs as protection against infiltrating water.

8.6. CONCLUSION

There is no doubt that great advances in tunnelling have taken place since the 19th century and these have obviated older techniques, such as timbering, which as a result are becoming lost arts. Machinery has reduced the risks greatly as well as improving the working conditions. All rock removal is now mechanized. For instance blasting in hard rock tunnels is followed by mechanical shovel debris collection with later loading on to conveyor belts. Similarly the slurry pumping plant is used for disposal and in gravelly soil, sifting may be accomplished by machine, the smallest fragments of the product being utilizable in association with grouting. Such tiny pieces are compressed air-injected into the spaces betwen the rock wall and the tunnel lining through holes in the latter. The quantities of cement needed for subsequent grouting are thereby reduced and so the costs diminish. There can be little doubt that tunnelling machinery will become increasingly elegant and it may well be that grouting will too, eventually becoming part of a continuous automated process also involving excavation, mucking and, if required, lining.

REFERENCES

1. Olivier-Martin, M. and Kobilinski, M., 1955. L'execution d'un grand souterrain pour l'amenagement hydroelectrique d'Isere-Arc. *Construction*, **X**(4), 145–56.
2. Bowen, R., 1981. *Grouting in Engineering Practice*, 2nd edn, Applied Science Publishers Ltd, London; Halsted Press, New York.
3. Zaruba, Q. and Mencl, V., 1976. Engineering geology. *Developments in Geotechnical Engineering, 10,* Elsevier Scientific Publishing Company, Amsterdam.
4. Zaruba, Q. and Andrusov, D., 1939. *Srovnani geologickych profilu Harmaneckym tunelem na draze Banska Bystrica-Diviaky*, Techn. obzor. Praha.
5. Dowding, C. H. and Rozen, A., 1978. Damage to rock tunnels from earthquake shaking. *Proc. Am. Soc. Civ. Engrs, J. Geotech. Eng. Division*, **104**(GT2), 175–91.

6. Borecki, M. and Kidybinski, A., 1966. Problems of stress measurements in rocks taken in the Polish coal industry. *Proc. 1st Int. Cong. Rock Mechs*, **II**, 9–16.
7. Langefors, U. and Kihlström, B., 1978. *The Modern Technique of Rock Blasting*, 3rd edn, John Wiley, New York.
8. Terzaghi, K., 1946. Introduction to tunnel geology. In: *Rock Tunnelling and Steel Supports*, eds R. V. Proctor and T. L. White, The Commercial Shearing and Stamping Company, Youngstown, Ohio, USA.
9. Mann, F. A. W., 1972. *Railway Bridge Construction: Some Recent Developments*, Hutchinson Educational, London.
10. ASCE, 1979. Washington Metro: a people's eye view. *Civil Engineering*, June.
11. Lizzi, F., 1982. *The Static Restoration of Monuments*, Sagep Publishers, Genoa, Italy.
12. Caron, C., Delisle, J. P. and Godden, W. H., 1963. Resin grouting with special reference to the treatment of the silty fine sand of the Woolwich and Reading beds at the new Blackwall tunnel. In: *Grouts and Drilling Muds in Engineering Practice*, Butterworths, London, pp. 142–5.
13. Lauffer, H. and Seeber, G., 1966. Die Messung die Felsnachgiebigkeit mit der TIWAG Radialpresse und ihre Kontrolle durch Dehnungsmessungen an der Druckschachtpanzerung des Kaunertallekraftwerkes. *1st Congress of the International Society of Rock Mechanics*, Lisbon, Vol. II, pp. 347–56.
14. Benson, W. N., 1946. Landslides and their relation to engineering in the Dunedin district. *Economic Geology*.
15. Carati, F., 1958. *L'Impianto Idroelettrico di Premadio*, L'Energia Elettrica, Milano.
16. Müller, L., 1953. *Setzungen von Bauwerken auf Felsuntergrund,*. Vorträge der Baugrund Tagung in Hannove, Hamburg.
17. Bowen, R., 1982. *Sci. Prog., Oxf.*, **68,** 97–125.

CHAPTER 9

Large Scale Hydraulic Structures: Dams

9.1. SOME HISTORICAL CONSIDERATIONS

Large scale hydraulic structures include the biggest engineering accomplishments of Man, namely dams, of various types, which constitute potentially the most dangerous human constructions, as Grunner has pointed out.[1] Their erection is intimately associated with the geological characteristics of a site, this comprising not just the dam but also its reservoir.

The significance of the local and regional geology was not recognized in pre-Christian times and the consequences may be illustrated by reference to the oldest known dam, the Sadd el-Kafara. This was discovered in 1885 by a German archaeologist, Schweinfurth, and built in Egypt during the 3rd or 4th Dynasty, i.e. between 2950 and 2750 BC. The associated reservoir was impounded in the Wadi el-Garawi probably to provide water for drinking and stone extraction at alabaster quarries located just over 3 km to the east. Although most ancient dams were intended for irrigation, the lack of agricultural activity in the Wadi el-Garawi in those days rules out the possibility of this usage here. The dam was built across an appropriately narrow point and it had a crest length of 104·4 m, a base length of 79·5 m and a maximum height of 11·1 m above the valley floor. It was extremely thick and the central section has been washed away so that the internal structure can be examined. Upstream and downstream sections are composed of separate rubble masonry walls each 23·4 m thick basally and attaining the total dam height. Approximately 22 000 m^3 of masonry weighing some 40 000 tons were quarried, transported to and emplaced in the dam. Additionally the 35·4 m space which separates the two rubble masonry walls basally across the entire length of the dam was infilled with about 60 000 tons of gravel from the bed of the Wadi as well as with stones from the nearby hills. The sloping water face of the structure was covered by roughly dressed limestone blocks arranged in rows of steps about 28 cm high. As the Ancient Egyptians probably only utilized mortar as a lubricant, in order to place stones or level the top of a building course prior to commencing the next and not at all as a binding agent, the fitted masonry facing was uncemented and hence permeable. However it did provide resistance against erosion and wave action. Such a technology was feasible for two reasons. The first was the favourable climate, hot and dry with only rare rainfall. The annual Nile flood led in ancient times to a basin irrigation system for land fertilization persisting until increased agricultural demand in the last century necessitated the construction of new dams, actually the first made after the Sadd el-Kafara. The second reason relates to the massive size of this structure which in itself tended to promote stability.

The fate of the Sadd el-Kafara, the Dam of the Pagans, may be considered because it was determined by lack of understanding of environmental hydrology and engineering geology together with low standards of workmanship. Site investigations show that the crest sloped towards the centre perhaps, as Murray has suggested,[2] through settlement of the essentially rather loose structure. Alternatively it could have been an intentional design feature incorporated in order to restrict overflow from the reservoir to the dam centre and just above the Wadi bed. This explanation cannot be verified because of the washing away of almost 45 m of the dam.[3] This also prevents determination of whether or not a spillway was present initially. It seems certain that the dam was utilized only for a very short period of time, an inference derived from lack of any evidence of siltation behind its remains. The cause of this was probably hydrogeological. As the climate around Helwan has not altered appreciably since the construction, the hydrology of the area around the Wadi el-Garawi as now known facilitates an estimate of the volume of water originally impounded by the dam. The catchment area is about 186 km^2 in extent and there is no runoff unless the individual falls of rain exceed 8 mm, a likely event. Records for the first half of this century demonstrate that in each 3 years out of every 4, the catchment area probably received a rainfall exceeding 10 mm daily. Once in every 4 years the catchment area received a rainfall greater than 20 mm. Hence the dam would have impounded a valuable quantity of water after rain. The capacity of the reservoir has been calculated as 560×10^3 m^3 and 20 mm of rainfall would have sufficed to fill it, if it were empty. Consequently several such rainfalls could have caused overtopping of the dam if they occurred within a few months. This together with the lack of sedimentation alluded to above could imply that the short period of use of the dam resulted from overflow-induced failure. The dam had no watertight face and was of loose construction so that water could percolate through it easily and cause throughflow erosion which reached the sand and gravel core. If there was a spillway, it was probably inadequate. As far as can be seen there was no cut-off trench which contributed to the insecurity of the dam. Although the dam must be considered as unsuccessful, this may have had something to do with undue haste as well as with a lack of appreciation of the geology and design failures. Lack of experience must have played a part also because up to that date nobody had tried to construct a dam of such a size in Egypt or indeed anywhere else.

Later the Romans built many dams and those in North Africa, with the exception of the water supply dam for Leptis Magna, were either for flood control, water retention or soil conservation.[3] In Persia the captured army of the Emperor Valerian constructed a bridge dam on the River Karun at Shushtar, first diverting this to facilitate working in dry conditions.

Geology was neglected still, however, by the Mongols who, between 1281 and 1284 AD, constructed an irrigation dam near Tehran, the Saveh Dam. This has survived undamaged as a result of the fact that it was never used.[3] The reason is that the limestones of the valley sides were considered sound, but the basal foundation alluvium was not. It was comprised of sand and gravel with bedrock about 27 m down. Infilling of the reservoir pressured water through the foundation alluvium to a permanent outlet. This was especially unfortunate since otherwise the geology was very favourable. There was a narrow gorge and the river carried enough water, particularly in the winter, to ensure a full reservoir during the summer drought. Modern attempts, both before and after the Second World War, to revive the project by means of a rockfill dam 69 m high with a reservoir capacity of a stated $25\,000 \times 10^6$ gallons (95×10^6 m^3) have been plagued by foundation problems.[4]

The Moslems were active dam builders and in 960 AD under Adud-ad-Dawlah the Band-i-Amir Dam was constructed on the River Kur in Fars province. It was made of masonry blocks throughout, these being connected by iron bars set in lead. The stones were mortared so that the dam was watertight and in its day it was highly successful as a diversion dam; it had no storage capacity.

After mediaeval times Spanish dam building excelled so that by the 17th century it was unrivalled. Thereafter economic decline reduced dam building activity but towards the end of the reign of Charles III three dam building projects were in progress and one of the dams, the Puentes, later failed, the first serious disaster in modern times. The Puentes Dam was 50 m high, 282 m long, 11 m thick at the crest and 44·25 m thick at the base. It was a gravity dam of polygonal outline, a shape determined by the site topography. The water face was vertical throughout its length and the uppermost 16·7 m of the air face comprised four steps, each 4·175 m high and 3·34 m wide. Below this, the face inclined uniformly down to the base. The structure was composed of a core of rubble masonry set in mortar and faced with large cut stones. The crest was completed with an ornamental parapet and overflow was intended to discharge over it. There may have been a separate spillway because on the left bank near the dam there are the remains of a channel which can be so interpreted. The site appears very suitable. The valley of the Guadalentin is narrow with hard rock on either side and the reservoir would have been 4 km long and contained over 12×10^6 m^3 of water. The geological defect consisted of a deep earth and gravel pocket encountered during construction and located near the centre. The engineers attempted to combat this using foundation piles. Hundreds of these, each 6 m long and 0·61 m square, were driven into the pocket and connected and braced at their caps using a network of horizontal beams at right angles. Into this grillage the lower courses of the dam masonry were constructed to a depth of just over 2 m. The pocket sediments extended downstream from the airface base, precisely where the scouring gallery and one of the outlet tunnels were supposed to discharge. To prevent undermining, the piled grillage was extended 39 m downstream and covered with over 2 m of masonry together with a layer of planks to prevent erosion. The approach was unfortunate. Although it succeeded from 1791 to 1802 AD, this was because the reservoir was not completely filled up during this period—the water level never exceeded 24·6 m. At the beginning of 1802 AD, demand had slackened and the water level had risen to 46·2 m by 30 April at which point the foundations could no longer take the strain. It would have been possible to obviate this if excavation through the sediments in the pocket down to bedrock had been effected, i.e. if the dam had been founded entirely on rock. At about 2.30 p.m. on the date in question, reddish water was observed bubbling out on the downstream side; half an hour later there was an explosion in the discharge wells built into the dam from top to bottom. Shortly after there was a second explosion and a vast mass of water enveloped piles, beams and other pieces of wood, forcing them upwards. Thereafter there was another explosion and the two gates closing the scouring gallery and the intermediate pier fell in. This was accompanied by a huge water escape, again of red colour due to mud or reflection of the sun. The volume involved was great enough to empty the reservoir within 1 h. The town of Lorca was inundated, with the loss of 608 lives. 82 years later a second Puentes Dam had been erected and stood 71 m high. Like its predecessor in its time, this was then the highest dam in the world. The first case demonstrates that although geological factors had become recognized as significant, their treatment when adverse was not always optimal.

9.2. APPRAISAL OF GEOLOGY

This should be effected in the first phase reconnaissance of a project. One or more engineering geologists should collect and collate relevant data with a view to preparing a report embodying site information on outcrops and exposures with particular reference to such deleterious factors as fracturation due to jointing and/or faulting, erosion and weathering either by water, wind or temperature change as well as the occurrence of associated drift deposits. Such a study will incorporate previous knowledge as well as that derived by personal inspection and investigation. Hence, an examination should be made of any available geological, hydrogeological and geomorphological maps, well and boring records, information on cuttings and appropriate publications.

Naturally, dams and other projects are carried out in the underdeveloped countries as well as in the advanced ones and here the problems encountered may be formidable. This is because no geological maps may be available or, if they are, the scale may be unsuitable for evaluation. For instance much of Sierra Leone is unmapped and this is true also for parts of Sri Lanka, two countries with which the author has been involved. The result is that in such places basic mapping and areal analyses must be effected, the former, for reasons of economy and speed, concentrating upon relevant parameters such as the dip and strike of strata, their tectonic state, folding, faulting, jointing, deformation, shattering, the depth and amount of weathering and so on. If all the products of such an endeavour are put together on to maps and into reports, it is feasible to utilize them in order to:

(i) select topographically suitable sites for dams;
(ii) assess the probable watertightness of areas to be used for impounding water in reservoirs;
(iii) predict the likely stability of banks of reservoirs to be flooded;
(iv) derive a useful guesstimate for the rate of siltation in the planned reservoir and from this extrapolate so as to arrive at an estimate of its probable lifespan.

As well as dams and reservoirs, the engineering geologist may become involved with other large scale hydraulic structures such as drainage systems. However the same approach as that discussed above is applicable, i.e. special maps should be constructed as a basis for the preparation of appropriate plans for water management. In dealing with river valleys, for instance, these may be divided into reaches possessing similarity of topography and geology in so far as, for example, the erection of a dam is concerned. Obviously some sections of a river are more suited for this purpose than others and data maps of the right kind facilitate the picking out of optimum ones. Up to this point therefore the initial studies for hydraulic structures may be grouped into the following categories:

(I) Geomorphological and geological investigation of the actual site accompanied by a less detailed survey of the wider environmental setting with particular reference to the location of 'borrow' areas from which appropriate building or grouting materials may be obtained economically.
(II) Subsurface investigation of the site and if this is done for a dam, its associated impoundment area. This will be very valuable if carried out properly because it will enable the depth to bedrock and thus the thickness of the weathering zone to be determined. It will facilitate also the accumulation of pertinent information on the presence or absence of fracturation in the rocks. Normally this does occur and adequate maps and sections will be very helpful in determining and delineating fault zones and joint systems.

It may not be taken for granted that these two sets of activities will be effected as a matter of course. Sometimes projects which are geologically practicable and entirely suitable from an engineering standpoint are abandoned for financial, political or other reasons.

Occasionally unexpected results may spring from investigations undertaken in an existing facility area. An excellent instance is given by blind mineralization which was discovered under 100 m of sediments at Olympic Dam which is located on Roxby Downs Station 90 km north of Woomera in South Australia. In 1975 drilling was initiated and by 1981 what may prove to be one of the world's largest orebodies had been outlined; it contains copper, uranium, gold and rare earth minerals. All are subhorizontally layered in a granitic and richly haematitic breccia pile thought to be 1 000 m thick. The first area of about 20 km^2 has been paralleled by discoveries in the adjacent region. Actually this discovery resulted from the application of sound geological reasoning coupled with geophysical investigations and regional lineament analysis. It began with the concept of mineralization within an Upper Proterozoic regional geological unit, namely the Stuart Shelf, a stable platform of older crystalline basement overlaid by horizontal Adelaidean sediments. Theories of ore genesis and the study of known copper and uranium occurrences combined to provide a basic approach suitable for the selection of appropriate target areas which were confirmed by lineament analysis of Landsat data identifying altered basaltic piles in the basement as source rocks for sedimentary copper deposits. Five drill holes provided stratigraphic data and one (RD1) at 353 m depth intersected the 38 m point at 1·05% copper and trace uranium. This work has been described by Haynes[5] and Davis.[6]

9.3. GEOPHYSICAL METHODOLOGY

This is very useful in reconnaissance investigations because the instrumentation required is usually portable and measurements can be made in the field both quickly and cheaply. Perhaps the most widely used are the techniques of resistivity and seismic investigations. Where such reconnaissance work is carried out it is often important also to effect chemical studies on the composition of surface and groundwater. After all preliminary investigations have been performed, the engineering geologist is in a position to make a detailed report on the results obtained and this will provide as complete a geological analysis of the site and adjacent areas as possible and include sections on stratigraphy, tectonics, morphogenetics, etc., together with an assessment of the suitability of the site with supporting data such as an engineering geology map, geological sections, boring logs, water pressure test records, pump test data, etc. A complete list of relevant previous literature with extracts where applicable should accompany the report as also should specialist reports on the hydrogeology, soil and rock mechanics, etc. After all preliminary studies are completed and their impact taken into account, a second phase of highly detailed geological investigation may begin.

9.4. DETAILED GEOLOGICAL INVESTIGATION

This is effected on a much larger scale than the preliminary reconnaissance work. An appropriate one is 1:1000, but for a reservoir site a less detailed scale such as 1:5000 may prove to be quite adequate. The engineering geology investigation will be geared now to the requirements of the plans for the actual hydraulic structure and involve the use of pits, trenches, adits, drifts and rotary and/or percussive borings. The valley floor at the site of a dam may have to be cleared to bedrock. Throughout, modification of the programme of investigation may have to be made in order to adjust the work to the data accumulated during its execution. Open excavations are almost always advisable because they are the best means of determining the extent of deep and extensive mechanical weathering of rocks. Core boring technology is normally employed in the case of solid rocks and the very solidity of these is advantageous in that it allows for drilling in various directions. Good recovery is essential in any orientation. Rock permeabilities are determined using water pressure tests and also by trial grouting. This latter can also show the interconnection of single joints and sets or families of joints. Obviously all such work must be carried out by experienced personnel and the process must be subjected to continuous supervision. All bore samples must be preserved (this is usually done on site), the cores being boxed in a dry room with appropriate lighting and complete labelling. Concomitantly all details of boreholes must be recorded and these include position and elevation.

The engineering geologist is deeply involved in all of the foregoing activities and therefore bears a heavy responsibility to the construction company. It is all the heavier because of the interpretive and subjective nature of a good deal of the work. For instance in joining up individual borehole records on a geological section, an informed imagination must be utilized and if this is inadequate then misinterpretation of data is quite possible with correspondingly adverse results. In order to avoid this human error factor, or at least to minimize its effects, continuous monitoring of the subsurface geology is usually undertaken during the actual construction work. This monitoring process necessitates that the engineering geologist or geological engineer should be present on site practically all the time and this is also true where, as is usually the case, grouting operations have to be carried out. The main tasks may be summarized as follows:

(a) Cooperation with site engineers with regard to the excavation of foundations, a matter often requiring the removal of weathered material and almost always involving the application of explosives for blasting. If highly weathered rock occurs, much of this may be removable without recourse to blasting. After clearance of the actual foundation site, the engineering geologist will proceed to make a geological survey of the site on a detailed scale thereby producing an appropriate geological map which may well be copiously supported by a report containing sketches and photographs.
(b) Continuous presence of the geologist and/or engineer facilitates the recording and evaluation of every temporary exposure and borehole from which samples may be obtained for laboratory testing. It is essential also that close collaboration exists between engineering geologist and site engineers in grouting during which inspection of grout holes (often totalling hundreds of thousands of metres) is effected and water pressure tests for permeability determinations are carried out. From the data derived from the latter it is possible to ascertain whether the proposed grout curtain depth is sufficient or not.
(c) After the construction work is finished the engineering geologist can make a final report which will contain data apropos the foundation and grouting procedures as well as the results of *in situ* and laboratory tests of naturally occurring building materials together with a very detailed geological map of the foundation area.

Each of these matters will be examined in further detail below.

9.5. GROUTING

This process, for damsites and elsewhere, has been discussed by the author in detail in *Grouting in Engineering Practice.*[7] Here it is only necessary to say that with dams grout curtains may have to be constructed. Ischy[8] has given a very good instance of this. Referring to the de Castillon Dam on the River Verdon in the French Alps, he mentioned that the site for this was eminently satisfactory from a geomorphological point of view as an arch dam was envisaged. However the geology was adverse, the gorge being cut in jointed limestones of Jurassic age. Their dip is approximately 20° on the left flank where they are massive, but on the right flank the dip is towards the valley and the beds are anticlinally folded (see Fig. 9.1). Fracturation occurs

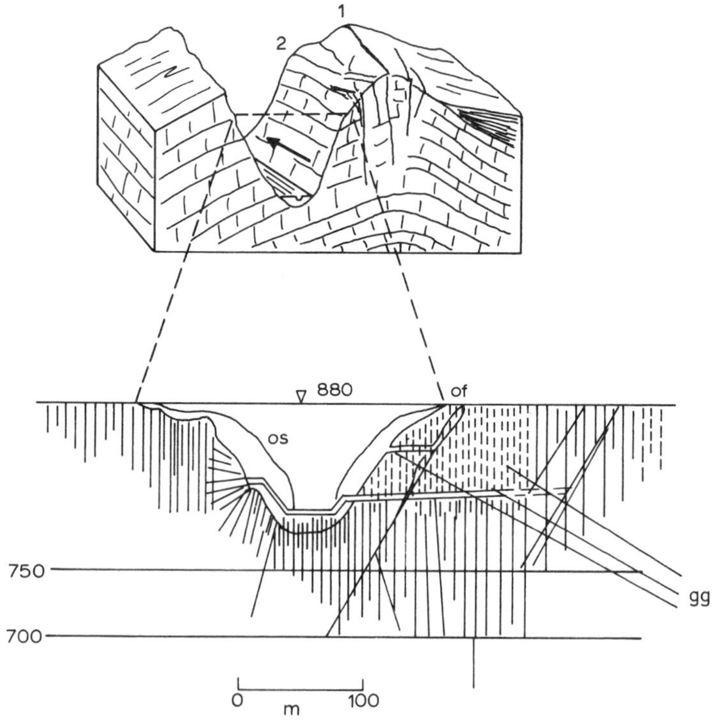

FIG. 9.1. De Castillon damsite, Verdon River, French Alps. Top: major fracturation of limestones in the right valley slope at Castillon. Bottom: arrangement of grout curtain at damsite with os, original surface; of, open fracture; gg, grout galleries and oblique boreholes. (From ref. 12.)

in three directions and fissures are sometimes rather wide (up to several metres) and of tectonic origin, actually a result of the anticlinal folding process. As a result of these geological considerations, the arch dam could not be keyed into the fractured rock with any success unless reliable support was provided. This was achieved by consolidation by concrete of the cleaned fissures. Subsequently the massif behind the toe of the dam was grouted, a huge operation because the relevant block volume was about 200 000 m^3. In addition, all loose limestone blocks at the foot of the incline were reinforced using concrete supports secured by means of pre-stressed bolts. The grouting exercise was extremely difficult because it was necessary to confine the work to a relatively small part of the entire rock mass (economics dictated this), ensuring that cement would not be transmitted through the fissure system to distances in excess of the chosen limits. The solution used was to drill peripheral holes set 2–5 m apart from one another in several different directions from three galleries. The very process of drilling was hard because of the fissuring of the limestone, but the holes were eventually drilled and grouting effected using a mixture of sufficient viscosity to prevent loss by flowage through open fissures. In fact, this mixture consisted of cement, bentonite and crushed stone and had the consistency of dry mortar. However it could still be injected utilizing a normal machine for grouting. Bentonite is of course thixotropic and this promoted consolidation very soon after injection, a strength of 30–40 kp cm^{-2} being attained after a week which doubled after a month. Fissures having widths as much as tens of centimetres were satisfactorily grouted with this mixture. Boreholes in the massif were so planned that they could be drilled following the construction work and in total their length reached 15 000 m. The rock mass of 200 000 m^3 was consolidated with 6000 m^3 of the grouting mixture and this represents grouting of a mere 3% of its total volume; however this proved to be sufficient and thus represents a very skillful approach to the problem.

In some terrains it is almost impossible to erect a grout curtain. This is true of limestone areas which are often cavernous. Nevertheless circumstances may arise which impose such a task on a contractor against geological advice. Not surprisingly such cases are rare and indeed very few dams have been built in karstic regions of the world. In these there is not only the problem of the grout curtain, but also of uplift, the total value of which must be calculated and taken into account.

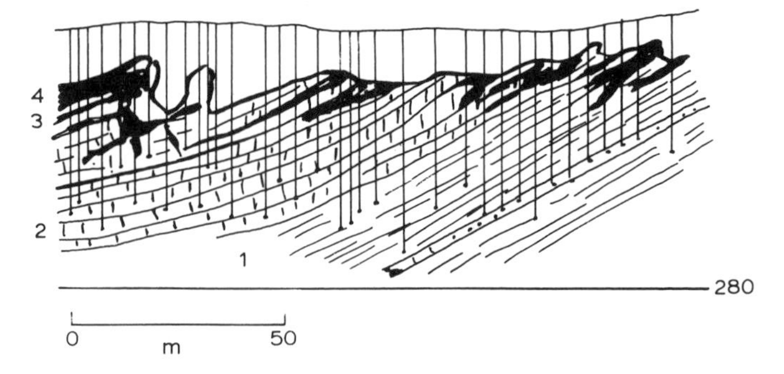

FIG. 9.2. Cherokee Dam, Holston River, showing extensive karst cavities under the valley floor. 1, Limestones and calcareous shales, Middle Cambrian in age; 2, limestones and calcareous shales; 3, dolomites and dolomitic limestones, Upper Cambrian in age; 4, zone of karstification, showing caverns at depth.

Figure 9.2 illustrates an instance of successful construction on karstified limestones, namely the Cherokee Dam on the Holston River (Tennessee Valley Authority) in the USA. Here the karstification occurred in dolomites and dolomitic limestones of Cambrian age and since many were formed prior to later subsidence, large caverns are found deep below the floor of the valley. Another instance is afforded by the Camarasa Dam in Catalonia, Spain, which stands upon Jurassic limestones and dolomites. After the impoundment of water in its associated reservoir spring leakages appeared downstream; these had yields as high as 11 $m^3 s^{-1}$ and were distributed widely. Remedial work involved the construction of a grout curtain on both flanks for a length of 1 029m using 224 grout holes with a maximum depth of 394 m as described by Lugeon.[9] For comparison, Fig. 9.3 shows a grout curtain below the Beauregard Dam, an arch dam in the Italian Alps based upon competent and disturbed mica schists.

The case of an earth dam is now considered—the Dillon Dam, Summit County, Colorado, USA, discussed by Wahlstrom and Hornback.[10] This was constructed on a foundation of faulted and highly jointed sedimentary rocks of Mesozoic age and its appurtenant features include a diversion tunnel and spillway shaft. The area is highly complex geologically and a major feature is the Williams Range thrust fault outcropping about 3 km east of the damsite and dipping northeast. This resulted from a probable 3 or 4 km displacement of Precambrian igneous and metamorphic rocks over Mesozoic and older

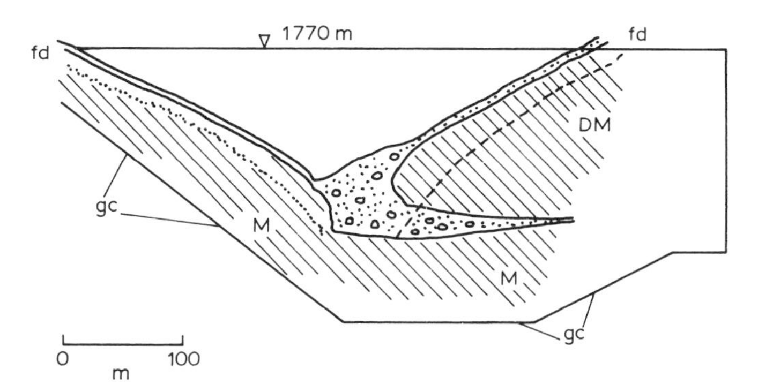

FIG. 9.3. Illustration of the grout curtain below the Beauregard arch dam in the Italian Alps. fd, foundation depth; gc, grout curtain; M, competent mica schists; DM, disturbed mica schists; [symbol] sands and gravels of glaciofluvial origin.

sedimentary rocks, a movement accompanied by extensive folding, faulting and jointing. These phenomena affected both the thrust sheet and the underlying rocks. Bedrock at the damsite comprises sandstone, sedimentary quartzite, shale, siltstone, mudstone and calcareous mudstone. Dykes and sills composed of felsite porphyry also occur. Drilling and grouting of the foundation, the tunnel and the spillway shaft had to be undertaken. Geological investigations were effected during construction and as this progressed, detailed maps and sections on a scale of 20 ft to the inch were made of the dam foundation, the diversion tunnel and the spillway shaft. Geological data were used to determine the depth and spacing of the holes for the cement grout in the dam foundation and the amount of excavation necessary for obtaining suitable foundation rock. Figure 9.4 shows the geological map and section of the damsite area and also the various formations encountered, their respective thicknesses and the fault system. Some interesting observations may be listed:

(i) Brittle fractured rocks such as sandstones and quartzites in the Dakota Group, the Lykins Formation, the Entrada Sandstone and the Maroon Formation took large quantities of grout.

(ii) Large quantities of grout were taken up by the Morrison Formation and the shales of the Dakota Group, a surprise because it was at first believed that these rocks could not maintain open and interconnected fractures.

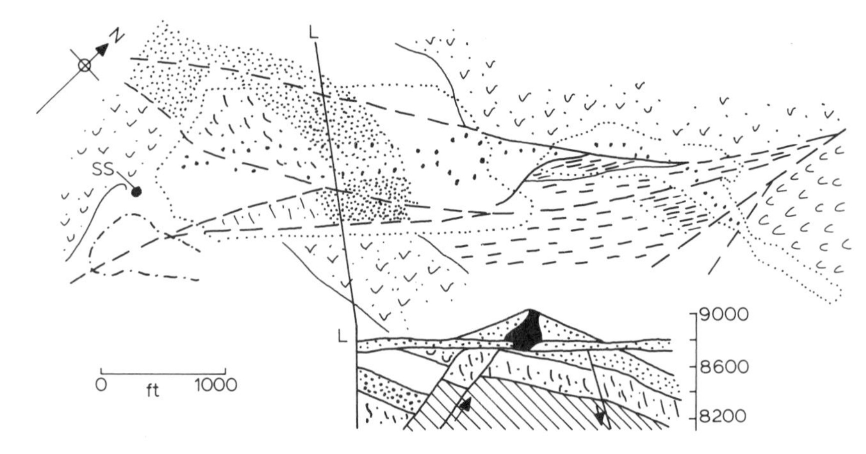

FIG. 9.4. Geological plan and section of the Dillon Dam, spillway shaft (SS) and diversion tunnel, Summit County, Colorado, USA. (- - - -) Fault; (L) line of section; (· · · · ·) dam outline. (From ref. 10.)

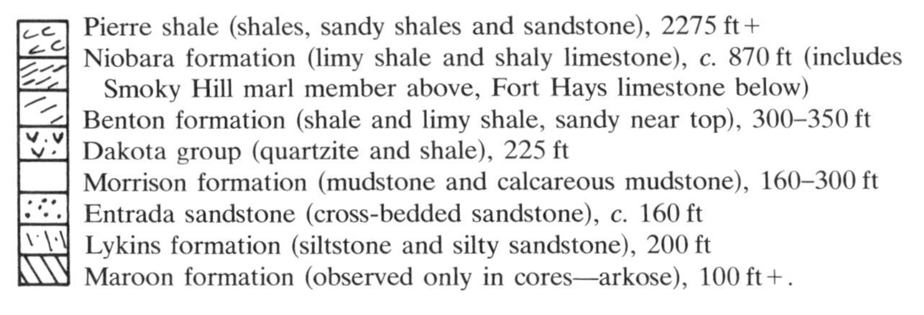

(iii) Large quantities of grout were necessary in order to seal off faults and fractured rocks associated with them.

A total of 630 grout holes was drilled and they added up to an aggregate depth of 17 363·1 m. Two core holes were drilled, one at Station 15 + 15 to a depth of 43·5 m and the other at Station 21 + 25 to a depth of 71·4 m. This was done in order to test the rock at depth after the grouting work was completed. A total of 175 957 sacks (4926 m^3) of cement were emplaced in the holes. Despite every precaution, the complexity of the fracture pattern in the foundation rocks is such that it would not have been surprising if leakage had occurred after the reservoir was filled. It was anticipated, however, that the dam load would tend to close a lot of fractures which were not penetrated by grout.

It is important to realize that the location of the Dillon Dam was primarily determined by geographical and topographical factors which were adverse to the geological considerations. These might have militated against siting the dam if fully taken into account since they are clearly unfavourable and include:

(I) The sedimentary bedrock is extensively fault-intersected and the brittle rocks such as sandstones and quartzites usually contain sets of interconnecting joints.
(II) The bedrock is deeply and extensively covered by fluvioglacial deposits, terrace gravels, slide rock and talus.
(III) Earth slides located upslope from the diversion tunnels intake structure need considerable excavation and a protective dam above this had to be made.
(IV) Excessive rainfall or reservoir filling or drawdown can trigger additional movement in the slide rock.

9.6. DAM TYPES

These comprise the following:[11]

(a) embankment dams, homogeneous or zoned;
(b) concrete, either arch and dome or multiple arch and dome;
(c) concrete gravity and gravity arch;
(d) concrete slab and buttress;
(e) combined.

The Dillon Dam belongs to type (a) and the extensive investigation of foundation geology entailed in its construction has been discussed above.

Similar investigations are requisite for the construction of almost any type of dam. However, it is pertinent to bear in mind that embankment dams may be constructed in a greater variety of foundaticn material than other types and these may range from loose and incompetent fluvioglacial deposits to very strong sedimentary rocks as well as crystalline igneous and metamorphic rocks. The major advantage of embankment dams over concrete ones is related to this as the bearing capacity of the foundation beds can be accepted even if low. As an embankment can adjust to minor dislocations without failure, small scale settlements owing to load stresses during and after construction do not constitute a serious problem. This does not of course mean that some foundation materials are to be preferred to others and among the less desirable kinds are argillaceous sediments such as clayey shales; the term includes consolidated pelitic sediments originally deposited as clayey or silty alluvium which was compacted subsequently by compressive forces due to superincumbent load as well as perhaps becoming cemented through percolation and/or recrystallization. The rate of consolidation of clayey shales depends upon:

(i) size and nature of particles;
(ii) presence or absence of sand layers (paths for the escape of water);
(iii) thickness and superincumbent load;
(iv) duration of loading;
(v) orogenic effects upon the rock mass.

9.7. DAM FOUNDATIONS

Rock and soil mechanics theory and practice are applicable and one of the advantages vis-à-vis tunnels is the comparatively small areas involved. There are two main approaches to the stress–strain analysis of dams, namely:

(I) Scale models which are subjected to stress can yield valuable data.
(II) The computer may be employed in connection with the application of analytical theory in the development of the finite element method in structural and continuum mechanics.[12]

Stress and strain are familiar concepts in engineering and their relationships are related to orthogonal x, y and z reference axes or to polar coordinates. In general, materials may be regarded as either elastic or non-elastic as regards their behaviour under stress and the perfectly elastic state may be contrasted with the perfectly plastic state. A material exhibiting the latter will deform permanently without limit at some critical stress and will not support a stress exceeding this in amount. Strengths of materials are tested uniaxially or triaxially. The relevance of all this in geology is that failure fractures which develop in semi-elastic materials are usually either of the shear or lateral extension type. The former develop at from a few to 45° to the direction of uniaxial stress in homogeneous materials. The latter develop roughly parallel to the direction of applied stress. They often occur in brittle sandstone layers between beds of shale. Load of a dam on bedrock or loose materials augmented by the weight of water in a reservoir on its underlying rocks have an effect related to the total magnitude of both and the porosity and/or permeability of the materials stressed. Usually, confining pressures produced by loading cause an increase in strength, but pore pressures arising from interstitial liquids have a tendency to offset the effective confining pressure and therefore reduce the strength. Most bedrocks possess primary permeabilities sufficiently low that pressures exerted by pore fluids are negligible whereas loose deposits (particularly those with interconnected pores occupying more than 5% volume) do not show a marked increase in strength with confining pressures.

Unfractured bedrock possessing low primary permeability increases in strength with the confining pressure, in fact much more so than is the case with bedrock which is intersected by interconnected and water-filled fractures. The subject has been examined by means of short duration triaxial compression tests and for the interpretation of these it is of course necessary to employ assumptions implicit in Mohr's circles of stress and strain. These matters are well known in engineering practice and it is unnecessary to discuss them in detail. However it may be useful to indicate that the shearing resistance of an isotropic substance to failure is taken as the sum of its cohesive strength and a term expressing frictional resistance to dislocation along a potential plane of failure, this latter being the product of the effective normal stress across such a plane and the coefficient of internal friction, $n = \tan \phi$, where ϕ is an angle analogous to the angle of ordinary sliding friction. Thus if the shearing resistance is written τ_θ, the cohesive strength τ_0, then:

$$\tau_\theta = \tau_0 + \tau_\theta \tan \phi$$

Some pertinent data are given in Table 9.1. The values in the table are for room temperature and the ultimate strengths recorded are taken from the maximum points on stress–strain curves for specified confining pressures, these constituting the maximum stress differences which the materials mentioned can withstand under the conditions of measurement.

TABLE 9.1

RESULTS TAKEN FROM SHORT DURATION TRIAXIAL COMPRESSION TESTS

Rock	*Confining pressure (bars)*	*Ultimate strength (bars)*	*Fault angle (degrees)*
Basalt	0	2 620	21
	690	4 620	29
	1 030	5 510	21
Granite	0	1 670	15
	490	4 710	20
	980	8 340	15
Limestone	0	530	30
	510	4 150	30
	1 020	3 130	27
Limestone	0	1 340	27
	230	2 010	29
	490	2 620	32
Quartzite	0	3 590	—
	1 010	10 790	25
	2 020	12 940	—
Sandstone	0	590	23
	350	1 610	36
	690	2 190	41
Sandstone	0	680	19
	280	2 000	35
	550	2 530	37
Shale	0	640	5
	260	980	—
	510	1 610	10
Siltstone	0	480	20
	500	1 470	30
	1 010	2 260	—

The dam and its impounded water exert enormous load stresses on the floor and sides of a valley which were not present prior to them. The appropriate site should, optimally, have the following characteristics:

(a) location in a narrow valley with steep hard rock slopes;
(b) absence of fracturation both at the damsite and in the reservoir area;

(c) absence of weathered rock at the foundation area of the dam;
(d) satisfactory slope stability at the damsite and above the proposed dam crest so that excavation at the future abutments will not disturb anything;
(e) tectonic stability;
(f) surface and groundwater of such a chemical nature that they will not attack the dam or its foundations;
(g) workable rocks in the area;
(h) adequate local supply of building materials.

Such a perfect combination of factors is almost never encountered and site engineers and geologists have to reckon with nature as normally manifested so that dams are often constructed in unsuitable areas.

As regards tectonic stability, man-made dams and reservoirs may promote instability; reservoir-induced seismicity was recorded at Lake Mead in Colorado as long ago as 1945. Later instances occurred when Lake Kremasta in Greece was filled, at Lake Kariba, southern Africa and at Koyna in India where the shock so produced had a magnitude of 6·5 on the Richter scale and caused over 200 deaths. The latest instance has been described by Adams[14] and concerns the Aswan Dam in Egypt. This case is particularly interesting because Lake Nasser is only 72 m deep at its maximum and the 100 or so known reservoir-induced earthquakes suggest that such events are much more likely to happen where the reservoir depth exceeds 100 m. Nevertheless one did take place at Lake Nasser, actually near the edge of the lake 65 km to the southwest of the dam on 14 November 1981. Its local magnitude was 5·6 (Helwan), body-wave magnitude about 5·25 (NEIS and ISC) and it was felt as far away as Khartoum, 900 km to the south. It also caused some damage at Aswan, but this did not affect the dam or installations. The event occurred well away from the deepest part of the lake and also a long time after its impoundment (6 years after reaching maximum depth). This might imply that the earthquake may not have been induced and certainly no instrumentally determined earthquakes have been located before in the Upper Nile Valley. However, it has been suggested by Simpson that induced earthquakes such as those mentioned above may have first events at some distance from the deepest part of a reservoir and also take place some time after impounding when the inducing effect of increased pore pressure first overcomes the inhibiting effect of loading. It is very probable therefore that the Aswan Dam incident with its concomitant long continuing aftershock sequence (another normal feature of reservoir-induced seismicity) really indicated an induced origin for the event.

Turning again to pressure arising from dam and reservoirs, it may be stated that the types of stresses imposed arising from a dam and acting upon its foundation relate to its shape and also to the materials utilized in its construction. Thus masonry or concrete dams behave as cohesive and rigid structures, but movement may take place along construction joints. Here the stress exerted on the foundation is a function of the dam's gross weight. With earth and rockfill dams, semi-plastic behaviour occurs and the pressure on the foundation at a point is dependent upon the dam thickness above that point. This latter kind of pressure parallels to some extent the pressure of water in the reservoir, but of course the distribution of pressure is influenced by the fact that the materials used in the construction process possess inherent strength and so do not fail until a critical stress is exceeded. On the other hand, water exerts pressures of an hydrostatic type which maximize linearly with depth. Such pressures are exerted on the upstream face of a dam as well as on the sides and floor of the reservoir. As regards the vertical face of a concrete dam, this is taken as acting like a rigid body and where P is the hydrostatic pressure per unit area, the change in pressure with depth y is given by

$$\mathrm{d}P/\mathrm{d}y = \rho g$$

ρ being the density of water (a temperature-dependent variable) and g the acceleration due to gravity.

In practice, foundation failures may arise from a variety of causes. In the case of embankment dams with almost horizontally layered materials, failure usually results from inadequate handling of seepage problems in or under them.

Concrete dams have more concentrated loads and shear dislocations due to them are much more significant. As regards the geology, the following conditions influence failure of the foundations of concrete dams:

(i) Fractured sandstones overlying weak shales dipping upstream.
(ii) Fractured crystalline rocks overlying a shallow fault zone containing sheared materials of very low strength.
(iii) Horizontally layered limestones overlying weak shales extending downstream to a steep slope in the floor of the valley.

(iv) Folded rocks containing thin layers of shale, an incompetent material.
(v) Sedimentary strata dipping downstream and intersected by a fault dipping upstream and containing materials of low strength.
(vi) Intersecting strong conjugate joints with orientations promoting mass shear dislocations. Such conditions were the cause of the disastrous failure of the Malpasset Dam in France in 1959 as described by Talobre.[15]

Slope failures may take place towards the dam axis and dislocate the abutments, but these are rare.

In concrete dams where slopes in the areas of abutments stay stable during foundation excavation work, there is little possibility of downslope movement along surfaces intersecting the dam foundation since added stability is given through the actual weight and strength of the structure. Nevertheless there is always a potential for failure of slopes above the dam in deep valleys with accompanying burial of surface structures under associated debris.

In the case of embankment dams, slopes may fail along surfaces intersecting the abutment parts, this occurring both from below and above such structures where sufficiently steep inclines are weakened through the infiltration of groundwater.

From this discussion, it is apparent that the testing of foundation materials is an essential prerequisite to the design and construction of a dam. There are several things to bear in mind. One is that laboratory determinations often give misleading results, e.g. by showing the strength of a rock as greater than is actually the case *in situ* and in the mass. This is because planes of weakness in the latter are not represented in test specimens. Then it is important to remember that in nature one is not dealing with isotropic materials. On the contrary, all natural materials (rocks and soils) are anisotropic to some degree. The most significant parameters are Young's modulus, E, the modulus of elasticity; Poisson's ratio, ν; and unit weight or density, ρ. All can be determined using cylindrical samples in the laboratory or by various methods directly in the field. Stress is denoted σ and strain ε.

The modulus of elasticity (see Table 9.2) may be defined as the ratio of longitudinal stress to longitudinal strain, i.e. σ/ε, and it is a measure of the ability of a material to resist directed stress. In the usual test, the modulus of elasticity is determined by measuring the shortening of a test cylinder under a given applied stress at the ends of the said

TABLE 9.2

MODULI OF ELASTICITY OF VARIOUS ROCKS[13]

Rock	*Elasticity, kN m*$^{-2}$($\times 10^6$)	*Bars* ($\times 10^5$)
Igneous		
Granite	10·35–82·11	1·0–8·2
Diorite	24·84–42·09	2·5–4·2
Andesite	32·43–47·61	3·3–4·8
Basalt	40·71–85·56	4·1–8·5
Diabase	70·38–95·91	7·0–9·6
Sedimentary		
Sandstone	4·14–55·2	0·4–8·0
Limestone	2·76–97·29	0·3–9·7
Shale	2·07–68·31	0·2–6·8
Siltstone	6·90–64·17	0·7–6·4
Metamorphic		
Quartzite	8·28–44·16	0·8–4·4
Gneiss	24·15–104·19	2·4–10·4
Greenstone	23·46–104·88	2·3–10·5

cylinder. When a cylinder is compressed, the shortening is accompanied by a lateral extension at right angles to the direction of the applied stress. Measurement of this lateral extension facilitates the calculation of Poisson's ratio. If the direction of applied stress is parallel to the x-axis and lateral extension is measured parallel to the y-axis of rectangular coordinates, then Poisson's ratio is given by

$$\nu = \varepsilon_y E/\sigma_x$$

As noted earlier, laboratory measurements of the physical properties and elastic constants of the materials of dam foundations can be made involving test cylinders of sample rock which are not totally representative of the relevant rock mass. *In situ* it is possible to note primary anisotropism in the latter due to the origin of the material and secondary anisotropism arising from the development of fracturation and folding, i.e. structural geological considerations. There are also the effects of solutional alteration of shallow or deep origin causing weaknesses usually unrepresented in laboratory testing.

Existing stresses in the foundation materials before dam construction and the impoundment of reservoirs fall into two categories:

(a) Those connected with the weight of material in slopes and the floor of the valley as modified through the slope configuration and internal structures such as fracturation and layering.

(b) Those of a residual or tectonic nature—residual stresses can be considered to be unbalanced and preserved in elastic or semi-elastic rocks, having been generated in the geological past when these were deformed; tectonic stresses are also unbalanced and related to the elastic strain of rocks by forces acting today, e.g. seismic forces accumulated in intensity along faults in currently active seismic regions where earthquakes occur.

Of course a number of residual stresses probably originated as tectonic ones and, after the relaxation of regional forces formerly acting when dislocation took place, remained because total relaxation of the relevant rocks did not occur subsequently.

All these stresses may be regarded as virgin stresses, i.e. stresses existing prior to the commencement of engineering works. Subsequent and man-made stresses resulting from the application of artificial loads may be termed induced stresses.

It is possible to subdivide virgin stresses into gravitational and latent stresses also. Gravitational stress derives from the effects of overburden on rock elements at depth. Latent stress may depend upon topography, high ground increasing stress in adjacent low ground, or tectonics; for example where sedimentary rocks in a basin have been uplifted, high lateral stresses may become locked into them. Release of tectonic stress (whether it is residual or not) occurs during quarrying where 'popping' takes place or in tunnelling where rock bursting may be encountered.

Obviously *in situ* measurement of the elastic properties of either elastic or semi-elastic rocks for abutments or foundations of dams is valuable in assessing the possible presence of residual or tectonic stresses.

The mass strength of foundations and reservoir slopes must be estimated and in order to do this it is necessary to derive data from shafts, adits and tunnels using equipment suitable for the determination of the elastic constants of rocks in boreholes. Rupture strengths may be ascertained by hydraulic fracturing by appropriate fluids introduced at high pressures between properly placed packers in boreholes. The state of stress can be obtained from the velocities of induced seismic waves through rock masses.

As well as facilitating the estimation of the strength of foundations, shafts and tunnels driven into the rock there and in the abutments permit visual examination of relevant rock features such as lithology, fracturation state and degree of weathering. The object of the whole exercise is to guesstimate the response of the materials of the foundation to later loading at the surface. In the case of embankment dams, it is anticipated that there will be some adjustment of the foundation materials as they are constructed and this is not regarded as a serious matter. However in the case of concrete structures, any settlement of foundations or dislocation of abutment rocks could induce failure and constitute a threat to stability. Non-elastic and uneven foundation adjustments will be of much greater threat to a dam's integrity than any systematically distributed and elastic dislocations. It is important that all measurements of bulk elastic constants and strengths of foundations as well as test borehole locations and results from work in excavations should be correlated with the local geological conditions as recorded on maps and sections to appropriate scales. Especially significant are the proper recording of the positions and geometry of planes of weakness and the location of rock bodies possessing low inherent strength.

In recent years more and more dams have been erected and there are more than 10 000 in the USA alone with many more thousands scattered around the rest of the World. Although these are mostly of the earth and rockfill types, there are also many concrete and masonry ones. Remote sensing Landsat satellite imagery has been particularly useful in locating many in the USA which appear not to have been listed by state geological surveys or other information sources. Their primary purpose is connected with the supply of energy (although dams impound reservoirs also for recreational purposes, etc.) so that their proliferation may be related to the rising demand for this. Concomitantly it is often necessary to place these structures in unsuitable geological sites for socioeconomic reasons.

It must be borne in mind that with large scale hydraulic structures the influence of water is paramount—beds liable to contain this fluid show a diminishing bearing capacity with increasing water content. This is true of shales which, with their relatively low moduli of elasticity (see Table 9.2) demonstrate plastic behaviour so that, under great load, they can be squeezed out from beneath foundations of buildings. Shales of this type slake when contacting water after partial drying. Another type of shale exists which is resistant to weathering and so does not slake. Shales belonging to this type have a greater bearing capacity and are stronger. Where a concrete dam is concerned, contact between this man-made material and shale which has been excavated is likely to be rather uneven which results in a greater shear resistance between rock and concrete than the actual shear strength of the shale. The shear strength of the shales must be determined, therefore, along those surfaces upon which it is anticipated that shearing might take place. Obviously the shear strength along strata or planes of schistosity will be less than that occurring where bedding and other planes are transgressed by the shear plane. This latter may not be regular also and where it cuts across beds, it will follow planes of weakness somewhere in the rock mass. In practice it has been found that the shear resistance as well as the bearing capacity of clayey-shales are too small to permit the erection of concrete dams and earth or rockfill dams are usually substituted.

Argillaceous materials have one great advantage, their impermeability. As they are plastic they do not form fissures when subjected to superincumbent loading and indeed any existing fracturation will be partially or wholly removed by this. Unless there are permeable inclusions of water-transmitting beds such as sandstones, a perfect foundation seal is established. However the percolation of water through such permeable beds if they are present may disturb shales at great depths, depths far below any susceptible to the effects of surficial weathering. Of course the above remarks apply to the type of shales which slake. Those which do not, i.e. compact shales, may well become interpenetrated by fissures and joints as well as by shattered zones which may have to be grouted in order to achieve watertightness. Shales of any type are subject to alteration if their water content alters and this renders them, in principle, unsuitable foundation bed materials. Nevertheless there are instances of successful projects on such foundations. One of the most adverse shale conditions is that of the Alpine and Carpathian flysch in which such beds alternate with thin layers of sandstone. In some places flysch is rather impermeable, with the alternating beds lying almost horizontally and extending over considerable distances. Mostly, however, the flysch is strongly folded, thrust and faulted, i.e. it has been subjected to intense orogenic stresses and as a consequence shale beds may be disharmonically and tightly folded with accompanying heavy fracturation of the sandstones. Steeply inclined beds may occur.

None of these manifestations can be considered favourable to dam building, but the Ben Metir Dam was founded upon a disturbed shale and sandstone flysch complex of Oligocene age as described by Stucky.[16] This dam, details of which are shown in Fig. 9.5, was built so as to impound water for supplying the city of Tunis. Its foundation is on thin sandstone strata, highly permeable and heavily fractured, separated by clayey shale beds which would normally be quite impermeable but which are found to be heavily weathered and plastic near the sandstones. On the left bank the flysch is covered by Pontian lacustrine clays and arenaceous sediments having an almost horizontal orientation. At the actual damsite the rocks are variable and hence non-uniform as regards compressibility. When these formidable geological conditions were considered in relation to the construction work, it was decided to design a rockfill dam at the left abutment with a buttress dam on the flysch rocks. The buttresses were so made that the resultant pressure imposed on the foundation materials would not exceed 8·5 kPa cm^{-2}. The sealing of inter-buttress joints allowed for a total settlement of 10 cm and up to 5 cm of differential settlement. Some low blocks on the right flank stood upon solid sandstones and were constructed as gravity blocks. The reservoir was successful and a subsequent settlement of up to 58 mm was measured with the differences in the settlements between individual buttresses not exceeding 1 mm. Afterwards, two drainage galleries were inserted under the dam in order to accelerate the consolidation of clayey beds and reduce the uplift. From these galleries, drainage borings in a fan-like arrangement were drilled under the foundation of the dam. These observations refer to the problems arising where dams are sited in sediments of unsuitable types.

In the case of crystalline rocks such as plutonics like granite, gabbro and diorite, hypabyssals like porphyry and its allies as well as volcanics like rhyolite, andesite and basalt, the situation is different because they, like metamorphic materials such as gneisses and schists, possess a much higher innate strength; in fact they are stronger than concrete so that they usually have a perfectly satisfactory bearing capacity. The main problems with them arise through alteration.

Shatter zones for instance may occur in granites as well as widespread zones of thermal decomposition. The geological history of the particular granites must be determined as it has a considerable bearing on the matter. Thus older granites will probably be more tectonically altered than young ones since more stages of folding will have occurred during their existences. Granites may be altered strongly along tectonic lines and also may be deeply weathered. In Europe this is often the

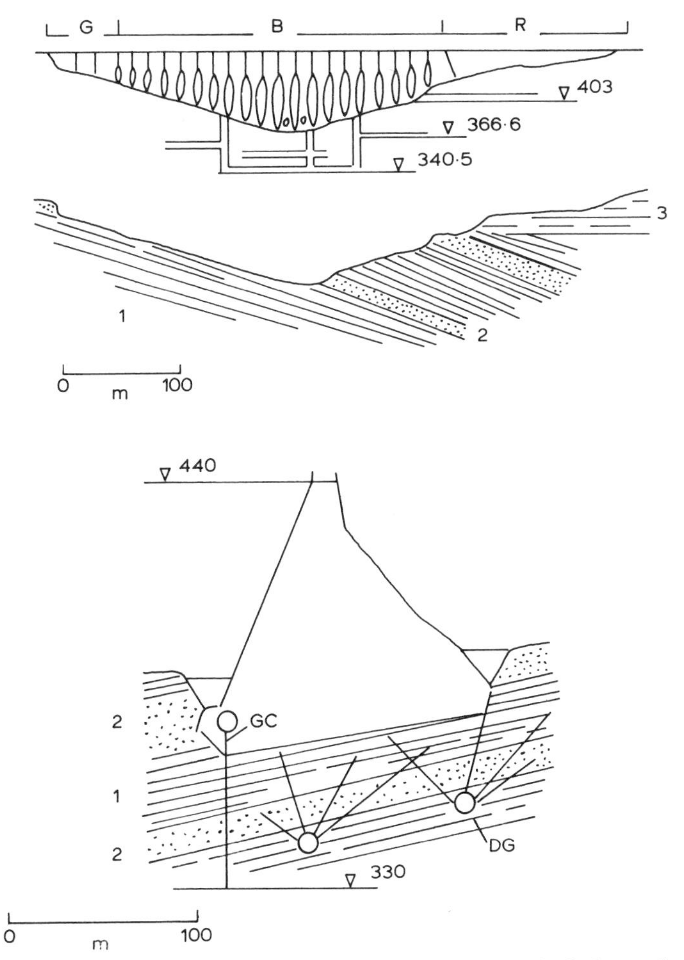

FIG. 9.5. Ben-Metir Dam (Tunisia): founded on disturbed shale–sandstone flysch. G, gravity blocks; B, buttress dam; R, rockfill dam; GC, grout curtain; DG, drainage gallery with drainage borings; 1, Oligocene clay shales; 2, sandstones; 3, Pontian lacustrine beds.

case because granites occur beneath Tertiary or Cretaceous sediments and prior to their deposition underwent weathering of an almost tropical nature with of course irregular depths of weathering reflecting variations in climatic and topographic conditions. Sometimes weathering extends deep down along joints and fracture zones. Hence, although at first sight granites seem to be almost perfect foundation materials, this cannot by any means be taken for granted. An instance of an abandoned damsite illustrates the point. At the Fork Site Dam, San Gabriel River, California, a wide shatter zone was revealed on the right bank and in this, the granite was heavily fractured and in part also decomposed by hydrothermal processes. Excavation caused a big landslide along this area of the proposed damsite. The downslipped material actually filled the whole area of the foundation and caused the abandonment of the project, as Lugeon[9] has indicated.

Of course weathering can affect any kind of rock. For example it adversely affected the construction work in micaceous gneiss, a metamorphic rock, for the Prettyboy Dam near Baltimore. This was originally planned as a gravity dam of 45 m height for supplying water to that city. Excavation on the right bank demonstrated that the initial geological investigation was inadequate. It was discovered that rock along a fault plane was decomposed to a depth of 50 m. This necessitated considerable deepening of the foundation and extension of the dam by means of a concrete wall on the right bank. Actually the work enlarged the planned volume of excavation by a factor of six with all that this entailed in terms of delay and added cost.

In rocks of this type the load of a gravity dam will produce yielding and the deformation properties of the rocks must be known in order to predict the results of compression on abutments and foundation. This is true for gravity and buttress dams, but with arch dams not only these features but also the rocks in the valley slopes are involved as well. Here laboratory testing is insufficient because, since an assumption of homogeneity is made, this gives unrealistically high values. Rock masses are almost invariably inhomogeneous so that the actual values are lowered by rock joints, bedding planes and other weakness features. Stiny[17] has noted that diabase at the site of the Zitterbach Dam gave a value of 800 000 kPa cm^{-2} in a sample as compared with only 92 000 kPa cm^{-2} on site.

9.8. ADVERSE GEOLOGY IN CONSTRUCTION

The optimum characteristics of a geologically appropriate damsite have been indicated already and these include location in a narrow valley with steep, hard rock slopes. However steep slopes of this type may conceal the presence of rockfall material since the hardest rock is

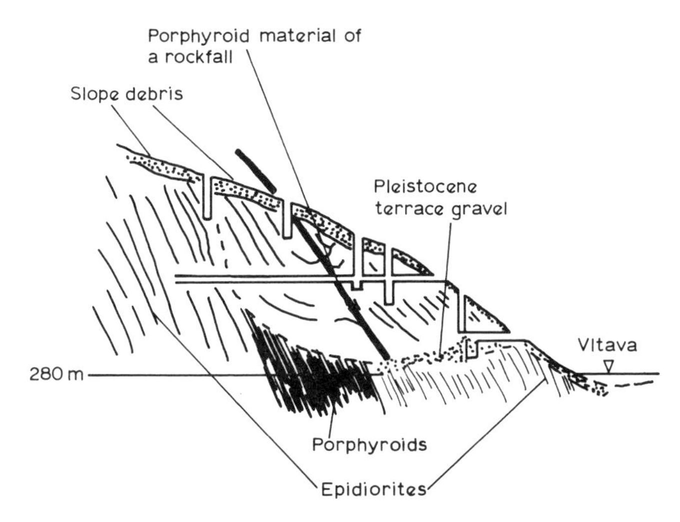

FIG. 9.6. Geological section of left bank of Vltava River (Bohemia) near Zlakovice (original site for Orlik Dam studied by Zoubek—see ref. 18). (——) Slope surface prior to rockfall.

subject to erosion. A case in point has been cited by Zoubek,[18] namely the Orlik damsite on the River Vltava in Bohemia (see Fig. 9.6) where there proved to be a sizeable rockfall on the left side of the valley with porphyroidal rocks showing loosening and rather irregular depositional characteristics. As a result of suspicions about the site, a number of test pits was inserted from an exploratory drift at the foot of the slope and gravel of the terrace of the river found to occur below the downslipped porphyroid. It proved to be necessary to abandon this and indeed several sites on the river.

There is no doubt that similar impediments may be expected to occur in many of the deep river valleys in central Europe because of Pleistocene climatic conditions, especially freeze–thaw action which broadened fractures and loosened the valley slopes. Even where rock sliding did not take place, the stability of the rocks was reduced and of course any human activities such as excavation would trigger a rockfall in many cases.

Another instance is afforded by the projected damsite in the upper part of the River Romanche, French Alps, which Gignoux and Barbier have described.[19] In this instance the left flank of the valley comprises granite together with sandstones and limestones, the last two being of

Triassic age. However it was the right flank, composed of apparently stable Liassic shales which are greatly inclined, which caused the trouble. The impermeability of such strata would seem to be ideal, but a detailed geological study showed that the valley had been narrowed by an ancient landslide. A deep borehole inserted from the proposed dam crest demonstrated that water-bearing gravels exist to a depth of 63 m. The old river valley in fact had been dammed by a landslide to an elevation of 40 m, this having moved the river channel about 70 m in the direction of the left bank. The inference was made that the supposedly satisfactory Liassic shales are not now *in situ*, but rather slid on to highly permeable alluvial beds of the old river valley. As a consequence it was decided to abandon the damsite. The situation is shown in Fig. 9.7.

Where the geology is right, narrow valleys are excellent. Arch dams are perfect for canyon-like valleys with appropriate rocks as the example of the Cancano Dam in the Italian Alps shows. Here the abutments are keyed in solid, competent limestones in a valley of classical shape with stable flanks.

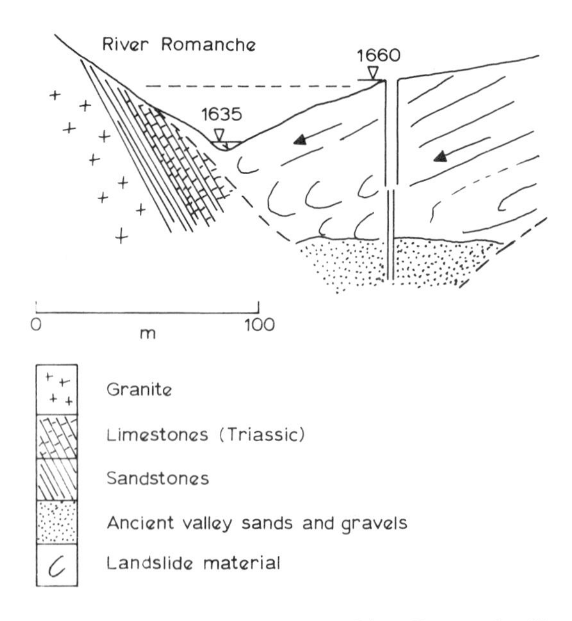

Fig. 9.7. Abandoned damsite on the upper River Romanche (French Alps).

Where the valley is not quite so ideal, appropriate design adjustments have to be made and this is exemplified in the case of the Pieve di Cadore Dam in Italy, as described by Zaruba and Mencl.[20] The relevant valley has a trapezoidal cross-section 50 m in elevation and 300 m wide having, on the right bank, a 55 m deep erosion gorge cut by the River Pieve in Triassic dolomitic limestones. The solution of this particular problem took the form of constructing two arch dams separated by an expansion joint.

Clearly erosion can change the shapes of valleys and weaken the slope stabilities of their flanks as also can other agencies such as earthquakes. The stability question is of major importance because dam construction involves quite deep excavations and these will disturb the natural stabilities in any event. It is axiomatic that gentle slopes are better than steep ones. Tectonic fracturing is a hazard because it means that induced planes of weakness can slip, e.g. with slates. This may occur also with normal sedimentary strata having bedding planes if the rocks dip unfavourably, i.e. towards the river. Where schistose beds occur, these metamorphic rocks develop similar planes and if these dip at the same angle as the slope of the valley side, a proposed dam foundation may become buried in debris when excavation begins. This has happened often—one case is described by Reuter.[21,22] Construction of the Rappbode Dam in East Germany required painstaking examination of the site and very careful excavation in order to prevent a major landslide from occurring.

As noted earlier one of the great hazards to dam construction in Europe and also in North America and other parts of the world arises because such regions have been in part subjected to several stages of Pleistocene glaciation, a phenomenon which deepened valleys through superincumbent loading by glaciers or ice sheets. These may be hidden by subsequently deposited, i.e. younger, sediments—Walters[23] has described an instance of this. He cited the Pitlochry Dam in Scotland which is approximately 20 m high and founded on competent crystalline schists offering optimum characteristics of impermeability and safety. However a cover of fluvioglacial sands and gravels infilled an antique pre-glacial valley of which the bottom is below the existing river floor. In consequence an expensive extension of the dam by means of a concrete wall had to be carried out for 280 m. This acted as a sealing screen.

A similar case has been noted by Gignoux and Barbier.[19] This was that of the dam on the River Isère (Savoy, France) erected on a vault.

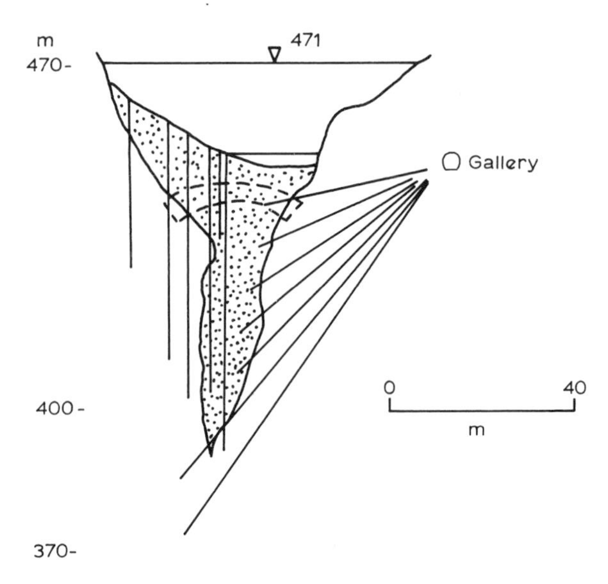

FIG. 9.8. Dam on River Isère (Savoy, France) erected on a vault. (From ref. 19.)

Borings in the area showed the shape and depth of the canyon and it was necessary to remove glaciofluvial infill material and replace it with concrete during construction (see Fig. 9.8).

If it is not practicable to remove all such glaciofluvial deposits from an area, then not merely the dam but also the reservoir and its impounded water may be affected in a deleterious manner. For instance the Sautet Reservoir, France, loses water along gravels occupying the pre-glacial valley, as illustrated in Fig. 9.9.

Many rivers follow reasonably straight courses, but others tend to wander and this wandering becomes more marked with age until eventually the river may develop a series of meander bends. If rejuvenation follows, i.e. if there is an isostatic elevation of the landscape incorporating the river, the meanders may remain as a series of alternating ox-bow lakes on the old flood plain, these possessing crescentic shapes. If the river is in the meander state, this can be useful. The Wallsee Dam in Austria was constructed on a meander of the River Danube now beyond the existing river bed. After completion in 1968 of appropriately geologically sited power plants, weirs and navigation locks, the river was diverted into a newly excavated channel (see Fig. 9.10).

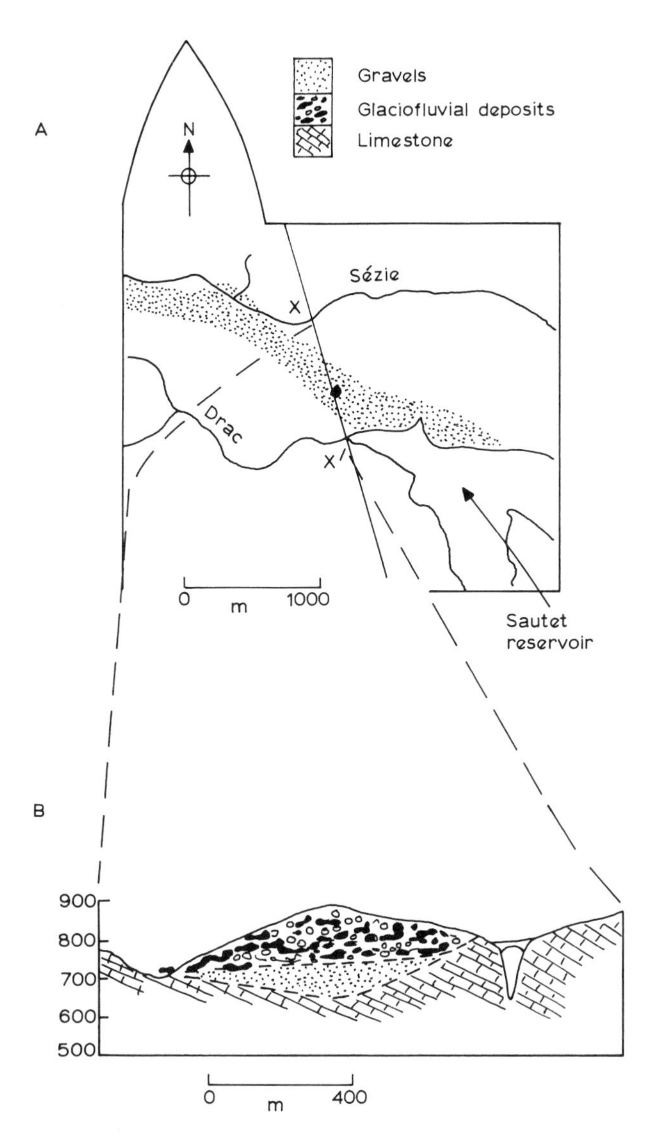

FIG. 9.9. The Sautet Reservoir, France, which loses water along the gravels occupying the preglacial valley (stippled). A, Plan of area; B, geological section XX′.

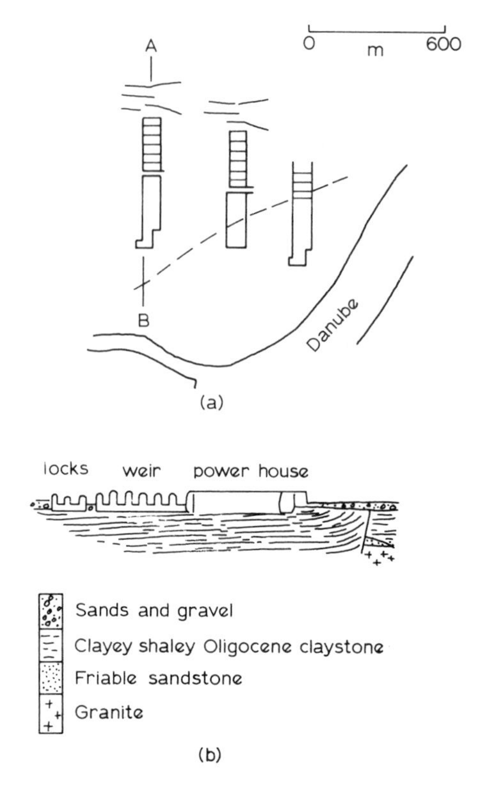

FIG. 9.10. Wallsee hydraulic works (sites 1–3, Danube). (a) Plan; (b) section AB.

Adverse geology must be overcome if a dam, sited for economic or other reasons on it, is to be successful. A case in point was the Spruce Run Dam and its associated reservoir located near Clinton, New Jersey, actually in Hunterdon County on Spruce Run which is a tributary of South Branch, Raritan River. The dam was intended to retain and supply excess flow for the off-river storage reservoir at Round Valley near Lebanon, New Jersey and raw water was to be supplied for sale to cities within the state. The dam is an earthfill dam. The areal geology involves both Spruce Run and its incised valley in the Cambro-Ordovician Kittatinny limestone outcrop which at the site has a northwesterly strike with concomitant southwesterly dips averaging 50°, but ranging from 15° to 70°. There is no evidence of folding, but isoclinal folds may occur if anticlinal crests have been eroded away and synclinal troughs buried. The geology is shown in Fig. 9.11. That at the site is far from optimal. There is a great deal of surface evidence of faulting in the region and this is paralleled by much subsurface evidence as well. However, nearly all the potential foundation weaknesses arose from the variable but consistent solubility of the limestone underlying all of the dam and much of the adjacent reservoir rim. This is augmented by bedding planes and cross-cutting intersecting joints, all planes of weakness. During the work extreme solutional action and weathering was encountered along the cutoff axis due to multiple faulting and massive brecciation. These very weak zones did not usually necessitate excavation down to rock underlying the pump-lowered water table, but they did require a triple line of grout holes for the desired cut-off. The preliminary subsurface exploration programme

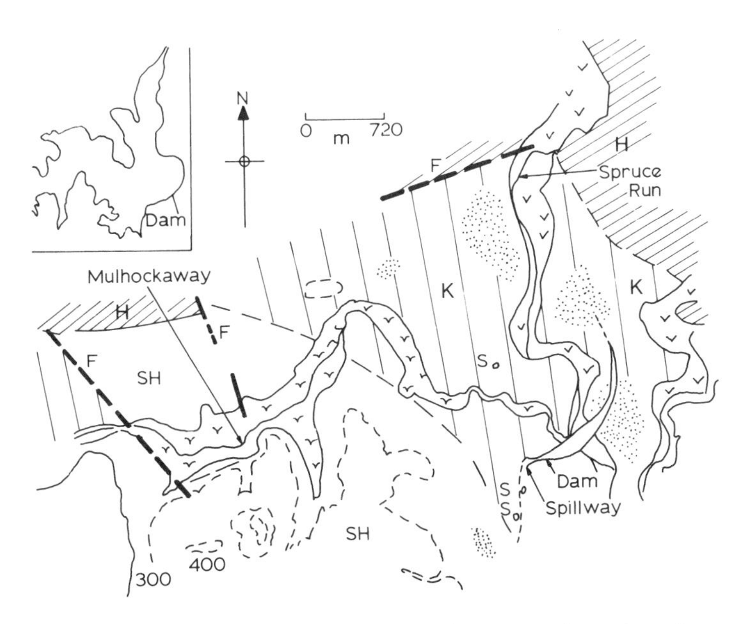

FIG. 9.11. Geology of the Spruce Run damsite, reservoir and environment showing spillway and sinkholes. F, Faults; SH, Martinsburg shale (Ordovician); K, Kittatinny limestone (Cambro-Ordovician); H, Hardystone quartzite (Cambrian); (— —) formation boundaries; S, sinkhole; [dotted box] residual soils and glacial drift; [v v box] alluvium (gravels, sands, silts). Inset: Spruce Run Lake reservoir.

did not delimit all the deeply weathered and dissolved zones accurately, but it did show the extent of several of the major zones, which information allowed design and specifications to cover adequately treatment methodology and also to foresee most specialized treatment. During construction, the rock surface weathered zone was excavated to a greater depth and extent and more concrete grout cap was employed than was expected. This resulted in use of less total grout than anticipated because of less grout leakage and also produced a tighter grout cutoff. Locally available materials were utilized for the impervious core, fill and riprap. This latter comprised not only limestone, but also gneiss and quartzite located in nearby deposits. As regards the construction of the grout curtain, this project utilized an unusually large number of well qualified and experienced engineering geologists and geological consultants in order to properly fulfil the requirements of a cutoff trench to 'rock' for the total length of the dam, relatively short sections of concrete grout cap, a single line of curtain grout holes and blanket grout holes under parts of the rolled fill. With the cutoff trench, much more weathered material and broken rock were excavated than expected and the grout cap had to be extended all along the core trench. All this tended to offset acceptance of grout. Leaks were reduced to a minimum by the grout cap as a result of the removal of a great deal of poor quality, groutable rock and this also made it easier to place the cutoff grout. As regards the grouting, the drilling consisted of about 2810 holes, 2057 in the curtain and 753 in the blanket areas. The drill length was about 61 530 m with 46 140 in the curtain and 15 390 in the blanket. The curtain took approximately 6478 m^3 of neat cement grout together with a small quantity of chemical grout. The blanket took approximately 9559 m^3 of cement-fly ash grout. Consequently the total grout quantity was about 16 037 m^3. Interestingly, the total length of drillage conformed reasonably well with the specifications, but the amount of grout utilized was appreciably lower than the estimate. The general situation is shown in the section through blanket and cutoff grout patterns in Fig. 9.12.

Sometimes it turns out to be impossible to overcome adverse geology; the famous landslide slippage into the reservoir of the arch dam at Vaiont with its resultant destruction of lives and property at Longarone, through the overtopping of the crest by flowage into the Piave River following excessive rains and failure to adjust the water level in the reservoir in October 1963, immediately comes to mind. If adequate attention had been given to previous geological reports, the

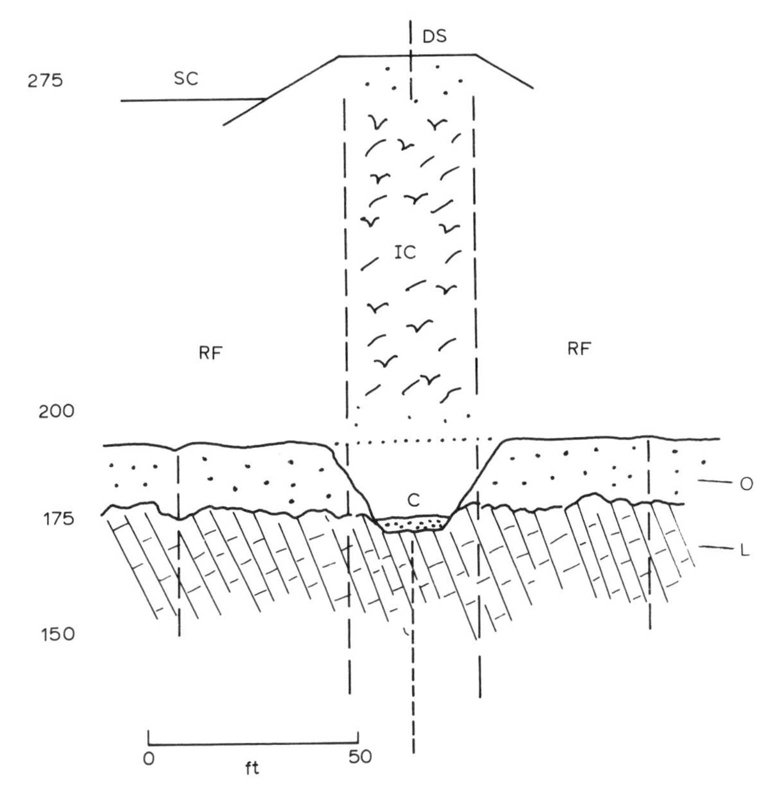

FIG. 9.12. Spruce Run earth-fill dam: section through blanket and cutoff grout patterns. DS, Dam summit, 283 ft; SC, crest of spillway at 273 ft; IC, impervious core of rolled material comprising appropriate borrow and other sediments; RF, random fill-rolled, weathered shale; C, cutoff trench with basal concrete cap ranging from 1 to 4 in. in thickness; O, alluvial overburden; L. Cambro-Ordovician limestone. (From ref. 44.)

disaster might have been averted and almost 3000 lives saved. As it was, this event in northern Italy occurred and 237×10^6 m^3 of rock and earth infilled part of the reservoir. As landsliding was known to have occurred here in pre-historic and early historic times, it was really inadvisable to make an impoundment in the first place.[24]

An unusual case of landsliding causing a natural dam to form comes from the USSR. In 1964 a rockfall took place on the Zeravshan River in the Uzbek SSR and this created a dam; it was therefore a natural one with a height of 200 m and a width of 400 m. It immediately began to impound water and as the site was accessible, engineers moved in and blasted, excavating a spillway directly through the slide itself. In this manner, danger to the old city of Samarkand which lies 160 km downstream was obviated.[25]

Dams and reservoirs are often associated with irrigation projects and the planning of these must include an assessment of possible socio-economic consequences. Large scale projects involve consideration of geological and geomorphological factors and should be considered as part of an integrated, multi-purpose water resources endeavour. This was the case with famous programmes such as the Tennessee Valley scheme and the Missouri River project in the USA as well as others such as the Orange River development in South Africa. The physical context of each is the actual drainage basin although in certain instances it may be necessary and justifiable to effect engineering work which transfers water across drainage divides as was the case with the Snowy Mountain scheme in Australia.

In the USA such matters are the responsibility of the US Bureau of Reclamation which handled construction of the Teton Dam in east Idaho; this failed on 5 June 1976 despite extensive design precautions taking into account the high state of fracture and jointing of the associated rocks including rhyolite. The affair has been discussed in detail in *Grouting in Engineering Practice.*[7] However the case is so important that Figs 9.13 and 9.14 here illustrate the location of the dam

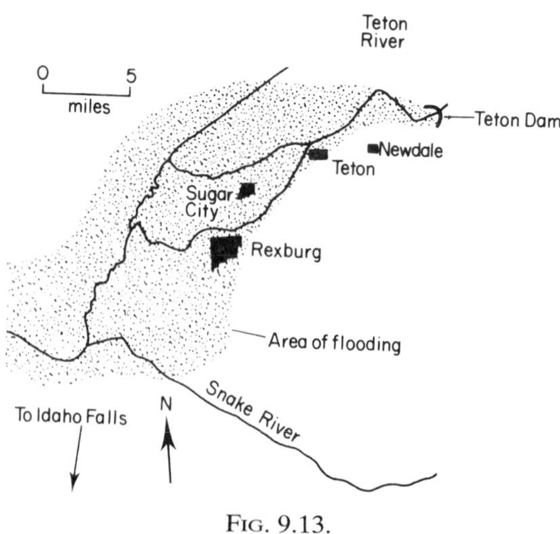

FIG. 9.13.

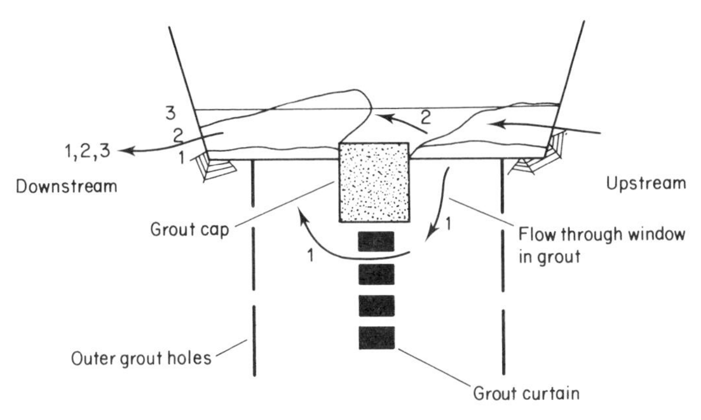

FIG. 9.14.

and the area of flooding as well as a possible failure mechanism (according to Schuster,[26] however, the actual mechanism may never be known fully). The geology of the site has been examined by Fecker.[27] One thing clearly emerged; that is that the intense jointing facilitates water movement except in local situations where grouting was effectively carried out. Sometimes this was not the case and for additional information reference may be made to the two most significant reports—see refs 28 and 29.

In Europe there are more stringent requirements as regards the watertightness of grout curtains than in the USA because safety against erosion is also an objective, not simply leakage prevention, and Fecker opines that leakage is not regarded as a particularly important matter in the USA. There can be little doubt therefore that the imperfect grouting of the rock was an unfortunate contribution to the ultimate failure of the Teton Dam.

It is apposite to refer here to guidelines adopted by the Association of Engineering Geologists for the design, construction and operation of dams and reservoirs. These may be divided into two categories, those pertaining to new dams and those relevant to existing dams.

(*a*) *New dams*: Firstly, the site and regional geology, seismicity and other significant factors must be evaluated and the optimum continuity in developing and understanding geological conditions are best attained by a single team of engineering geologists and engineers working in close association through the stages of site investigation, site selection, final design modifications, specifications, periodic inspections of construction work and performance evaluation of the dam when the reservoir is initially filled as well as thereafter. Subsequently a peer-group review of geologists and engineers normally not associated with the owner of the facility should be set up for all structures which might involve a risk to the public. Such a group should review periodically the various geological, seismic and engineering decisions made during the design and construction activities. During the construction, there ought to be an open communication between all key personnel, i.e. those from the construction side and those comprising the engineering geologist/engineer design team. Field inspections should be carried out often in order to determine whether the existing geological complexities revealed are being understood and appropriately treated.

(*b*) *Existing dams*: Firstly, all existing dams, the failure of which could endanger life or property, should be examined periodically by safety inspectors and reviews made by a team of engineering geologists and engineers not usually in the employ of the owners of such facilities. The intervals between inspections must be short enough to ensure detection of safety-related changes in the structures themselves or their environments. The actual safety reviews must take into account state-of-the-art procedures or design, new engineering, engineering geology and seismic data. As regards structures with higher risk, inspection frequency is recommended not to exceed two years. In certain cases where there is insufficient geological information, it may be advisable to effect subsurface studies. The necessity for such studies ought to be recognized by the engineering geologist/engineer team and of course it will be based upon the hazard potential as well as the history of performance for the facility together with the overall geology of the area.

An instance of the applications is now cited: the proposed Auburn Dam of 1979. This Californian project involved a dam capable of withstanding a fault displacement of just over 23 cm and public hearings were held regarding it on 17 May 1979 in Sacramento. The then Secretary of the Interior Cecil D. Andrus approved minimum criteria for the design of the dam recommended by commissioner Keith Higginson of the Bureau of Reclamation (BurRec) and agreed to by the state. The BurRec is said to have more seismic data regarding this site than anywhere else in the world and indeed the Auburn investigations are believed to have advanced the state-of-the-art of seismic analysis probably more than any prior work. The Woodward-Clyde Consultants Co. made a regional seismic study and the US Geological Survey's recommendations were based on this and were for a design allowing almost 1 m of slippage on the foundation faults. The US Geological Survey has not made detailed analyses of data regarding the faults in the dam foundation area however and their requirement was coincidental with the metre of historic movement estimated for the major fault, F-1. Their criteria were based upon a single seismic event whereas the foundation area movement occurred over 120 million years. Interestingly many observers disagreed with the conclusions of the US Geological Survey. The criteria adopted were as follows:

(a) The maximum believable earthquake would have a Richter scale magnitude of 6·5 with the epicentre just over 3 km away and a focal depth of 8 km.
(b) The dam foundation would be able to withstand a displacement of 23 cm from one seismic event despite the opinion of BurRec that the expected movement would be a mere 3–5 cm. On site geological work has established that faults in the foundation area have moved by anything up to 1 m over the past 120 million years (cf. the San Andreas Fault which has been displaced by at least 320 km in the past 10 million years).
(c) The design of the dam must ensure that the structure withstands ground motion according to an acceleration curve or response spectrum curve recommended by the State.

The curve is, in fact, somewhat different from that of Woodward-Clyde's and one standard deviation higher than the mean response spectrum of the BurRec. The US Geological Survey did not prepare a curve, instead recommending an upward scaling of the spectral acceleration two or three times that recommended by Woodward-Clyde at the 1·0 s interval (already one standard deviation higher than the mean of the acceptable strong-motion recordings of 6·5 magnitude earthquakes).

Adverse geology on damsites may be paralleled by adverse geomorphology causing floods, and fortunately it is possible to predict reasonably accurately probable magnitudes for flood flows in any river basin over a stated number of years (say for a 100-year flood). To reduce the magnitude of peak floods, strategically located large scale dams may be utilized for water impoundment. Complementary measures involving geology more directly include the construction of little dams and various soil conservation activities as well as the adoption of better agricultural practices, e.g. contour ploughing on inclined land. With large scale dams, grout curtains are made; this subject has been extensively treated in *Grouting in Engineering Practice.*[7] Their primary objective is to provide maximum watertightness and this is specified by various permeability standards the attainment of which depends mainly upon hole spacing and hole depths. The deeper the latter the looser is the standard obtained so that the more superficial stage may be constructed to a tighter standard which is therefore more satisfactory. The actual assessment of watertightness is a very important matter and the units employed may be related either to velocity ($cm\,s^{-1}$, for instance) or to take (quantity of water taken in holes during pressure tests).

During grouting injection it is necessary to check the effectiveness of the grout curtain therefore and the German Standard DIN 19700 prescribes that

> 'At dams the grout curtain should be placed under the core, preferably in conjunction with a control gallery from where its effectiveness can be checked and improved if necessary. The reduction of foundation water pressure at the downstream toe of the dam caused by injections must be controlled in boreholes downstream of the grout curtain'.

Such checking is effected by means of water pressure tests but if these are inconclusive so that the coefficient of permeability is hard to determine, then the limits of water losses given in Table 9.3, taken from Fecker, can constitute a standard for the effectiveness of the grout curtain. As a result of the fact that the Lugeon criterion of 1 litre $min^{-1}\,m^{-1}$ at 10 $kg\,cm^{-2}$ pressure for dams higher than 30 m is very strict and seldom achieved, the criterion of Terzaghi and Water Research Commission represent an upper limit. Some typical Lugeon permeability tests for the Sarrans Dam in France are shown in Fig. 9.15.

It is interesting to add some typical data (see Table 9.4) from such water pressure tests. The examples in the table are from the Bumbuna damsite, Sierra Leone, and they were taken during a feasibility study in 1980. They were effected whenever possible in 5 m stretches with pressure steps of 3–6–10–6–3 $kg\,m^{-2}$, each for 10 min, using both the 'successive stage' and 'from bottom upwards stage' methods, as necessary. The single example given illustrates typical data and it is taken

TABLE 9.3

RECOMMENDED ACCEPTABLE WATER LOSS IN WATER PRESSURE TESTS

Source	Satisfactory water loss	
	Water pressure tests, litres min^{-1} m^{-1}	*Pressure (kg cm^{-2})*
Ref. 9		
Height = 30 m	1	10
Height = 30 m	3	10
Heitfeld (1964)		
Height = 100 m	3	10
Height = 50 m	4	10
Height = 20 m	4·8	10
Terzaghi (1929)	5	10
Water Research Commission Standard	7	10
International Teton Dam Review Group	18	10

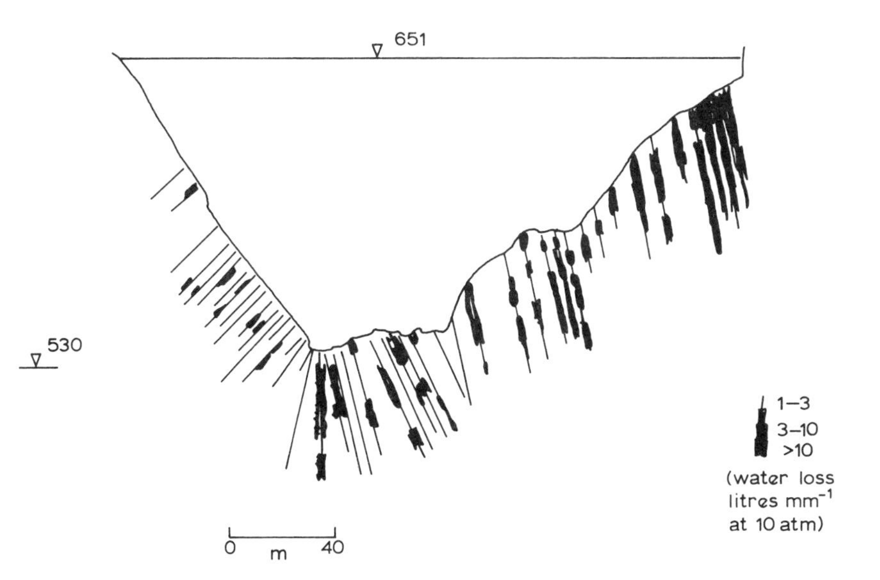

Fig. 9.15. Sarrans Dam, France: Lugeon permeability tests (water pressure tests).

TABLE 9.4

ABSORPTION CHART

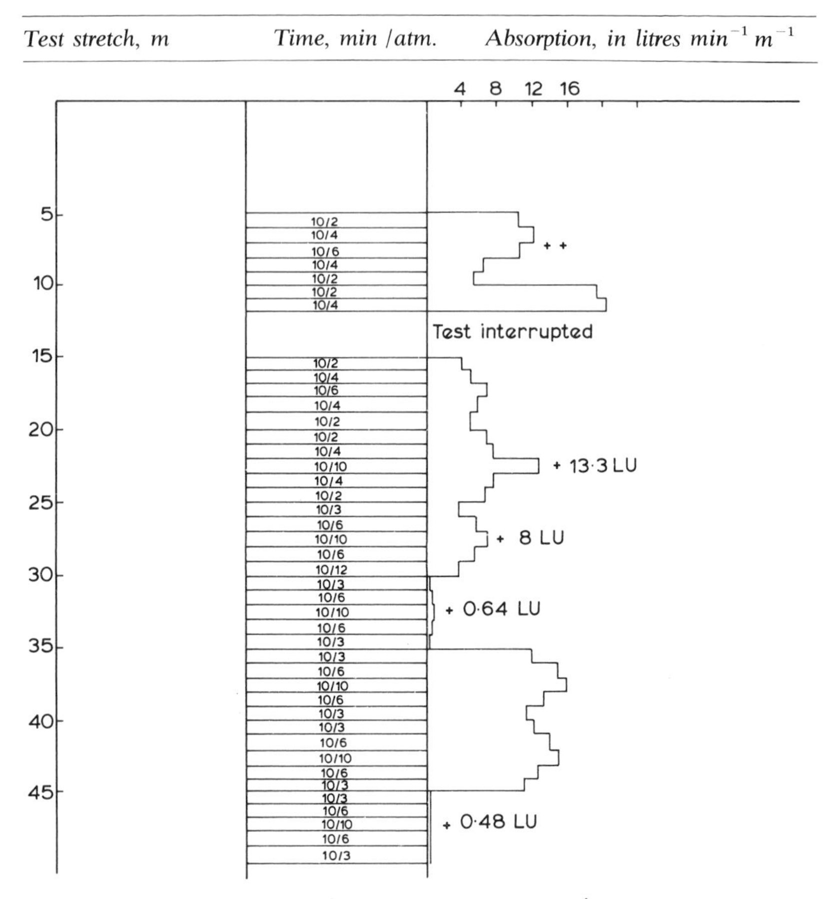

from borehole 340/S1 on the left bank at the damsite at which drilling commenced on 9 December 1978 and was completed on 8 January 1979, the angle from vertical being nil, the method rotary coring, the ground elevation 214·87 m and the total depth 50·01 m. The associated borehole log 340/S1 is given in Table 9.5. There were no special indications such as water inflows and the core recovery was around 80% below 27 m depth, negligible above this level.

TABLE 9.5
BOREHOLE LOG 340/S1

Elevation, m	*Depth, m*	*Drill diameter, mm*	*Description of strata and other rocks encountered*
213·55	0	Cemented	
210			No recovery
208·27	5		Friable yellow brown silty clay soil (amphibolite fragments)
202·88	10	76	Similar, of granitic origin
200·72	15		Friable yellow brown silty clay soil again with amphibolite fragments and fragments of quartz at 24–24·33 m
	20		
188·05	25	66	Fine grained foliated dark grey amphibolite
	30		Fractures: 8 per m at 60–80°, to 27·99 m; 6·5 per m at 45°, to 45·01 m; 3·5 per m at 60° until the end of the borehole
	35		
	40	56	
	45		
163·54			End of the borehole

Final reference to the Teton Dam disaster will now be made. As well as the damage to life and property already indicated, the nearby towns of Rexburg and Sugar City together with much farmland were subjected to major setbacks. Thus, between 16 000 and 20 000 head of livestock were lost and 51 km of railway destroyed or damaged. About 100 000 acres (400 km^2) of prime agricultural land were inundated and up to 16 March 1977, the Bureau of Reclamation had received 5616 claims totalling over US$250 million and paid out US$138 million or more on 4938 claims. The final total of claims was expected to reach US$400 million, excluding damage to the dam and appurtenant structures. Of US$86 million expended on the original Teton Dam, i.e. to 1977, about US$40 million in facilities were lost or damaged. Rexburg initiated a rapid reconstruction programme and an emergency volunteer effort organized by the Mormon church began in the weeks following the disaster. Much food was transported from church warehouses in Utah and Mormon sources put about US$$\frac{1}{2}$ million into relief work accompanied by many volunteers. Mennonites were involved also. The total number of volunteers reached 50 000. More data are available from ref. 30.

The effects of adverse geology coupled with seismic hazards are admirably demonstrated by the failure of Baldwin Hills Reservoir, Los Angeles, California, which has been described by James.[31] The hills in question are situated in Los Angeles County about 6·5 km northeast of LA International Airport and the reservoir was located on their north slope at the head of a small ravine over 100 m above the plain which was inundated through the failure. Construction was carried out between 1947 and 1951 and the objective was to meet an inadequacy of supply to peak demand in the expanding area of Los Angeles. Some adverse geological conditions were recognized, but no alternative site could be found. Hence specifications included the capability of emptying the reservoir in a 24 h period if an emergency arose. From 1951 to 14 December 1963, minor cracking in the concrete walls of the inspection gallery constructed below the reservoir floor had been noted as well as in the parapet wall of the main dam. Additionally, subsidence was seen at the reservoir and its immediate surroundings and there had been perceptible horizontal shifting of triangulation stations nearby and movement of survey stations along the perimeter of the reservoir. A slight lengthening of the northeast–southwest diagonal of the reservoir and a stretching of the dam crest were observed. At about 11.15 in the morning of Saturday 14 December 1963, a caretaker noticed an unusually loud sound of running water at the spillway discharge pipe which was connected to the elaborate underdrainage system of the reservoir.

This led to inspection of a chamber where three underdrain pipes were observed to be blowing like firehoses and discharging muddy water at a high rate. After the authorities were notified, at 12.20 p.m. the drainage of the reservoir began. At about 1.00 p.m., muddy water was seen to emerge from the east abutment some 25 m below the dam crest elevation. Drainage was accelerated and reached 450 ft^3 (12·6 m^3) per second. Half an hour later the police were requested to evacuate the area below the main dam. Up to 3.30 p.m., a detention basin near the downstream toe of the dam accommodated all the leakage, but by 3.38 p.m. failure was complete with overflowing of this basin and flooding of the residential zone below the reservoir.

The geologic setting may be examined in relation to the above disaster. The Baldwin Hills represent the most northerly topographic expression of the Newport–Inglewood uplift, a chain of structural domes and saddles extending 65 km northwest–southeast between Beverly Hills and Newport Beach. They reflect the structure of a subterranean faulted dome which is a fruitful source of both oil and gas tapped by wells in the Inglewood oil field. The deepest wells in Baldwin Hills pass through 3600 m of Tertiary marine sediments and terminate in schists supposedly of Jurassic age. Lateral movement along a fault in this basement caused a series of echelon folds to arise in the overlying sediments, one of which constitutes the Baldwin Hills. These events are connected to the creation of the Newport–Inglewood uplift one of the most significant features of which is the Inglewood fault. This passes through Baldwin Hills about 150 m west of the reservoir. It is approximately 1414 km in length and is interrupted by cross faults by which it is offset and divided into seven sections. Near the reservoir vertical displacement of 81 m and right lateral movement of 450 m have been measured (see ref. 32). Such deformations affected late Pleistocene formations and as a result it is evident that substantial tectonic movement has taken place during the last half million years. On the other hand the stream channels and fan deposits are not visibly offset when fault-crossed in Baldwin Hills which infer that movement on the faults concerned did not take place in recent geological times. The Newport–Inglewood uplift is seismically active and there are many epicentres of earthquakes in the Los Angeles area concentrated around Baldwin Hills. Tectonic events continue in the area. The abutments and foundation for the main dam are Tertiary sediments which are mostly of marine origin and the oldest exposed formation is the Pliocene Pico, silt, sand and clay beds with lenses of sand and gravel which are moderately consolidated with cementation in some beds. Overlying this is the Inglewood formation which is of Pleistocene age and comprises silts and sands. Near the rim of the reservoir is exposed the Palos Verdes formation, poorly consolidated silts, sands and gravels of Upper Pleistocene origin. Several faults were mapped near the reservoir both before and during construction. Many are now buried beneath the enbankments. Faults numbered as 1 and 5 are thought to have been the most significant because movement occurred along them during failure as is shown by displacement of the asphaltic lining of the reservoir, its drains and observation gallery. Actually fault 1 coincides with the axis of the breach through the main dam. The two faults in question were examined carefully after failure and 13 test pits, 2 shafts and almost 77 m of adit were excavated in order to do this properly. The most important findings were that:

(i) Both faults dip steeply to the west and towards the Inglewood fault.
(ii) During failure, both experienced normal dip–slip displacement amounting to as much as 18 cm on fault 1.
(iii) Tile drain pipes sheared by the faults are not offset laterally.
(iv) Loose sand seams up to over 6 cm thickness occur between the hanging and footwalls of fault 1 beneath the reservoir floor and also beneath the bottom of the breach; these occasionally contain voids and fragments of asphalt apparently originating from the lining of the reservoir.

All these factors imply that the faults opened during the process of failure, the gaps later being almost infilled with detrital material deposited by leakage from the ruptured reservoir.

Among recent deformations in the Baldwin Hills area may be cited lateral movement and subsidence. In fact a subsidence bowl has been defined there. It is elliptical and centres some 0·8 km west of the reservoir. Observations indicate about 2·9 m of subsidence from 1917 to 1963, about 0·9 m having occurred at the actual reservoir site in this period. Geodetic surveys effected in 1934, 1961 and 1963 demonstrated that first order triangulation stations in Baldwin Hills are in lateral movement, usually towards the trough of the bowl of subsidence, one, 'Baldwin Aux' located 180 m south of the rim of the reservoir, having shifted approximately 0·75 m during the 29 year period 1934–1963. Lateral movements of this type were paralleled by stretching of the surface of the Earth near the subsidence bowl perimeter. Thus the diagonal line connecting two stations through the reservoir was observed to have lengthened by 0·12 m from 1947 to 1962. As early as May 1957 several open earthcracks were noted and some were as much as 750 m long. Like those ultimately rupturing the reservoir along faults 1 and 5, they were generally parallel to pre-existing faults, were commonly orientated north–south, possessed steep dips, showed practically no horizontal displacement and occurred in areas where the rate of subsidence changes markedly in short distances. As regards the actual phenomenon of subsidence, its exact causes are not really known. No doubt oil field production plays a part, but slow tectonic activity is perhaps also significant. One cause can be ruled out and this is groundwater extraction (there are no aquifers in the underlying formations). Figure 9.16 shows the relationship between the Inglewood fault and the Baldwin Hills reservoir together with details of other faults, movements of some survey stations and the boundary of the Vickers zone of the Inglewood oil field. This latter, as may be seen, adjoins the reservoir site on the west and south. It was discovered in 1924 and has been very productive, production arising from two unconnected pools separated by the Inglewood fault. Regarded vertically, the field is divided into nine productive zones of

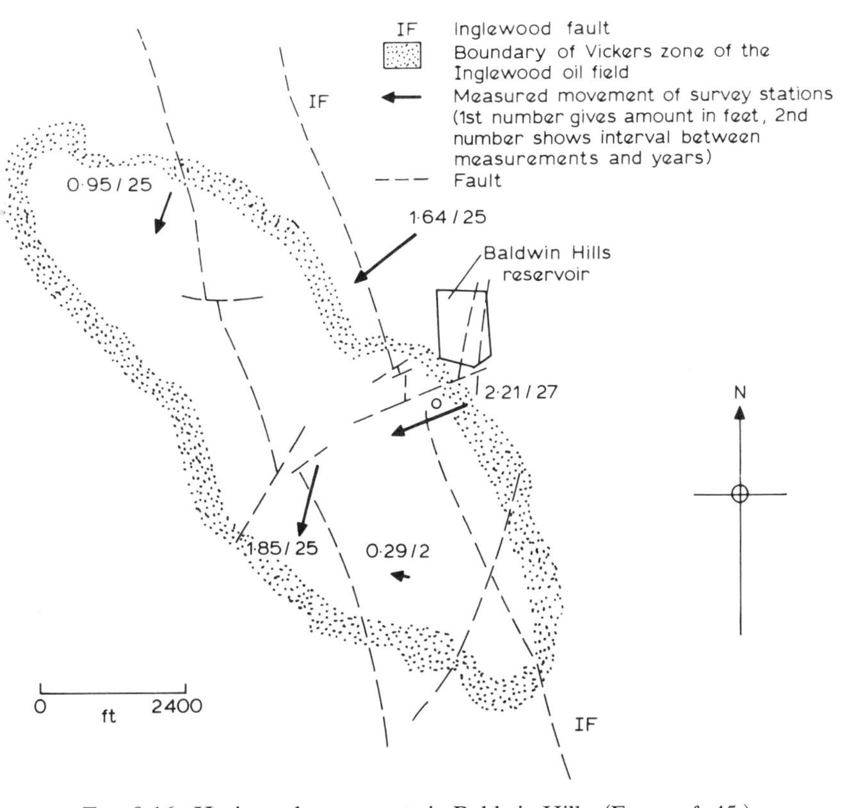

FIG. 9.16. Horizontal movements in Baldwin Hills. (From ref. 45.)

which the Vickers is one, the shallowest except for the Investment zone. Wells nearest to the reservoir site enter the zone at elevations between 270 and 300 m (cf. the reservoir floor elevation of approximately 126 m). The structural dome of the Inglewood oil field is very complexly faulted with many of the faults joining the Inglewood fault at depth. The plane of one of these was mapped by subsurface methods by petroleum geologists and found to incline towards the area of the reservoir. The plane is possibly an extension of fault 1 alluded to above.[31]

It may be seen from the examples cited above that dam construction and the infilling of reservoirs are always to some extent hazardous operations. Difficulties may arise during and after construction, many related to difficult geology. For instance Bishop *et al.*[33] have discussed the development of uplift pressures downstream of a grouted cutoff during the impounding of the Selset reservoir behind an earth dam erected in a valley containing boulder clay. Here a concrete filled cutoff trench extended by a cement grout curtain failed to prevent uplift pressure and relief wells had to be installed. The reservoir was constructed in a valley heavily and widely blanketed by the impermeable boulder clay. However artesian pressures are known to have existed beneath this prior to construction and the bedrock is Carboniferous limestone. The site is situated on the northern side of the Cotherstone syncline so that the rocks dip south at about 10° and comprise alternating shales, sandstones and thin limestones overlying massive limestone. The latter outcrops within the reservoir and unfortunately no attempt was made to extend the cutoff into it. The shales are laminated and often have fine fissures. The sandstones are thought to have bulk permeabilities of the order 10^{-3}–10^{-4} cm s^{-1} and the limestone bulk permeabilities will be larger by at least an order of magnitude. By contrast the bulk permeabilities of the boulder clay are much lower, the measured coefficient of permeability *in situ* being in the range 10^{-7}–10^{-8} cm s^{-1}. Groundwater flow is of course entirely restricted to rocks underlying the boulder clay. It proceeds to various rock exposures and springs, but there are insufficient of these, hence the build-up of artesian pressures over most of the area. It may be inferred that groundwater conditions are an important factor in determining the cutoff problem at any dam so that detailed hydrogeological investigations are, or should be, an important concomitant of its construction.

Earth dams may develop settlement problems if founded on loess which is an aeolian and mainly siliceous soil either of glacial or desert origin and well graded in nature. Loess consolidates rapidly when wetted, hence the settlement. The phenomenon is termed hydroconsolidation and may arise from one of two causes. In the USA the loess is associated with the Pleistocene glaciations and possesses clay films around the silt-sized particles comprising it. The addition of water lubricates these, thus promoting sliding of the particles with respect to each other.[34] In other parts of the world settlement may result from a combination of this lubrication effect and also a removal by the water of calcium carbonate cement from the loess.[35] As may be inferred, loess is extremely porous with high permeability, the latter being greater vertically than horizontally due to the presence of vertical tubing possibly originating from plant roots. Grouting copes with such settlement problems and is applied by pumping a loess–bentonite slurry into boreholes drilled into the foundation material as for instance in Nebraska where the loessal foundations of earth dams were consolidated using such a slurry prepared from local materials. The weight of loess increased by 20% after treatment.[36]

Uplift problems may occur also with concrete dams and a case in shown in Fig. 9.17. At Boulder Dam the amount decreased appreciably after the grout curtain was reconstructed. Water pressure tests may initiate upheaval of bedrock surfaces as well and an instance is given in Fig. 9.18 (this is taken from Czechoslovakia).

As mentioned above, filling the reservoir is usually a tense operation with dam installations and this may be illustrated with reference to the Moulay Youssef Dam described by Benzekri and Marchand.[37] A cross-section is shown in Fig. 9.19. The location is at the Ait Aadel site on the Oued Tessauot River in the Atlas Mountains about 60 km from Marrakech in Morocco. The fill volume is 5·3 million m^3 with a reservoir capacity of 200 million m^3. The foundation geology is horizontally bedded dolerite with several almost vertical faults which break it up into sections with differing characteristics. Three of these were selected with the expectation of minimal costs since there it was hoped that it would prove feasible to:

(a) minimize the extent of the grout curtain by increasing the spacing and reducing the depth of the holes;
(b) use an experienced specialized drilling and grouting company;
(c) keep the working areas of the main and specialized contractors separate by making the latter work from the foundation control

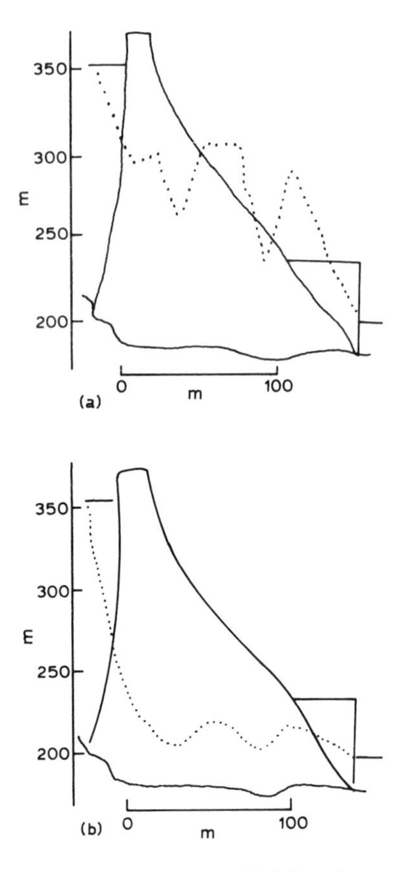

FIG. 9.17. Uplift under the Boulder Dam. (a) After the reservoir was filled; (b) after the reconstruction of the grout curtain.

tunnel; the main contractor did the blanket grouting from the ground surface and the sub-contractor was responsible for the deep grout curtain and its junction with the blanket grouting;
(d) put in drains and instal piezometers.

The drainage and piezometer systems were extensive, especially on the right bank. Details of the grouting have been discussed previously by the author.[7]

The reservoir was filled starting in April 1971 and nothing special took place until heavy discharge was observed from a right bank drain after attainment of about two-thirds of the projected maximum depth.

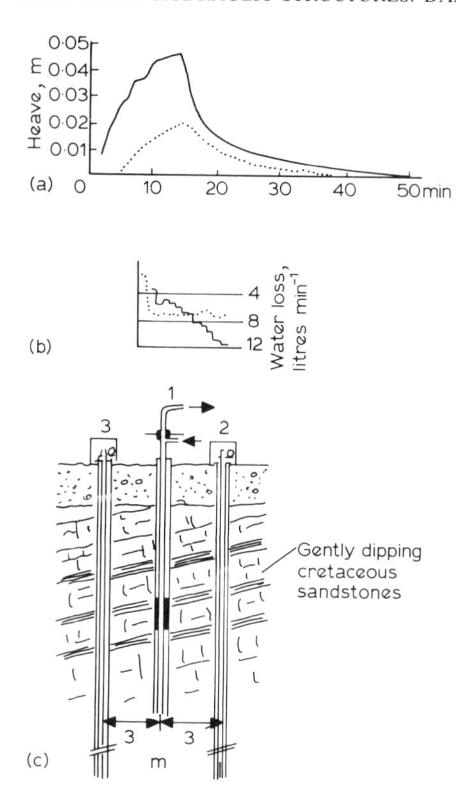

FIG. 9.18. Upheaval of bedrock surface resulting from a water-pressure test. (a) Deformation measured in boreholes 2 and 3: (– – –) borehole 2; (· · · ·) borehole 3. (b) Loss of water; (c) Upheaval near grouted borehole 1 is measured from fixed reference points set up in the boreholes 2 and 3. (From ref. 46.)

Shortly after, leaks appeared in the control tunnel and the pressures in the piezometers in the bank rose slightly. These events were not considered as constituting a danger to the dam, but flow was reduced in the affected drain (D 10) and the pressure rise controlled. As an interim step, extra drains were drilled and almost at once caused a drop in the discharge from the original drains and stabilization of pore pressures in the bank. Naturally the total discharge from all drains rose. Interpretation of the phenomena was rather difficult but the cause was probably geological. The most pervious rock is a basalt which underlies the weathered dolerite and this was not successfully

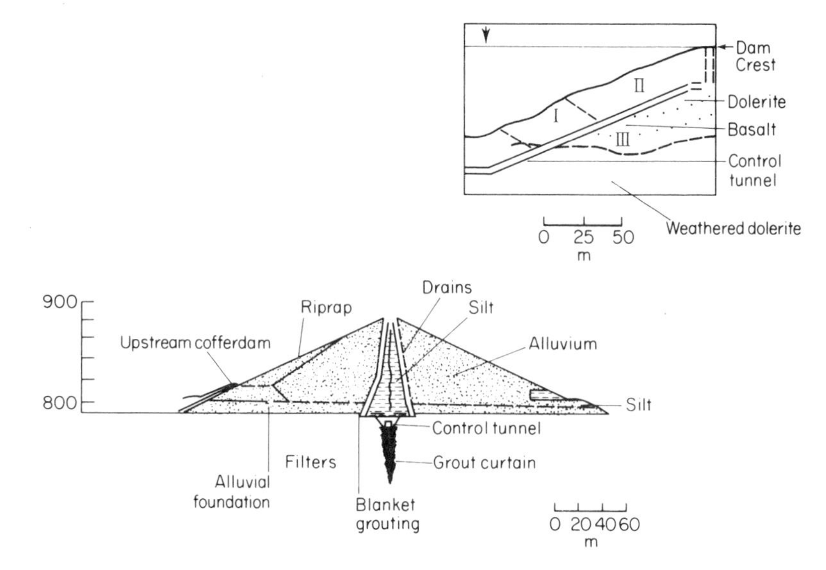

FIG. 9.19.

sealed by the original grouting operation—somehow the grout curtain holes had not intercepted the vertical paths of water movement. Consequently a second phase of grouting was carried out from the control tunnel and proved very successful. It remained successful through a subsequent monitoring period of 6 years.

Adverse geology was successfully overcome in a number of instances listed in the work of the Cementation Group of Companies, one being the Selset earth dam at Selset in Yorkshire alluded to earlier where the curtain penetrated to a maximum depth of 99 m. Another example is the Errochty Dam in Scotland where curtain and blanket grouting were carried out in mica schist below the dam and tunnels and shafts were treated. The clay seams in the foundation beds were jetted and replaced by sand–cement grout (54 000 m of drilling took place).

Sometimes even favourable rocks may have to be treated extensively as happened with heavily fissured quartzite outcroppings in the upper part of the right side of the Kariba Gorge constituting the boundary between Zambia and Zimbabwe. Here all erodible materials had to be removed (there were few) and the continuity of fissures broken. The main grout curtain was first completed and blanket grouting effected over the entire concrete to rock area. Jetting of the quartzite was carried out in cells, each comprising several roughly vertical holes drilled to a pattern giving equal distances between them. All were drilled to the same depth. Water was inserted into one hole under pressure and then air under pressure. These operations were continued on an alternating basis until air bubbles were observed at the head of a nearby hole filled with water.

All other holes were then capped and water pressured in until erodible material from the fissures washed out through the uncapped hole. The work continued until all fine material was removed and thereafter the capped holes were opened and the outlet one capped. The procedure was repeated between all the holes. Eventually no more fine material at all could be obtained and pressure grouting with cement or sand and cement was effected.[38] In the case of Kariba, fissured quartzites were being treated and the jetting and grouting approach proved to be suitable for these and other hard, jointed and fissured rocks with clays or other fine-grained sediment in the fractures, the continuity of which can be interrupted and the permeability reduced.

At Kariba the upper part of the right side of the gorge is composed of quartzite, the strike being parallel to the gorge and the dip between 90° and 70° towards the river. For the rest, the geology is different. The lower two-thirds of the right bank, the valley of the river and the entire left bank is gneiss. The contact between these two rock types is in the form a horizontal cylinder with its axis parallel to the river and it is generally weathered, the degree of weathering varying quite considerably from locality to locality. Large scale excavations were carried out in the treated zone in connection with a massive concrete abutment. Mass concrete which is tied to the dam by means of post-stressed cables provided extra weight, directing the thrust from the overlying horizontal arches of the dam into underground concrete buttresses. The additional weight of the latter and the rock between them direct the thrust on to massive unweathered gneiss. The treated rock is utilized in the design so as to constitute a seepage barrier between the grout curtain of the dam and the curtain provided at the intakes; it acts in conjunction with an extensive drainage system serving the dual objective of relieving lateral pressure and uplift at the abutment and preventing excessive saturation and weathering of the gneiss-quartzite contact. Figure 9.20 illustrates the constructional phases of the Kariba Dam.

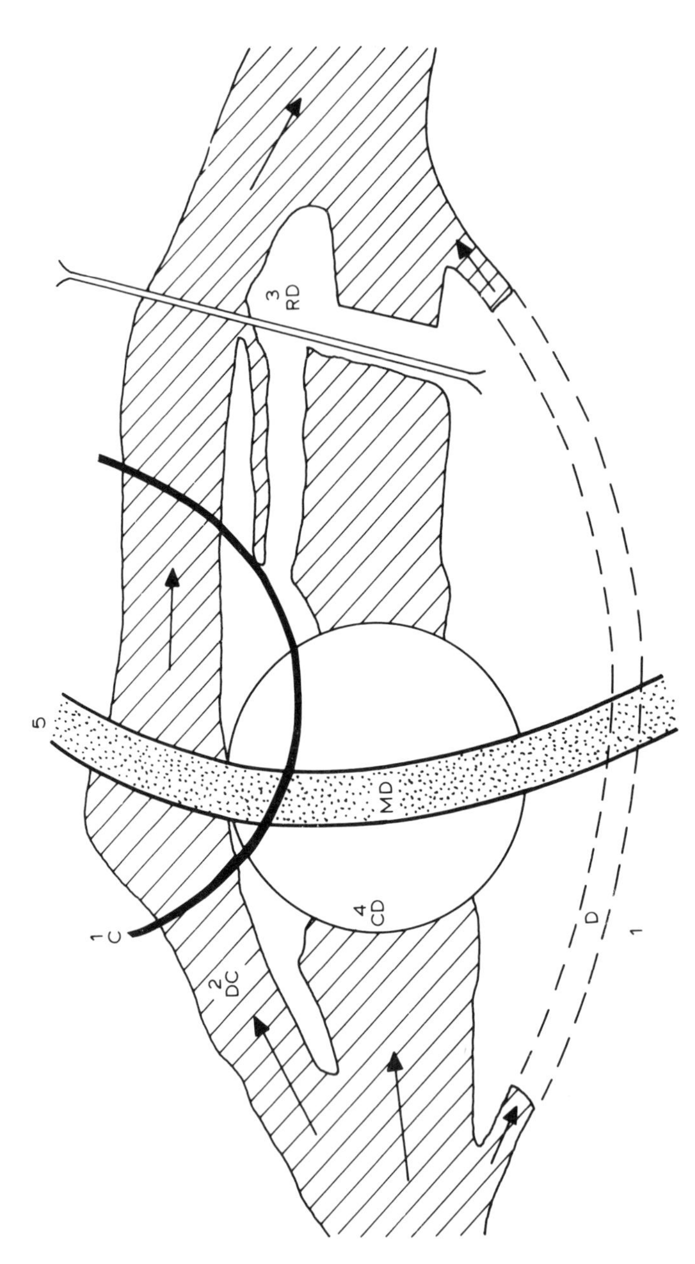

Fig. 9.20. Kariba Dam. Originally linking Rhodesia (now Zimbabwe) and Zambia, this was constructed in several stages: 1, Cofferdam, C, and diversion tunnel, D, were built simultaneously; 2, diversion channel DC was dug and main dam blocks laid in cofferdam; 3, river was dammed downstream with rockfill (RD) and walls of C dynamited to effect complete water diversion; 4, main cofferdam, CD, was constructed, pumped dry and main dam commenced; 5, diversion channel and tunnel were sealed and main dam (MD) completed.

Sometimes too enthusiastic grouting can be deleterious and even cause fracturing of the bedrock. This was the case at the Mangla Dam in Pakistan.[39] The foundation rocks at this site are extremely weak and the bedrock belongs to the Middle Siwalik Series of the Miocene, comprising beds of cemented sandstone interbedded with hard clays and siltstones. Only about 10% of the sandstones are well cemented. The more cemented parts often form hard lenses several scores of centimetres in thickness and up to 15 m in diameter. There is low resistance to crushing and the clays and siltstones are more resistant, frequently showing strengths up to 700 kN m^{-2} or so. There are two series of joints at right angles intersecting the beds and forming a 'saw tooth' pattern. In elevation the weathering has produced a succession of scarps marking the sandstone outcrops and flatter slopes with debris marking clays. The river has incised and left a number of alluvial terraces. The flood plain deposits are, in places, as much as 30 m deep and comprise coarse alluvium made up of sand, gravel and many cobbles. The dam was to be 2·4 km long. Prior to its construction and in order to look into the feasibility of providing a grouted cutoff of conventional type four tests were made at various locations so as to test the efficacy of grouting. From these it emerged that in a number of cases, high grouting pressures fractured the bedrock. It was inferred that therefore a low pressure criterion for grouting in soft bedrock is to be desired and used in such jobs. Attempts to limit pressures by observation of uplift gauges proved to be unsuccessful. Below about 20 m grouting was found to be reasonably effective as a tool for the sealing off of fissures. However even there at least one undesirable result was obtained. Various grout mixtures were utilized in the tests and included cement, clay–cement, bentonite and chemicals such as AM-9 (this latter now banned because of its toxicity). Plain cement injections seem to have been as successful as any others. Mechanical packers of soft rubber were successful. The optimum approach was stage and packer grouting. These have been discussed by the author elsewhere.[7] Figure 9.21 shows grouting test sites at the Mangla Dam construction project.

Bussey[40] has given data regarding the Priest Rapids Dam founded on the top of a basalt flow series. The basalt is columnar with practically vertical columns and the contact between the base of the flow bearing the dam foundation and the underlying flow is open; hence it had to be sealed by grouting. Holes were drilled to 3 m below the contact region which was grouted by injecting a thick sand–cement

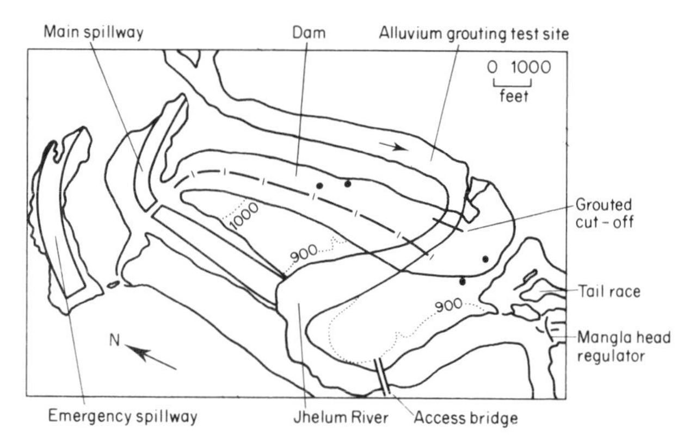

FIG. 9.21.

grout below a packer set 1·5 m above. Afterwards the rest of the hole was grouted using neat cement injected through a nipple. Subsequently the foundation of the powerhouse was excavated through the contact and seepage into it was negligible.

A number of other examples of adverse geology might be cited, but two more may suffice to emphasize the importance of thorough examination of the circumstances prior to construction. One, shown in Fig. 9.22, refers to gallery and borehole determination of the Jurassic

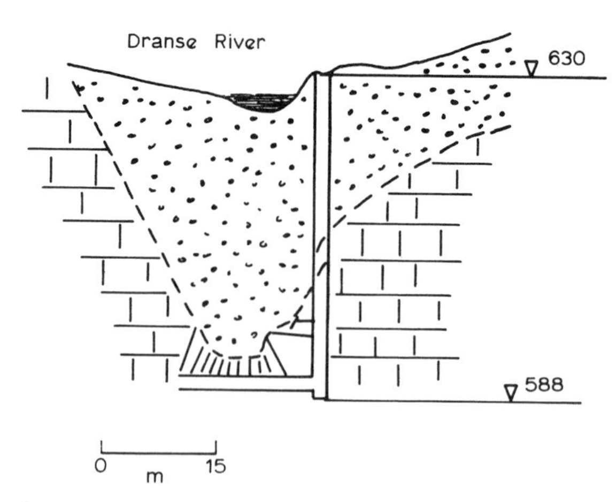

FIG. 9.22. Gallery and borehole determination of the bedrock surface under the floor of the valley of the River Dranse, Savoy (France). Stippled area: alluvial deposits of the river. Block deposits area: Jurassic limestones. (From ref. 19.)

limestone bedrock surface under the floor of the valley of the River Dranse, Savoy, France, and has been described by Gignoux and Barbier.[19] The second is the Serre-Poncon Dam, France, a preliminary examination of the over-deepened valley of the relevant River Durance being shown in Fig. 9.23. This has been described by Ischy and Haffen.[41]

The River Durance has an interesting connection with irrigation in the 12th century: in 1171 the Count of Toulouse permitted the Bishop of Cavaillon, as a matter of right, to divert water from the river in order to operate flour mills. The water, once having passed the water wheels, was utilized to fulfil local requirements for irrigation through the employment of a canal system.[3] A later irrigation dam was built on the same river in 1554 by Adam de Craponne. The Craponne Canal was the first large waterway constructed in France and it was not intended for navigation, but rather to convey Durance water to the fertile but dry lands of Provence in the valley of the Rhone. Together with several branches a total canal length of over 150 km was involved.

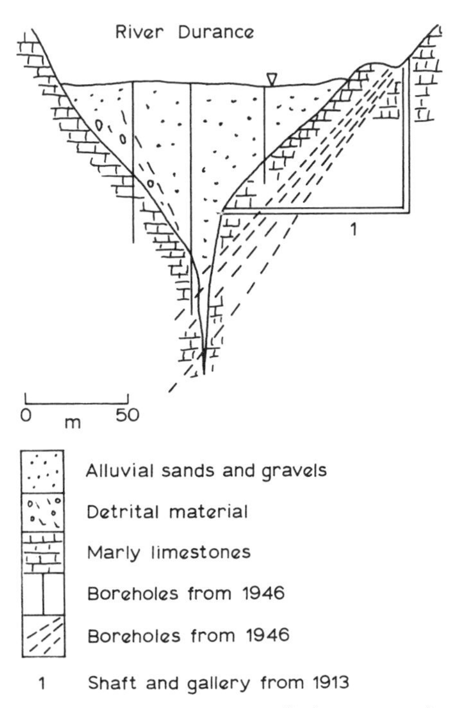

FIG. 9.23. Serre-Ponçon Dam, France: preliminary examination of the over-deepened valley of the River Durance. (From ref. 41.)

9.9. SOME RESULTS OF CONSTRUCTION

On 2 December 1959 a thin arch dam in southern France collapsed and the nearby town of Frejus was inundated with the loss of more than 300 lives. This was the Malpasset Dam alluded to earlier. The inquiry which followed ascribed the failure to weakness in the rocks on one side of the valley.[42] A side benefit was the incentive given to the study of rock mechanics thereafter, indicating its validity as a subject and also its general applicability. This is of continuing importance because as dam building accelerates the number of optimal sites diminishes. Hence the adoption of geologically unfavourable sites becomes ever commoner and nowadays many dams are constructed in earthquake prone areas such as Peru, India, Japan and New Zealand. Triggering of seismic shocks by reservoirs has been alluded to above. Another problem with these man-made lakes is siltation. This, together with salination, constitute major difficulties, a complete solution to which has not yet been found.

Figure 9.24 concerns the Pont du Loup reservoir in the French Alps. This is a masonry dam of height 30 m with an associated reservoir of which the original capacity was 1 750 000 m^3. The siltation which occurred between 1927 and 1930 has been described by Drouhin.[47] It was effected in connection with the Sautet Dam project and transverse profiles through the reservoir set 50 m apart were surveyed at regular intervals, the depth being measured at 20 m distances. In 1928, there was a flood with 750 $m^3 s^{-1}$ discharge which supplied 536 000 m^3

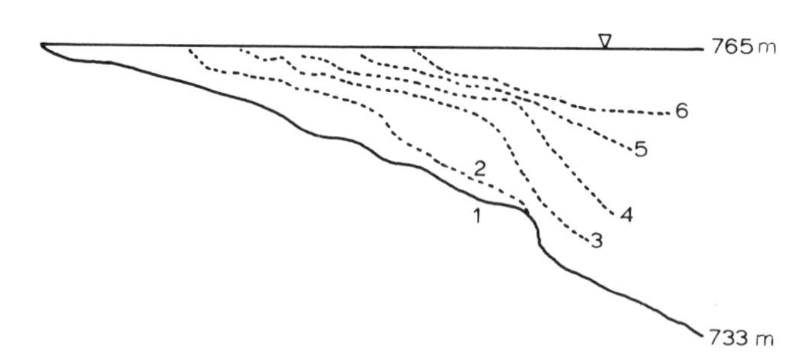

FIG. 9.24. Pont du Loup Reservoir, French Alps: siltation during a 3-year period (November 1927 to August 1930). 1, 13 November 1927; 2, 20 July 1928; 3, 12 October 1928; 4, 16 November 1928; 5, 20 July 1929; 6, 21 August 1930.

material (*c.* 550 m^3 km^{-2}). 464 000 m^3 extra came in during the same year by another flood with a discharge of 700 m^3 s^{-1}. In 1929, during rather dry weather, accretion only reached 230 000 m^3. The huge rise in bottom elevation consequent upon the 1928 floods can be seen in the figures. By 1932, the reservoir was silted up and, after several fruitless attempts to open the outlets, a tunnel was driven 23 m below water level. The final stage of this was blasted with 400 kg of dynamite and 5 min after firing the water started carrying away consolidated material; within 4 days, the part of the reservoir immediately upstream of the dam was cleared.

Of course as siltation proceeds, there is a concomitant loss of storage capacity. For instance it was estimated in 1949 that Lake Mead had lost about 2·7% of its capacity through siltation in its first 14 years of operation. With Lake Mangla, the rate of siltation is so great that one million acre feet of sediment is being deposited every two decades.

Evaporative loss is very high in reservoirs, indeed much higher than occurs with river water. Water so lost cannot be utilized for power and irrigation and in addition, an increase in salt concentration takes place.

An excellent instance of these deleterious developments is given by Lake Nasser where it has been suggested that as much as 40% of the total quantity of water may be lost through evaporation and this is accompanied by a truly Pharaonic problem of siltation. Difficulties have arisen because, while the first Aswan Dam was fitted with multiple low-level sluices facilitating the continuance of the ancient irrigation system using the natural fertility of the Nile silt, the High Aswan Dam is not so equipped. Many think that the second, Soviet-financed and technically aided, dam is both technically and also economically unsound with all sorts of undesirable effects on the ecology and agriculture of Egypt. As at Kariba, bilharzia is another hazard. The area of Lake Nasser is approximately 5120 km^2, that of Lake Kariba about 4400 km^2 and the very size of these reservoirs removes much land formerly available for agriculture. Not to mention the disruption to towns and villages around them—at least 50 000 people were involved at Kariba and 120 000 at Lake Nasser. This threat extends also to wildlife and, with Lake Nasser, to ancient temples and monuments such as Abu Simbel. It is still not certain that the removal and re-siting of this was the best way to tackle the problem, but at least it was not lost for ever as happened to many other archaeologically valuable monuments (such as the Temple of Philae which was flooded when the first Aswan Dam was heightened in 1912).

It is very disturbing to reflect upon the potential consequences if one of these vast dams were to be destroyed, either accidentally or deliberately (as happened for instance to the Möhne Dam in Germany, destroyed by the Royal Air Force, or the Dnjeprogues Dam in the USSR destroyed by the retreating Red Army, during the Second World War).

Loss of water through leakage from reservoirs has been alluded to already and here it is mentioned again as an adverse effect of impoundment. In this connection some appropriate data are given in Table 9.6.

As regards the determination of K in the table, this may be done directly or indirectly.

For materials with K exceeding 10^{-4} cm s^{-1}, soil may be tested through original, position pumping tests which, if effected by an experienced operator, give very reliable results. A constant head permeameter may be utilized in the laboratory.

For materials with lower values of K, such direct determinations are not feasible. Indirect determination may be effected in the range $K = 1{\cdot}0$–10^{-10} cm s^{-1} using falling head permeameters which are reliable except in the range 10^{-4}–10^{-6}.

TABLE 9.6

RELATIVE PERMEABILITIES, K, OF UNCONSOLIDATED MATERIALS

(*I*) *Low*:
- (a) Alluvial clays, muds and silts, K ranging from 10^{-5} to 10^{-8} cm s^{-1}
- (b) Altered volcanic ash (this may display swelling properties)
- (c) Clean sand and gravel mixed with compact clay matrices
- (d) Buried mudflows and landslides

(*II*) *Moderate*:
- (a) Alluvial sands or gravel with some clay matrix, K ranging from 10^{-3} down
- (b) Fine grained sands, K between 10^{-4} and 10^{-5} cm s^{-1}
- (c) Unaltered volcanic ash
- (d) Glacial till with clay or rock flour, K around 10^{-6} cm s^{-1}
- (e) Loess and landslides with rock fragments and interconnecting voids

(*III*) *High*:
- (a) Clean gravel. K ranging from $1{\cdot}0$ to 10^{2} cm s^{-1}
- (b) Clean sands, K ranging from 10^{-4} to $1{\cdot}0$ cm s^{-1}
- (c) Glacial till with piping which has removed fine grained matrix locally
- (d) Buried sand dunes
- (e) Landslides comprising angular rock fragments having interconnecting interstitial voids

As the coefficient of permeability decreases, the amount of operating experience required increases.

Materials of the above kind may cover the slopes of valleys or canyons to be used for reservoir impoundment and the effects of groundwater penetration and fluctuation within them may create hazards. Drawdown of a reservoir causes complex and continually changing movement of groundwater. Movement of pore water from deeper to shallower parts of an accumulation of talus under noteworthy pressure differentials together with the actual pressure themselves will tend towards instability in such an accumulation. Prior to impoundment the shear stress within the accumulation will be insufficient to trigger failure along one or several slip surfaces and will be lower than a critical value expressed:

$$\tau_{\text{critical}} = (\sigma_{\text{n}} - P) \tan \phi$$

where τ_{critical} is the critical value, σ_{n} is the normal stress across a potential slip surface tending to prevent slipping as a result of frictional resistance, P is the pressure exerted by interstitial water (and in opposition to σ_{n}) and ϕ is the angle with the horizontal made by a potential slip surface.

If during the process of drawdown the flow of pore water from an accumulation does not accord with diminishing hydrostatic pressures in the reservoir, then unbalanced residual pressures arise. This is the opposite case from equilibrium where the infilling of the reservoir involves increasing pore pressures in the talus which are counterbalanced by the hydrostatic pressures within the reservoir. In the former, non-equilibrium, case the critical stress for failure along slips may be expressed:

$$\tau_{\text{critical}} = (\sigma_{\text{n}} - P') \tan \phi$$

in which P' represents the sum of the initial pore pressure P together with the residual pressure from reservoir infilling. Clearly since $P' > P$, the critical shear stress which existed before impoundment has been reduced and failure along slip surfaces becomes very possible. The mechanism is that the adding of residual pressures to initial ones has effectively reduced the resistance of the talus accumulation to shear failure.

An especially dangerous situation may occur if a slide develops in shales with a typical lower curved surface of shear associated with low average permeability and high average porosity. Internal dislocation will promote increase in the permeability. Processes analogous to these can produce instability in types of rocks other than shales and perhaps the ones most likely to suffer pore pressure increase during impounding of water in a reservoir are weathered and/or fractured materials. Slope failures during drawdown are quite frequent, but as this goes on they become less and less of a hazard until they are manageable at the stage where the reservoir is almost empty.

Construction of affiliated structures associated with dams depends to a great extent upon similar factors. Hence a diversion tunnel is feasible only if the valley sides are of the appropriate form and the rocks involved are solid. In the Bumbuna Dam area the rocks are competent basement material so that in this Sierra Leone project, two diversions are planned. If rocks occur which are shattered or incompetent, it may not be possible economically to drive diversion tunnels (see Fig. 9.25).

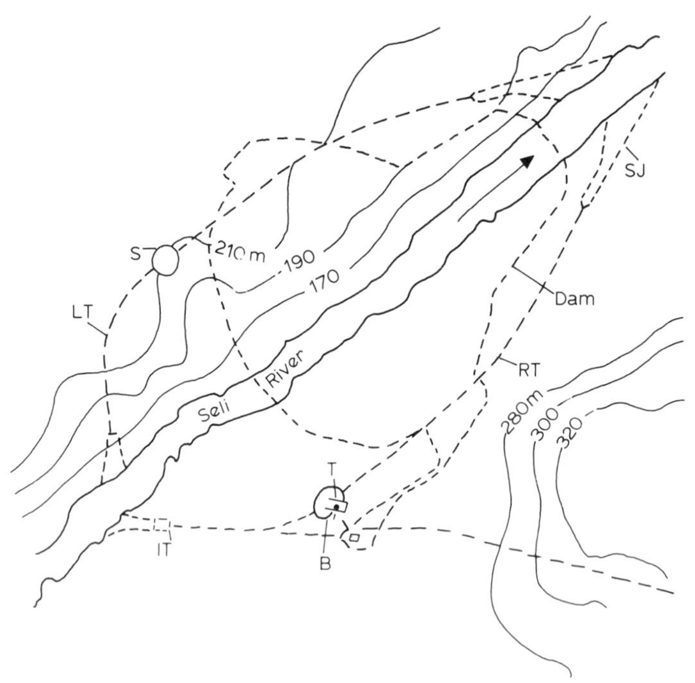

Fig. 9.25. Bumbuna damsite, Sierra Leone, showing proposed structures. Dam, Bumbuna rockfill dam; S, spillway: LT, RT, left and right tunnels for diversion; T, trench; B, borehole; IT, intake tower; SJ, ski jump (energy dissipator).

As for cofferdams, the construction of these temporary water-diverting dams entails determination of any probable groundwater flow to the foundation pit. This is best accomplished by an hydrogeologist who will analyse the solid and drift geology in the valley itself as well as make the necessary hydrological assessments. In some embankment dams the cofferdam may be later incorporated into the final, larger structure.

Many undesirable results can be obviated or minimized by adequate prior borehole investigations. The major objectives are as follows:

1. Unconsolidated foundations, embankment dams:
   (a) horizontal and vertical variations in mineral composition and physical properties leading to assignment of units to the Unified Soil or other classification;
   (b) recording of engineering properties (density, shear strength, compressibility, consequences of changes in moisture content, effects of structural loading);
   (c) horizontal and vertical variations in permeability (which could localize impermissible subsurface seepage);
   (d) depth to bedrock if a concomitant cutoff is under consideration.
2. Bedrock or consolidated foundations; concrete or embankment dams:
   (a) depths and contours on the surface of the bedrock under unconsolidated surficial materials;
   (b) lithology of the units of the bedrock which covers structural characteristics of anisotropic rocks such as planes of bedding, foliation and schistosity as well as the flow structures of igneous rocks;
   (c) assessment of weathering and other alteration;
   (d) assessment of fracturation (faults and joints);
   (e) identification of seepage zones (especially where limestone, dolomite and gypsum are concerned);
   (f) geophysical logging;
   (g) assessment of engineering properties.

As regards the various types of borehole logs, these may be categorized as follows:

A. Geological from core drilling and percolation testing, data:
   (i) results of percolation tests;

(ii) core loss or recovery as a percentage;
(iii) fracture spacing;
(iv) log of rock types with descriptions;
(v) rock quality designation (RQD rating), the percentage of hard pieces of core 10·3 cm long or longer in a stated interval of the core;[43]
(vi) special notes, e.g. regarding the water table, lost drill water, casing, weathering, etc.

B. Geophysical from instrumentation lowered into the borehole, data:
(i) investigations into the apparent electrical resistivity;
(ii) neutron probe moisture measurements;
(iii) gamma attenuation measurements of density;
(iv) gamma ray identification of rock types by the measurement of natural radioactivity;
(v) induced compressional and shear waves measurements of elasticity and Poisson's ratio;
(vi) methods for measuring the potential difference between a borehole electrode and a surface electrode;
(vii) temperature measurements for establishing the thermal gradient;
(viii) magnetic measurements;
(ix) sonic method of measuring the velocities of compressional waves as a function of depth.

C. Soil profile from conversion of drilling with geophysical measurements, data:
(i) dimensions and descriptions of test samples;
(ii) size classification with symbols through screening;
(iii) engineering properties;
(iv) elevation of water table if one exists.

D. Unconsolidated materials from drive tubes with penetration resistances obtained simultaneously with the sampling, data:
(i) methods used, moisture contents and penetrations;
(ii) description and classification of materials;
(iii) engineering properties;
(iv) elevation of water table if one exists.

E. Drilling data, including the mechanical details of the operation:
(i) equipment;
(ii) core size and replacement of bits;
(iii) drilling time;

(iv) loss of drilling water or the intersection of groundwater under head;
(v) special procedures such as cementing of the borehole, insertion of casing, altering the direction of the borehole, etc.

REFERENCES

1. Grunner, E., 1963. *Dam Disasters*, Inst. Civ. Engrs, London.
2. Murray, G. W., 1955. Water from the desert: some ancient Egyptian achievements. *Geog. J.*, **121,** 171–81.
3. Smith, N., 1971. *A History of Dams*, Peter Davies, London.
4. Goblot, H., 1965. Kebar en Iran sans doute le plus ancien des barrages-voutes: l'an 1300 environ. In: *Arts et Manufactures*, **154,** 43–9.
5. Haynes, D. W., 1979. Geological technology in mineral resource exploration. In: *Mineral Resources of Australia*, eds D. F. Kelsall and J. T. Woodcock, Australian Academy of Technological Sciences, Parkville, Victoria, pp. 75–95.
6. Davis, G. R., 1981. Geologists in the minerals industry. *Presidential Address to the Institution of Mining and Metallurgy, 21 May*, Steven Austin and Sons Ltd.
7. Bowen, R., 1981. *Grouting in Engineering Practice*, 2nd edn, Applied Science Publishers Ltd, London; Halsted Press, New York.
8. Ischy, E., 1948. Barrage de Castillon. Lutte contre les erosions souterraines. *3e Congres des Grands Barrages*, Stockholm.
9. Lugeon, M., 1933. *Barrages et Geologie*, Librarie de l'Université de Lausanne.
10. Wahlstrom, E. E. and Hornback, V. Q., 1968. Engineering geology of Dillon Dam, spillway shaft and diversion tunnel, Summit County, Colorado. In: *Eng. Geol. Case Histories*, No. 6, ed. G. H. Kiersch, Division on Engineering Geology, Geol. Soc. Am., pp. 13–21.
11. Wahlstrom, E. E., 1974. Dams, dam foundations, and reservoir sites. *Developments in Geotechnical Engineering—6*, Elsevier Scientific Publishing Company, Amsterdam.
12. Zienkiewicz, O. C. and Cheung, Y. K., 1967. *The Finite Element Method in Structural and Continuum Mechanics*, McGraw Hill, London.
13. Handin, J., 1966. Strength and ductility. In: *Handbook of Physical Constants*, ed. S. P. Clarke, Jr, Geol. Soc. Am. Mem. No. 97, pp. 223–89.
14. Adams, R. D., 1983. Incident at the Aswan Dam. *Nature*, **301,** 14.
15. Talobre, J. A., 1967. *La Mechanique des Roches*, Dunod, Paris.
16. Stucky, A., 1956. Ben Metir Dam. *Water Power.*
17. Stiny, J., 1953. *Tunnelbaugeologie*, Springer-Verlag, Wien.
18. Zoubek, V., 1953. Geologicke podklady k projektu udolni u Orlickych Zlakovic.

19. GIGNOUX, M. and BARBIER, R., 1955. *Geologie des Barrages et des Amenagements Hydrauliques*, Masson et Cie, Paris.
20. ZARUBA, Q. and MENCL, V., 1976. Engineering geology. *Developments in Geotechnical Engineering—10*, Elsevier Scientific Publishing Company, Amsterdam.
21. REUTER, F., 1958. Die Wasserdurchlässigkeitsprüfungen und Baugrundvergütung als ingenieurgeologische Untersuchungsmethode bei Talsperrenbauten. *Zeitschr. für angew. Geologie.*
22. REUTER, F., 1958. Hangrutschungen bei den Baustellen des Bodeswerkes. *Zeitschr. für angew. Geologie*, **2/3,** 94–8.
23. WALTERS, R. C. S., 1971. *Dam Geology*, Butterworth, London.
24. KIERSCH, G. A., 1964. Vaiont reservoir disaster. *Civ. Eng.*, **34,** 32.
25. Anon., 1964. Russians blast through landslide dam. *Eng. News-Record*, **172,** 24 (7 May).
26. SCHUSTER, R. L., 1980. Discussion on geological causes of dam incidents. *I. A. Eng. Geologists, Bull.*, **21,** 231–2.
27. FECKER, E., 1980. The influence of jointing on failure of Teton Dam—a review and commentary. *I. A. Eng. Geologists, Bull.*, **21,** 232–8.
28. Independent Panel (to review cause of Teton Dam failure), 1976. *Failure of Teton Dam, Idaho Falls.*
29. Review Group, 1977. *Teton Dam Failure, a Report of Findings*, US Department of the Interior.
30. Anon., 1977. *Civil Engrg, ASCE*, 55–61.
31. JAMES, L. B., 1968. Failure of Baldwin Hills reservoir, Los Angeles, California. In: *Eng. Geol. Case Histories*, No. 6, ed. G. H. Kiersch, Division on Engineering Geology, Geol. Soc. Am., pp. 1–11.
32. DRIVER, H. L., 1943. *Inglewood Oil Field*, Division on Oil and Gas, Calif. Department of Natural Resources, Bull. 118.
33. BISHOP, A. W., KENNARD, M. F. and VAUGHAN, P. R., 1963. The development of uplift pressures downstream of a grouted cutoff during the impoundment of the Selset reservoir. In: *Grouts and Drilling Muds in Engineering Practice*, Butterworths, London, pp. 98–105.
34. KRYNINE, D. and JUDD, W. R., 1957. *Principles of Engineering Geology and Geotechnics*, McGraw Hill, New York.
35. SCHIEDIG, A., 1934. *Der Loess und seine technische Eigenschaften*, T. Steinkopf, Leipzig.
36. JOHNSON, G. E., 1953. Stabilization of soil by silt injection method. *Proc. Am. Soc. Civ. Engrs, J. Soil Mechs and Foundations Divn*, **79** (323), 1–18.
37. BENZEKRI, M. and MARCHAND, R. J., 1978. Foundation grouting at Moulay Youssef Dam. *Proc. Am. Soc. Civ. Engrs, J. Geotech. Eng. Divn*, **104** (GT9), 1169–81.
38. LANE, R. G. T., 1964. The jetting and grouting of fissured quartzite at Kariba. In: *Grouts and Drilling Muds in Engineering Practice*, Butterworths, London, pp. 85–90.
39. LITTLE, A. L., STEWART, J. C. and FOOKES, P. J., 1964. Bedrock grouting tests at Mangla Dam, West Pakistan. In: *Grouts and Drilling Muds in Engineering Practice*, Butterworths, London, pp. 91–7.

40. Bussey, W. H., 1964. Some rock grouting experiences. In: *Grouts and Drilling Muds in Engineering Practice*, Butterworths, London, pp. 65–9.
41. Ischy, E. and Haffen, M., 1955. Barrage de Serre-Ponçon. Campagne de reconnaissance. *5e Congrès de Grands Barrages*, Paris.
42. Jaeger, C., 1963. The Malpasset report. *Water Power*, **15,** 55–61.
43. Deere, D. U., 1968. Geological considerations. In: *Rock Mechanics in Engineering Practice*, eds K. G. Stagg and O. C. Zienkiewicz, John Wiley, New York.
44. McGavock, C. B., Jr., 1968. Engineering geology of Spruce Run Dam and reservoir. New Jersey. In: *Eng. Geol. Case Histories*, No. 6, ed. G. H. Kiersch, Division on Engineering Geology, Geol. Soc. Am., pp. 23–32.
45. James, L. B., 1968. Failure of Baldwin Hills reservoir. In: *Eng. Geol. Case Histories*, No. 6, ed. G. H. Kiersch, Division on Engineering Geology, Geol. Soc. Am. pp. 1–11.
46. Verfel, J., 1957. Provadeni vodnich tlakovych a injekcnich zkousek. Brno.
47. Drouhin, G., 1951. Sedimentation des reservoirs et problemes connexes. *4e Congres des Grands Barrages*, New Delhi.

CHAPTER 10

Remote Sensing

10.1. INTRODUCTION

Ever-increasing applications of engineering geology to construction projects have resulted in a rising need for special engineering geology mapping methods and other procedures and a number of papers have dealt with this in recent years including those of Dearman and Fookes[1] and the Commission on Engineering Geological Maps of the International Association of Engineering Geology.[2] Maps of this type must contain data regarding the distribution of rock and soil units in the relevant region, their types and the hydrogeological conditions with especial reference to factors such as depths-to-water. They should show also the geomorphology and contain information on geodynamic phenomena such as erosion, deposition, mass movements such as solifluction, karstic terrains, active faults, joint systems and so on. Of course there will be varying emphases according to the actual projects involved so that engineering geology maps may be made for specific projects—this is not the usual practice for normal geological maps. However such 'site investigation' work is accompanied by wider 'regional studies' and with regard to these, various phases of activity must be considered. (These have been discused by Matula.[3])

10.2. PHASES OF REMOTE SENSING

10.2.1. Acquisition, Storage and Retrieval of Primary Data

These aspects must be developed in parallel with the techniques of geology, geophysics, remote sensing, etc., if the results obtained by the use of these methods of investigation are to be made available and utilized with optimum efficiency. Two essentials have been indicated, namely the preparation of a precise terminology coupled with exact evaluation in connection with phenomena recorded on engineering geology maps and the preparation of quantitative classifications for their main attributes. Appropriate engineering geology data banks can provide outputs tailored to special requirements such as geological sections, tables of necessary data, etc.

10.2.2. Quantitative Classification of Data on Engineering Geology Maps

This must be regarded as very valuable in increasing the exactness of regional engineering geological work and the IAEG mapping commission has undertaken to prepare a proposal for such a quantitative classification of the various characteristics of engineering rocks and soils.

10.2.3. Transformation of Engineering Geological and Geotechnical Maps

Multipurpose engineering geological maps may be made for an area by collecting comprehensive information from data banks. Transformation procedures can be utilized in order to produce special purpose engineering geological and geotechnical maps.

The UNESCO–IAEG guide to the preparation of engineering geological maps issued in 1976 indicated that two types of multipurpose map are desirable.[2] One is a map of the engineering geological conditions containing classified rock and soil units, groundwater types, landforms, geodynamic phenomena, their distribution and variability as well as properties over the entire area of the map. The other is a map of engineering geology zoning. By evaluation of the spatial and functional interrelationships among the basic geoenvironmental components (i.e. rocks, water, landforms, geodynamic phenomena), it is feasible to delineate individual territorial units in terms of engineering geological zoning. The precision of such zoning operations is improved by introducing the principles of typology. Matula[3] has cited an example which may be given here also:

Quaternary soils

g Gravelly soils
p Sandy soils

Pre-Quaternary rocks and soils

S Solid rocks
B Semisolid rocks

Quaternary soils

- n Alternation of gravelly and sandy soils
- h Cohesive soils
- k Combined cohesive and non-cohesive soils
- s Loessic soils
- o Organic soils
- b Bouldery soils

Pre-Quaternary rocks and soils

- F Alternation of hard and weak rocks (S and B)
- Z Highly weathered rocks
- G Gravelly soils
- N Alternation of gravels and sands
- I Cohesive soils
- G Combined cohesive and non-cohesive soils

Strata thickness index

- 1 <2 m
- 2 2–5 m
- 3 >5 m

Depth to pre-Quaternary surface index

- 1 <5 m
- 2 5–10 m
- 3 >10 m

Thus the single symbol set h1g2S2 indicates a well quantified ground model in which cohesive soils of thickness less than 2 m at the surface are underlain by gravels of thickness 2–5 m and at a depth of more than 5 m by hard rocks of the pre-Quaternary basement.

The above maps form the basis for preparation of various types of terrain evaluation and special purpose zoning maps wherein different selected attributes are applied for a gradual transformation of more generalized multipurpose models of engineering geological situations into very specialized ones.

Four basic types of zoning maps can be prepared at various scales for various purposes. The first includes maps for the protection and rational exploitation of groundwater and other resources while the second comprises maps for regional planning, land development and construction. The third consists of maps for the delimiting of endangered areas together with details regarding warning systems and suggested measures against active or potential geological risks (hazard maps). Finally the fourth relates to maps for the protection of vulnerable geological environments prone to undesirable changes caused by development.

From such work geotechnical maps can be made and these conform to the requisite of how to adopt and/or adapt particular engineering geologically defined environments by specific engineering, i.e. geotechnical, activities to different types of technical use (perhaps the exploitation and development of manmade works without significant adverse effects from detrimental natural processes).

10.2.4. Optimization Procedures

These are applicable both to regional and urban geology situations and comprise:

(a) The estimation of optimum land use potentials for various zoning units or sites in the relevant area covered by the map.
(b) The selection of optimum sites or zones for any particular land use within the area of the map such as for residential districts or industrial zones.
(c) The decision on optimum utilization of an area or its method of development or construction on a particular site.

Factors taken into account include foundation situations, excavation requirements, drainage, the water table, corrosivity of environmental water, slopes, weathering, seismic considerations, mass movements, flooding possibilities which relate to the engineering geological suitability. Undesirable alterations to land stability, erosion, the water regimen and the properties of the rocks in a site increase geoenvironmental vulnerability and must be avoided or minimized. Local georesources must be protected and these include fertile soils, water, materials of construction, etc. Appropriate rating values may be normalized to a unitless measure and assigned to each classification step of the individual factors, such values being calculated in correlation with real conditions in individual map units. Factors may be weighted according to their relative importance.

A somewhat similar quantified approach has been utilized by Rodriguez Ortiz and Prieto[4] regarding terrain evaluation for highway construction in Spain. They stated that an overall terrain evaluation for road construction must involve:

(a) the geomorphological features of the terrain, γ;
(b) the engineering geological properties of the ground, i.e. hydrogeological and drainage conditions, δ, bearing capacity, β, ease of excavation, ρ, value as borrow material, π, slope and overall stability, η.

The mapping scale used is between 1:50 000 and 1:25 000. Categories (a) and (b) may seem to overlap because geomorphology determines stability conditions and drainage, but the effects of it depend upon the geometric features of a new road. Consequently one requiring major construction (say 40 m in width) entails stringent excavation requirements and also strict requirements for bearing capacity and drainage. In the case where a secondary road is involved, all these are lessened and such a simple two-lane road could be adapted easily to an undulating topography.

All the above-listed factors can be presented in a final evaluation showing both the geomorphological evaluation and the geotechnical feasibility.

A geotechnical feasibility index was proposed and this follows the expression:

$$\alpha = a_1\delta + a_2\beta + a_3\rho + a_4\pi + a_5\eta$$

where α = the index and the values δ, β, ρ, π, and η are the geotechnical parameters mentioned above. a_1 to a_5 are weighting factors denoting the relative importance of each parameter on the road works. Rodriquez Ortiz and Prieto listed various values for the respective geotechnical parameters and noted that the following range of values for the index α may be anticipated:

α	*Conditions*
0·2–0·5	Very bad
0·5–0·7	Bad
0·7–1·0	Poor
1·0–1·4	Mean
1·4–1·7	Good
1·7–2·0	Very good

Of course these values apply to the situation in which they worked in southern Spain, a semi-arid region. Application elsewhere is possible.

10.3. REMOTE SENSING

The method of multispectral satellite imagery and other remote sensing techniques can now be utilized to effect surveys over large areas of land so that, for instance, fuller use may be made of locally available materials in order to reduce the costs of road construction. Actually such resources are not used as extensively as they might be in many developing countries to some degree because of the adaptation of specifications for road building materials made in industrialized countries with completely different climatic regimens and for that matter geology also.

It is appropriate to make a state-of-the-art survey of this subject: Reeves[5] has edited a manual of remote sensing from which may be observed the truly tremendous advances which have been made in recent years (cf. for instance Hempenius[6]). In a later paper, Hempenius[7] reviewed and criticized the technology.

The limitations of remote sensing techniques in general may be summarized as follows:

(a) Each technique employs a specific range of wavelengths of the electromagnetic spectrum and each of these wavelengths yields exclusive data about a specific property of the planetary surface, e.g. colour, temperature, relief, etc.
(b) Differences in the various approaches to collecting information result in images of different scales or resolution as well as variable geometrical quality.
(c) Cost, depending upon flying and satellite expenses and the price of processing the images (including rectification), may be high.

Significant remote sensing techniques now available include those given below.

10.3.1. Photography

This applies optical systems to imaging on negative film data from the visible part of the electromagnetic spectrum coupled with a small extension into the invisible infra-red and ultra-violet parts. Figure 10.1 shows the electromagnetic spectrum which in fact comprises three windows, namely:

(i) Optical window—radiation in the visible range of the spectrum, a small adjacent range of the near ultraviolet and an adjacent range of the near infrared.
(ii) Heat window—the intermediate infra-red. This depends upon the fact that a body at 300 K has a maximum radiation at a wavelength of 10 μm. As the Earth has this temperature, its characteristics as a radiator can be observed in the thermal window.
(iii) Microwave window—with natural radiation at a very low level so that it is suited to radar, i.e. sending and receiving of radio signals.

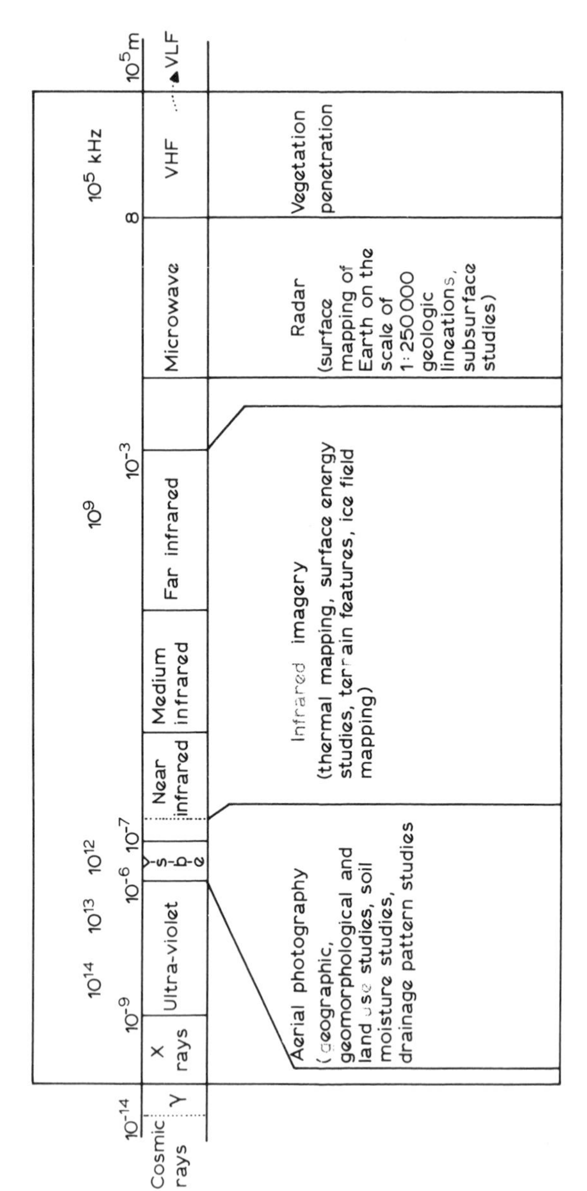

FIG. 10.1. The electromagnetic spectrum.

The use of different types of combination of films and filters enables the investigator to restrict or broaden the spectral range which is recorded either in black and white, in colour or in false colour. Actually, colour photographs frequently give better results because the tonal difference is more easily distinguishable than with black and white panchromatic photography. Photography is valuable because it provides an instantaneous synoptic picture of an entire region which is visible through the camera lens. A pair of photographs covering the same terrain but taken from slightly different positions of exposure can be placed adjacent to each other and viewed simultaneously with both eyes, this giving stereoscopic vision. The central perspective causes relief displacement towards or away from the centre of the image. Such relief displacement and related scale differences obviously disturb the image's geometry and become accentuated in dealing with terrains having considerable relief, especially when the survey is effected at low altitude. The consequence of all this is that transferring photographic data to a base map is usually time-consuming and almost always causes inaccuracies in the final version of the map.

A variant is orthophotography which is obtained from ordinary photographs by the optical conversion of the central projection to a parallel projection at uniform scale without relief displacement (see ref. 8). The transfer of orthophoto data to a base map is both easy and fast and stereo-orthophotography is simply a refinement of the method and one not used too much to date for projects in engineering.

The production and application of orthophotographs generally has been discussed by Visser and others.[9] Off-line as against on-line production has been examined, the former having been available since 1964, the year when the Orthophoto-projector GZ-1 of Zeiss Oberkochen was introduced. GZ-1 steering devices for automatic differential scaling taking into account terrain slopes above the slit are operable only by feeding them with height data which have been stored in an analogue form. This digital steering using digitally stored height data permits a greater flexibility in the design of alternative orthophoto production processes. In 1976 Wild presented the OR1, an especially popular, digitally steered, automatic, orthophoto printer and in 1980, Zeiss Oberkochen presented another one, the Orthocomp Z2.

Off-line methods have a number of advantages, among them the possibility of using height information from sources other than the photography utilized in the actual orthophoto production, for instance from any available Digital Height Model (DHM) in the form of say digitized contours and spot heights made from existing maps. Also, their quality is better (no gaps or double images) than with the on-line method although the latter is easier to employ and also the necessary equipment is much cheaper.

In practice, both approaches may be used as for example by the US Geological Survey. Computation of contours from three-dimensional profiles is feasible both in on-line and off-line processes bearing in mind the differential costs. The Wild OR1, including a Nova 3 computer, cost around US$225 000 in 1980 whereas the Zeiss Z2 including an HP 1000 cost slightly more, about US$250 000. By contrast the Galileo G6 plus Orthosimplex on-line equipment in 1980 cost some US$75 000. For optimum image quality, the following requirements must be met:

(i) The density range of the primary data must be lower than that for normal mapping.
(ii) Knowledge of the performance of the objective of the camera at certain apertures must be sound.
(iii) Involved with the second need is the effective aerial film speed, the type of emulsion used, the filter-factor and the processing.
(iv) Direction of illumination at the time of flight should be roughly the same for each photograph.
(v) Flying conditions should be optimum, i.e. without cloud shadow, haze, high solar angles, etc.
(vi) The optimum scale must be selected and it should be the largest feasible and consistent with accuracy of detail.
(vii) In terms of overall resolution, the final image quality of the orthophoto must be better than the resolving power of the human eye at its minimum distance of clear vision, say 20–25 cm.

Further information is provided by Eden[10] in a paper entitled 'The art of taking air photographs'.

Electronic correlators are utilized because vegetation and man-made constructions may make height readings unreliable and these permit faster profiling by a factor of three than with manual profiling, but with the same degree of accuracy.

The smallest scale for useful photomaps appears to be about 1:10 000 (except in arid terrains); hence the US Geological Survey orthophotomaps at a scale of 1:24 000 may be considered as constituting merely provisional substitutes for maps.

Turning to stereo-orthophoto pairs, the production of these involves extra computing time and if the Wild OR1 is used, this amounts to 25% of the time required for the conversion of the three-dimensional profile coordinates into image coordinates, the 1980 cost being estimated as about US$6 per model for computing. The extra OR1 time of about 30 min per model was then estimated as approximately US$12·25 per model.

In the case of a flat terrain, any error in height data acquired for differential rectification, e.g. an interpolation error dH (H = height) in the DHM as used in the off-line method, has no effect on the horizontal parallaxes in the resulting stereo-orthophoto pair. This is true only when the orthophoto and the stereomate are made from the same height data, i.e. in the on-line method, when both are produced simultaneously and in the off-line method, when both are produced using the same DHM.

In the case of a terrain which is not flat, small errors will be introduced into the x-parallaxes of the stereo pair, these being a function of the error dH (in the profiling or in the DHM) of the terrain slope and of the angles of the two projecting rays relative to the vertical. Finsterwalder[11] has shown that with terrain slopes of the order of 20°, the height deformations in the stereo-orthophoto pair are of the order of 10–20% of dH (only).

If large scales such as 1:10 000 are employed, then substantial systematic errors (dH) are to be expected in the electronic heighting of the terrain, especially near vertical objects such as buildings. This makes it unsurprising that the height accuracy of stereo-orthophoto pairs observed in an NRC Stereocompiler from original GPM orthophotos after photographic enlargement by a factor of four was only about 0·4% of the flying height.[9]

Photomaps have come more and more into general use and sometimes are preferred. While a scale of 1:10 000 may be regarded as the norm, for city areas this can be improved to 1:1000. On the other hand, in arid regions a scale of 1 : 100 000 is acceptable. Photomaps are mostly black and white, but sometimes they may be in colour. The US Geological Survey reports rather bad experiences with colour, however, pointing out that when high altitudes are involved (12 000 m or above) for 1:24 000 orthophotoquad production, poor contrast occurs due to atmospheric scattering of the shorter wavelengths. Some instances of photomap production for special applications are now cited:

(a) 1:24 000 scale mapping in the USA is country-wide in scope, usually covering 7000 square miles and with contours at intervals of 4 ft. Applications include flood control, i.e. for the design of approprite channels, roads, borrow pits, etc.

(b) 1:5000 scale orthophoto resource maps annotated with cadastral information have been set up for the province of Prince Edward Island in Canada.

(c) In Sweden, between 1966 and 1978, 60% of the country was covered by 1:10 000 photomaps with contours at intervals of 5 m. These were produced by differential rectification from 1:30 000 scale photography utilizing orthophoto printing system OR1.

(d) US Geological Survey 1:24 000 orthophotoquads are made at high altitudes (12 000 m) and a single photo of a triplet is centred over each 7·5 min quadrangle and completely covers this area. In 1978 the US Geological Survey spent US$4·5 million and 125 man-years to produce 4400 orthophotos. Orthophotoquads have only a small degree of cartographic enhancement and they are used as map substitutes in unmapped regions; for land use planning and inventorying of natural resources; in photo inspection and revision of existing maps (by overprinting the altered data in a distinctive colour); and in producing line maps by scribing a special scribecoated photo base.

Stereo-orthophoto pairs have been employed in a technical cooperation bilateral programme involving Canada and Colombia, the pilot project having involved the establishing of a multipurpose cadastre and data bank. A Gestalt photomapper is utilized to produce stereo-orthophotographs and four stereo-compilers equipped with digital recording devices are used to extract metric information. The pilot area is 2000 km^2 in extent and was covered previously by photography at scales of 1:54 000, 1:27 000 and 1:18 000. GPM stereo-orthophotographs will have been enlarged to 1 : 25 000 and 1:15 000 scales. Actually Colombia was the first country in the world to apply stereo-orthophotography in such a comprehensive survey.

The following interesting projects are or will be in progress if financed:

(a) 1:50 000 scale photomaps of Saudi Arabia (initiated in 1970) and Libya;

(b) 1:5000 scale photomap series in the Federal Republic of Germany;

(c) 1:2000 scale photomaps of all French cities;
(d) 1:2000 scale photomaps for flood control in Canada;
(e) 1:2500 scale photomaps for cities of Queensland, Australia;
(f) 1:5000 scale and 1:10 000 scale photomaps in Queensland, Australia;
(g) Floodplain maps as photomaps with conventional rectification in Victoria, Australia.

In orthophotography, resolution may be expressed as the size of the smallest object recognizable on the image by its salient characteristics and it will depend upon the scale of the photographs and the type of film and lens utilized. An approximate guide is 0·25 m at the scale 1:5000 and larger, 2·5 m on a scale of 1:50 000. In aerial photography the usual scale range is 1:5000 to 1:50 000. However smaller and larger scales are obtainable if the flying altitude and/or the focal length of the camera lens is varied. The so-called height resolution, i.e. the accuracy of height determination in stereophotogrammetry, is about 0·01% of the flying altitude for normal aerial and satellite photography. For high altitude photography this may be improved by a factor of two or three. In summary it may be stated that aerial photography is the only remote sensing technique which is valuable during engineering reconnaissance investigations as Rengers and Sosters[12] have pointed out.

Satellite as opposed to aerial photography is available in the data provided through manned space excursions in the Gemini, Apollo, Skylab and Soyuz programmes. However, the coverage of the surface of the Earth is haphazard and also small scale. These are disadvantages in terms of applications to engineering projects. No doubt future developments will remedy these defects. (Doyle[13] has introduced a programme of systematic satellite photography for use with a high performance cartographic camera yielding stereo photography on a scale of 1:1 000 000 with a resolution of 10–20 m—already a step in the right direction.)

Horizontal or slightly inclined terrestrial photography is widely used for photogrammetric or interpretation purposes as Reuss[14] has indicated.

Photography can provide invaluable information on the nature and distribution of soil and rock units visible through indirect evidence derived from the planet's surface morphology which is seen in stereophotography, the landscape and its morphology resulting from weathering, denudation and deposition processes acting upon the available geological materials composed of rock and drift overburden. The interaction is strongly influenced by physical properties such as strength, resistance to erosion and permeability. Details of the morphology, type and density of the drainage system as well as data on the types and abundance of vegetational cover are invaluable in assessing flood hazards, seepage, infiltration and saturation. The actual state of the geomorphology is determinable using quantitative information on the forms of slopes, their steepness, gradients of rivers, etc. All these can be investigated by the stereoscopic interpretation of aerial photographs in conjunction with quite simple photogrammetric procedures as outlined by Verstappen.[15]

Such information also permits an assessment to be made of the short and long term effects of future evolution of the landscape on an engineering project. Geodynamic phenomena such as processes of mass movement, erosion, etc., can be recognized from their expression in the morphology of a region as well as from their location in the landscape. The size of mappable features relates directly to the scale and resolution of the photography. Karstic phenomena for instance can be detected from details in the drainage system (undrained depressions) and from features such as sink holes.

Active faults can be detected where recent deposits are affected causing irregularities in the morphology of a region. A special technique, that relating to sun angle and alluded to by Cluff and Brogan,[16] is applicable. In this, patterns of linear shadows can reveal otherwise invisible morphological lineaments. However a limitation of the technique is that only lineaments striking approximately between northwest and southeast can be detected. A lineament is a large scale linear feature expressing on the surface some underlying structural characteristic; they include fault- or joint-controlled valleys, lines of isolated hills, ridges, straight coastlines and so on, reflecting such geological features as fault and joint zones, fold axes, linear igneous intrusions and so on.

Photography can be utilized during the various stages of an engineering project as has been indicated by Rengers.[17] This is illustrated in Table 10.1.

10.3.2. Multispectral Scanning

Multispectral optical–mechanical scanners are utilized in recording the electromagnetic radiant energy emitted or reflected by the planetary

TABLE 10.1
ENGINEERING GEOLOGY INVESTIGATIONS

Project stage	*Mapping type*	*Scale*	*Usage*	*Data types*			
				Rock and soil	*Hydrology*	*Geomorphology*	*Geodynamics*
Initial	Compilation of existing data and preparation of multi- or special purpose map	1:10 000 to 1:50 000	Base for plan of reconnaissance	General outlining and structural trends, faults	Drainage and potential flood	Land forms	Main zones
Reconnaissance	Engineering classification of soil and rock units by interpreting remote sensing imagery and field survey	1:1 000 to 1:10 000	For selecting the correct engineering sites	Zoning in quality units, e.g. grain size of soils	As above plus springs and seepage and infiltration	Details of landforms and slope inclination	Erosion and mass movement maps of karst and coasts
Main investigation	Mapping of soil and rock units using quantative data re their engineering behaviour; analytic or comprehensive maps prepared	1:100 to 1:5 000	Precise siting of project	Described re rock mass strength; soils described in terms of friction cohesion, bearing capacity, water content, etc.	As above plus depth-to-water plus data on permeability	Predictions on geomorphology development	Details
Construction investigation	Recording of all acquired data	1:100 to 1:1 000	Data storage	As above	As above plus potential changes in hydrological regime	Details re morphology after completion of works	Details re effects of construction on various dynamic processes (reactivation of erosion, mass movement, etc.)

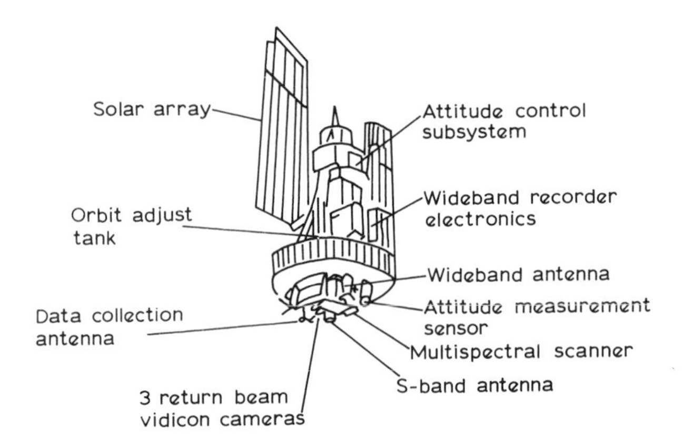

FIG. 10.2. The LANDSAT 1 earth resources satellite launched in July 1972.

surface in several or many precisely defined bands ('colours') of the electromagnetic spectrum between 0·3 and 3·0 μm. Data on radiation intensity are integrated over a certain region of the Earth's surface and the area of this is determined by the 'Instantaneous Field Of View' (the opening angle), termed the IFOV, of the scanner following the flying altitude. As regards scanners employed in aircraft, the individual picture element, the so-called 'pixel'-size is about 3×3 m of the planet's surface at an altitude of flight of some 2000 m with a scale of 1:15 000 fully enlarged. In the case of imagery from LANDSAT scanners, a pixel-size of 60×80 m can be attained using a much smaller angle (see Fig. 10.2). The equivalent scale is 1:3 500 000 from which enlargements up to 1:300 000 are feasible. Extremely detailed data on the radiation intensity are provided by the scanners and these are stored on tape after being electronically corrected for geometrical distortions arising from the actual scanning operation. Subsequently the data can be converted to imagery strips which show the radiation intensity of one spectral band in shades of black and white or alternatively two or three radiation ranges may be employed. These are then superimposed in different colours to make up a 'colour composite'. Multi-Spectral Scanning (MSS) will yield detailed information on the spectral signatures of areas, but will not show sharp boundaries as a result of the fact that the resolution is of the dimensions of an area covered by at least 6 to 9 pixels.

As noted above, LANDSATS 1, 2 and 3 satellite scanners possess small scanning angles and a small ratio of terrain relief to flying altitude; hence the imagery obtained can be corrected automatically rather simply to obtain a satisfactory geometry. With aircraft however MSS imagery entails great geometrical drawbacks because of flight instabilities and the wide angle nature of the scanner. Where irregular terrains are involved the effect is correspondingly worse. Of course correction is possible if there is sufficient geometrical ground control, but even then the process is very expensive and also takes a long time. In any event the results are inferior to those of LANDSAT images of which the pixels are larger by a factor of ten. MSS imagery displays relief displacement in the scanning direction perpendicular to the flight line. Overlapping portions of parallel strips of LANDSAT images are observable stereoscopically but MSS images from aircraft usually afford only a poor stereoscopic vision as a result of the fact that the lines of flight are not usually straight and parallel, but variable and in addition there is difficulty in matching the x- and y-scales of the images.

The sort of data which MSS imagery can provide differs from the information given by aerial photography. While the latter is a delineation-detection instrument, MSS imagery is an areal sensing technique. Morphology only becomes apparent where terrain relief is pronounced enough to provide shadows and/or affect the vegetational cover. Every pixel possesses a spectral signature which is a combination of the levels of radiation of the pixel in question in the different spectral bands of channels. Actually the quantity of data available on a single pixel exceeds that which can be made visible on a single image (there are more radiance levels than those perceptible to the human eye—no less than 128 in LANDSAT—together with, normally, more than three spectral bands). To overcome this problem, an interpreter–computer interactive system is employed to take an input of a limited quantity of ground truth as discussed by Donker and Mulder[18] and Donker and Meijerink.[19]

One advantage of satellite MSS imagery is its high temporal resolution. LANDSAT returns every 18 days for successive flights over the same lines and this facilitates a regular recording of alterations in planetary surface conditions over a wide area. This includes for instance the spread of floods and also of ice and snow in mountainous regions or at sea. The temporal resolution will depend upon weather conditions and a high temporal resolution is invaluable for the planning of roads, harbours, etc., in remote and uninhabited areas. It is also useful for predicting floods from snowmelt or excessive precipitation.

Specific uses may be listed:

(a) MSS recognizes units of vegetation, soil and rock on the bases of their spectral signatures. This is not so good for the Earth Sciences in that the vegetation has a strong influence and one undesirable in their discipline. In fact even where there is not much vegetative cover, there is only a poor correlation between the spectral signature and engineering matters. Hence most geological information from MSS refers to recognizing patterns of drainage, vegetation and so on. However, LANDSAT satellite data (and radar data) have been used to locate calcrete deposits (with poor results as regards reliability in quality of these) to be used for low cost, feeder road construction in Botswana (see ref. 20).
(b) In hydrology, MSS imagery is useful because of the marked difference between the spectral signatures of water and land. Consequently the delineation of drainage systems is easy.
(c) Geodynamic processes cannot be detected usually unless of considerable size, i.e. up to several hundreds of metres and geomorphological units are also not normally recognizable except by association with for instance the main drainage system of an area.

In summary therefore MSS imagery in engineering investigations has a limited value at certain stages. However aerial photography is certainly much more useful. On a regional basis large scale lineaments detectable on LANDSAT imagery can relate to fault patterns which may be accompanied by seismic events and hence influence future construction plans.

10.3.3. Radar

Side-Looking Airborne Radar (SLAR) and Synthetic Aperture Airborne Radar (SAAR) are important remote sensing techniques depending on the emission of pulses of radio from an aircraft, the return signals being recorded with an antenna. The range in the electromagnetic spectrum which is usable is from 8 mm to several metres, but the normal range is 0·8–15 cm. The techniques are usable both during the day and at night and with longer wavelengths (e.g. with SAAR) heavy rain showers can be penetrated (thus making the technique employable under practically any set of meteorological circumstances). The return signals intensity is transformed into grey tones and when corrected for geometry distortion can be utilized to form radar image strips of the areas on both sides of the aircraft. As there is very wide angle approach, the method cannot be applied to hilly terrain as Mekel[21] has pointed out.

Recently however high altitude flown SAAR has been found to give better image-geometry for hilly terrains. The optimum areas for investigation by this technique are extensive regions in the tropics which do not permit good aerial photography. The scales used vary from 1:100 000 to 1:400 000 whereas the resolution attains up to 30 × 30 m for SLAR and up to 10 × 10 m for SAAR. It is anticipated that pixels with sizes of 3 × 3 m will be feasible soon. The type of data available through the application of the technique relate to the reflective capacity of the planetary surface or its vegetation together with the slope of the terrain. Consequently, all the major morphological components of an area become visible in a radar image and large structural elements such as folds are recognizable if they obtrude morphologically to a sufficient degree (see ref. 22). An important problem must be mentioned and this is that linear elements of patterns show up better when perpendicular to the scanning direction than when parallel to it. Stereo is obtainable when parallel and overlapping strips are looked at microscopically, but the model is unreliable for deriving morphological conclusions of any soundness. As for soil and rock units, the poor resolution allows only a broad zonation of some regional aspects to be made and of course no inferences as to the engineering properties can be drawn because there is no relationship between these and the scattering and reflection processes which contribute to radar image formation.

Apropos hydrology and hydrogeology, large surface water bodies provide specular reflections and become apparent on radar images because of their strong contrast with the surrounding soils and rocks. Some small drainage lines may be visible also, but this is true only if they are related to the morphology of the region and even then those parallel with rather than perpendicular to the line of flight are better defined in the radar image. Slope changes may be visible in radar imagery so that from these some idea of the major features of the areal geomorphology may be derived.

Turning to geodynamic phenomena, some can be recognized in radar imagery and these include large scale regional patterns of faulting.

In sum the applications of airborne radar imagery to engineering projects lie mainly in the preliminary stages of investigation when the identification of regional fault patterns may be feasible. In seismic regions these together with epicentre data may enable an estimation of potential earthquake damage danger.

In 1981 some very interesting work was done using the Space Shuttle Columbia in the month of November. The mission was cut from 124 h to 54 h because of a faulty fuel cell, but the astronauts completed an experiment with the radar equipment nevertheless. They took a 50 km wide scan of the Sahara desert. Normally radar waves penetrate only a few centimetres into the Earth's surface, but in this dry, arid area they were able to pierce to depths of 5 m, reflecting from bedrock. The signals obtained were processed through image-enhancing computer techniques and the results analysed by the US Geological Survey, the Jet Propulsion Laboratory in Pasadena, the University of Arizona and the Egyptian Geological Survey and Mining Authority. A buried topography was revealed, one without trace at the surface. The images showed stream channels, broad flood plains and river valleys some of which were as wide as the Nile.* The trend of the ancient rivers is south and west, i.e. the opposite of existing movements and they may have joined up into one large basin of interior drainage comparable with the Caspian Sea today. Obviously the region was once sufficiently wet to support plant, animal and human life. It is known that rainy episodes, pluvials, occurred during the Pleistocene and these are thought to have taken place about 200 000, 60 000 and 10 000 years ago. Tools of *Homo erectus* have been located in the area as well, in fact at sites along the banks of the concealed river beds now revealed in the new radar-derived maps. From this work it seems likely that this type of radar scanning will be applicable to other arid regions for the detection of waterways not too far below the surface, possible sites being those where groundwater intersects the surface. As for

* For additional details regarding the revealing of previously unknown buried valleys, geological structures and possible Stone Age occupation sites by the Shuttle Imaging Radar (SIR-A) carried on the Space Shuttle Columbia in November 1981, the reader is advised to consult the paper on 'Subsurface valleys and geoarchaeology of the Eastern Sahara revealed by shuttle radar' by J. F. McCauley, G. G. Schaber, C. S. Breed, M. J. Grolier, C. V. Haynes, B. Issawi, C. Elachi and R. Blom which appeared in *Science*, **218,** 3 December 1982, pp. 1004–20.

archaeology, it may be possible to determine sites of early human habitation near former rivers and lakes. It is also possible that, by indicating such subsurface features, surveys for oil and minerals could be assisted. The interest generated was enough to obtain approval for a follow-up mission set for August 1984, one aim of which is to apply more sophisticated radar imaging to wider coverage of the Sahara region. It might be added that just such an approach has been valuable also in sub-cloud exploration of the planet Venus.

10.3.4. Thermography

This is the accumulation of data on the surficial temperatures of the planet and these may be obtained by making a record of the radiation in the middle (3–5 m) and far (8–13 m) infra-red using optical mechanical scanners as with MSS or making a record in the microwave range (8 mm to 20 cm) using a passive microwave radiometer with a scanning antenna, as in radar. From the data obtained during a scanning exercise, a thermal image can be derived and this will show different tones of grey or colours based upon the temperature differences of the various objects.

If the flying altitude is too high, a poor resolution image results. On the other hand, very poor geometry images are produced if the flying altitude is too low over irregular terrain. A number of parameters can affect images in thermographic work adversely and these may be air temperature, terrain inclination direction, angles of slopes in the terrain, wind velocity, atmospheric moisture content, vegetation and soil moisture. All will influence the absolute temperature data and also the contrast between different materials.

The drawbacks alluded to above naturally make thermography a remote sensing technique which does not as a rule offer much of value to the initial investigational stage of an engineering geology project. However in certain cases the approach may be useful, for instance where a particular aspect is correlated with appreciable ground temperature differences which are not disturbed or obscured by any or all of the above-mentioned factors. Cases in point might be the occurrence of fresh water springs in the sea or sources of geothermal energy or heat emission losses from subterranean pipe systems. However such uses will be rare and so the technique at present remains not generally applicable.

10.4. EXPENSES IN REMOTE SENSING

In the case of aircraft, there are three main components:

(a) Cost of arrival, i.e. flying the airplane together with the necessary equipment to the study region.
(b) Cost of flight, i.e. dependent upon the area to be covered, the scale requested and the overlap percentage of the imagery. Weather conditions are an important consideration.
(c) Cost of processing the images which will include geometrical correction, the development of films and the printing of positive or diapositive copies.

Satellite images are much too expensive for them to be made specifically for a particular project. Although satellite MSS (LANDSAT) imagery and photography (Skylab) can be purchased at a fixed price per frame, a specification of scale, overlap, spectral and temporal resolution cannot be adapted for the specific requirements of a particular project.

Ordinary terrestrial photographic costs relate to the time required and the area to be covered.

10.4.1. Expenses of Photography

As regards aerial photography, the cost of getting aircraft and equipment to the area to be surveyed is usually quite low if the camera and appropriate airplanes are available. Climatic conditions are important because in bad weather there may be an appreciable standby time for the airplane. A 1979 price quotation may be cited: it was about US$5–10 per km^2 for scales between 1 : 20 000 and 1 : 10 000 for areas exceeding 1000 km^2. Obviously greater detail costs more. In the case of regions for which appropriate scale photographs are available already, black and white prints or diapositives may cost from US$2–4 per copy of size 23×23 cm. Colour prints and diapositives cost about US$10 per frame.

Satellite photographs made during manned NASA space flights cover about 10% of the Earth's surface at a scale of about 1 : 1 000 000. Soyuz flights produced multispectral photography with a 10 m resolution for particular test areas.

As for terrestrial photography for interpretive uses, this is possible to make using rather simple photographic equipment such as a hand camera or a technical camera with a large frame size. In photogrammetric investigations, twin cameras with a fixed base length or a photo-theodolite can be employed. This latter is costlier however: whereas cameras may run up to US$2000 or US$3000, such an instrument can cost as much as US$30 000 or more.

10.4.2. Expenses of MSS Imagery

This is a more expensive technique than aerial photography because of the equipment price and also the geometrical rectification, which costs a lot. A 1979 cost figure is for satellite MSS imagery from LANDSAT and for one image of 19×19 cm at a scale of 1:1 000 000; for one channel, this was about US$10.

10.4.3. Expenses of Radar Imagery

This is certainly very expensive and anyhow the technique is too costly to use on a special flight for a particular engineering project. However in some parts of the tropics such as Brazil and Indonesia, coverage is extensive. The copies of resultant imagery bear a cost dependent upon the owner of copyright.

10.5. DEVELOPMENTS IN REMOTE SENSING

Clearly the acquisition of remote sensing data has benefited Mankind greatly, for instance in land usage classification, agricultural assessments, energy and resources evaluation and engineering geology as well as other aspects of human activities. However there has been a certain reluctance to place satellite systems on a normal operational basis because of the fear of lack of continuity of information assemblage coupled with continual alterations to the systems and some degree of dissatisfaction with the results. Almost certainly these objections will be overcome as satellite remote sensing improves; hence it is useful to consider work in progress and future anticipated developments.

The USA is the acknowledged leader in the field and among its most significant efforts is of course the Space Shuttle alluded to earlier. This has a main component, the orbiter vehicle, which is returned to Earth after each mission and consequently is re-usable. The first flight of the Shuttle was from 12 April to 14 April 1981 and it proved to be a great success despite troubles with the heat-resistant tiles. The orbiter vehicle is about the same size as a Douglas DC-9 aircraft and its back opens into a cargo bay 18·3 m long and 4·6 m in diameter. This bay is valuable for installing manned modules and instrument pallets operable while the Shuttle is on space missions. Launches have been east from Cape Canaveral covering latitudes 30°N to 30°S and the maximum orbital inclination available from this geographical location is 57° covering most of the populated regions of the planet. After 1984 polar orbits should become available after a Western test range becomes operational. Missions of the Shuttle will be designated STS (Space Transportation System) followed by the appropriate flight number.

Another important development in the USA is the Multi-mission Modular Spacecraft (MMS) comprising a central core on to which modules for power supply, attitude control, command and data handling can be attached. In-orbit manoeuvering is attained by a propulsion module and there is also an adaptor between the spacecraft and its payload. The assembled spacecraft can be fitted with solar panels and communication antennae and the MMS is specifically designed for launching and retrieval by the Shuttle employing a Remote Manipulator System (RMS), a long articulated arm. The MMS may comprise a principal vehicle for future unmanned missions in low orbits.

A third development is a Tracking and Data Relay Satellite System (TDRSS) which is made up of two communication satellites and a ground station located in White Sands, New Mexico, the missile test range. Commands originating on the ground to both manned and unmanned spacecraft will reach the TDRSS first and then the spacecraft. Data collected by the spacecraft will go to the TDRSS and then be re-transmitted to the ground station. Both TDRSS satellites will be located in geosynchronous orbit at longitudes 41°W and 171°W from which positions they almost completely cover the Earth with the exception of an exclusion zone in the Indian Ocean and extending into the USSR. These TDRSS spacecraft should be launched by the Shuttle and be in full operation with two spacecraft by the middle of 1983.

Fourthly there is a second generation multispectral scanner, the thematic mapper (TM), which will have six spectral bands in the visible and short wavelength infra-red with a 30 m IFOV and one band in the thermal infra-red with 120 m IFOV. It is intended for an altitude of about 700 km and should provide a swath width of 185 km (the same as LANDSAT 1 and also LANDSATs 2 and 3). Its data rate is 85 megabits per second.

Finally there is the Global Positioning System (GPS) to consist of an array of 18 separate satellites arranged in three different orbital planes. When this is in place, four spacecraft will always be visible from any place on Earth or in low Earth orbit.

All of the above advances are related to US policy decisions taken recently, with the following consequences:

(i) The shuttle is to replace all expendable vehicles.
(ii) All spacecraft data transmission is to use the TDRSS system so that the existing NASA ground receiving stations network is to be phased out.
(iii) Responsibility for operational Earth observation satellites is with the National Oceanic and Atmospheric Administration (NOAA) in the Department of Commerce.
(iv) Prices of satellite data for users will ensure the maximum recovery of the complete system costs rather than just the cost of data reproduction and distribution as now.
(v) The ultimate aim is commercial ownership and operation of Earth observation satellite systems.

These reflect the views of the Reagan administration, especially items (iv) and (v). A new LANDSAT, LANDSAT 4, is to be launched. The MMS will be launched by the Delta 3920 expendable vehicle into a 705 km near-polar, sun-synchronous orbit, the payload being the TM and an MSS which is almost the same as that on LANDSATs 1, 2 and 3. The spacecraft will be equipped with solar arrays for power together with an antenna for use with the GPS plus a steerable antenna of large size for communication with TDRSS satellites. There are no tape recorders on board for either the MSS or the TM. During operation, the two sensors record simultaneously, the former with an 80 m pixel, the latter with a 30 m pixel. In the range of a suitable ground station, i.e. one equipped to receive signals, data are transmitted directly. MSS data are on S-band at 15 megabits per second and TM data on X-band at 85 megabits per second. When a suitable ground station is not in range, data are to be relayed to the TDRSS and, via that, back to the ground station at White Sands. Subsequent to preliminary processing, the data will be relayed by domestic communication satellite to the Goddard Space Flight Center (GSFC) for further attention and the generation of a product.

It is important to remember that the current LANDSATs 2 and 3 are showing signs of failure and so the launch of LANDSAT 4 is most important.

MSS data are transmissible on S-band directly to GSFC, also to foreign receiving stations with tape recorders through which data is transmissible to GSFC for processing. GSFC can transmit High Density Digital Tape data (HDDT) through Domsat (domestic communication satellite) to the EROS Data Center at Sioux Falls, South Dakota, USA. Film and computer compatible tapes can be produced for distribution. The MSS ground processing system at GSFC can process 200 scenes per day for the present and when NOAA assumes full responsibility for command and control of the spacecraft and distribution this could increase (see ref. 23). At first TM data is transmitted on X-band directly to the GSFC ground station and no foreign stations are appropriately configured to receive and process the 85 megabit per second X-band data. The separate TM facility at GSFC is operated by NASA, initially on an experimental basis to produce one or two scenes daily, but it is anticipated that a rate of a dozen scenes per day should be attained by the first quarter of 1984 and full capacity, 50 scenes daily, by 1985. GSFC will produce the film and computer compatible tape products.

It is expected that LANDSAT D′, another and identical spacecraft, will provide additional wide band video tape recorders for the MSS and take over when LANDSAT 4(D) fails. The life of the TM is expected to be two years and that of the MSS three years so that they should provide information until 1988 by which time it is anticipated that NOAA will have made an agreement for the commercial operation of the Earth observation satellite system. It is not expected that LANDSAT D and LANDSAT D′ will be recovered and relaunched by the Space Shuttle (although they could be) because it is believed that the expected commercial operation might utilize a different system configuration.

As noted earlier, the first Space Shuttle flight was made in April 1981 and the second, STS-2, occurred in November of that year. On the initial flight an unexpectedly powerful shock wave from the blast of the solid fuel rockets caused the control flaps on the trailing edge of the delta wings of Columbia to flutter wildly and almost reach breaking point. This shock also bent and buckled several of the metal trusses linking Columbia to its big external fuel tank. To obviate this on the second flight, engineers had to overhaul the shock suppression system and deluge the flame pits on the launch pad beneath Columbia's solid fuel rockets with even more water than before. Another delay arose when a jammed valve caused a back-up of nitrogen tetroxide. This corrosive liquid (part of the mixture which powers 14 small manoeuvering rockets on the nose of the spacecraft) spilled down the orbiter's sides and loosened about 50 of the 31 000 heat-shielding tiles as well as damaging others. In all 379 tiles had to be detached, cleaned and re-glued. These troubles obviated the original plan that the shuttle would operate every two weeks and retarded the number of flights envisaged over the period to 1984 from 44 to 32. This second flight carried an Earth observation payload package (OSTA-1) developed by NASA, the major components being mounted on a pallet carried by the spacecraft in its cargo bay. The pallet is part of Spacelab, Europe's contribution to the shuttle programme. The instruments include the following:

(a) SMIRR, the Shuttle Multispectral Infra-Red Radiometer—It comprises a telescope, rotating filter wheel and detector to record surface reflectivity in 10 spectral bands between 0·5 and 2·5 μm with 100 m IFOV to evaluate effectiveness for geological classification. Two 16 mm film cameras (black and white and colour) record the scene and data correlated post-flight with field spectrometer data.
(b) MAPS, Measurement of Air Pollution from Satellites—A gas filter correlation radiometer for measuring carbon monoxide concentrations in the troposphere.
(c) OCE, the Ocean Colour Experiment—A rotating mirror imaging scanner with a 3 km IFOV 550 km swath, 8 spectral bands between 0·49 and 0·79 μm used to determine chlorophyll concentration as a measure of oceanic bioproductivity.
(d) FILE, Feature Identification and Location Experiment—Intended to develop the ultimate capacity of on-board classification terrain features. The output of two television cameras (one in the visible red, the other in the near infra-red) is to be ratioed in order to determine the scene classes. A 70 mm camera takes colour photographs for post-flight comparison with each TV frame.
(e) SIR-A, Shuttle Imaging Radar—A modification of the Seasat synthetic aperture radar to provide a 47° look angle, 56 km swath and 40 m resolution. Data recorded on film. The primary application is interpretation of geological structures; cf. the relevation of the Sahara's buried river system during the

November 1981 flight of the Space Shuttle Columbia, its second voyage, discussed above.

A separate experiment for the crew compartment was NOSL, a Night–day Optical Survey of Lightning and consisting of a hand-held 16 mm motion picture camera and a photocell for recording lightning flashes connected with major thunderstorms.

As noted earlier the STS-2 was projected for longer than it actually lasted in a 275 km orbit at 40·3° inclination but, after a flawless take-off, one of the Shuttle's three battery-like fuel cells was ruined thus reducing the electrical capability by one third. Actually the departure itself was delayed as well through clogged oil filters two weeks prior to lift-off. There is no doubt that the US$10 billon project has been plagued with difficulties and the latest flight, when the nose of the Columbia rose alarmingly, was no exception. Nonetheless the potential rewards in terms of the Earth Sciences remain very great. This is recognized; the US Geological Survey, for instance, has completed a study contract for a system called MAPSAT which would provide data for the compilation of 1:50 000 scale maps with 20 m contours in an orbit envisaged as the same as that of LANDSAT's 1, 2 and 3. Three multispectral linear array instruments would look in the vertical ±26° fore and aft in order to provide stereo data. Each array would have three spectral bands and a minimum 10 m IFOV, but the data could be clustered by on-board processing so as to provide lower resolution at multiples of the 10 m IFOV. On-board compression and selection could reduce the average data rate to 15 megabits per second which is compatible with existing worldwide network LANDSAT ground receiving stations. A stable spacecraft would permit the utilization of a simple algorithm in the ground data reduction for correlation of the stereo records in order to produce digital elevation data. Unfortunately to date MAPSAT is not approved and hence not funded after the completion of the feasibility study.

STEREOSAT is a rather similar idea and this again derives from geological requirements; vertical fore and aft looking linear arrays would be panchromatic with multispectral data to come from LANDSAT. STEREOSAT would utilize the MMS and be launched from the Western test range by the Shuttle into the same orbit as LANDSAT D. It would have a 15 m IFOV and a 61·4 km swath. Again, it has not yet been approved beyond the feasibility stage.

After the termination of Seasat, through failure after 3 months success, a follow-up system for oceanic surveillance is proposed in the form of NOSS, a National Oceanic Satellite System—once more unapproved to date.

An Ice and Climate EXperiment satellite, ICEX, has been proposed to monitor snow and ice conditions at the poles. If approved, it would have carried an X-band synthetic aperture radar capable of 100 m resolution over a 360 km swath.

A Shuttle payload sensor of great importance is the large format camera with a 30·5 cm focal length and 23 × 46 cm size. It has automatic exposure sensors and forward motion compensation allowing the employment of high resolution fine grain film, the magazine having a capacity of 2400 frames. A ground resolution of 10–15 m can be obtained from nominal Shuttle altitudes. This LFC is scheduled for flight on STS-26 in the middle of 1984. The mission is to launch three communication satellites, forcing the shuttle into a 28·5° inclination at an altitude of 296 km. Thereafter the shuttle will stay in orbit for 6 days so that the third observation payload package (OSTA-3) can be exercised.

As regards other countries, the USSR's manned space craft SALYUT 6 has been in a 260 km 65° inclination orbit for several years and among its instruments should be noted the MKF-6 multispectral camera system. Exposed film is recovered and new film brought by the Soyuz and Progress spacecraft. The system comprises six cameras of 125 mm focal length and 55 × 81 mm format equipped with appropriate filters. It gives a ground resolution of 20 m. The configuration is approximately comparable with the S-190A system of the US Skylab. Higher resolution imagery is obtained however because of the lower orbital altitude.

A joint European Space Agency (ESA)–NASA project is the Spacelab with a payload consisting of manned experimental modules and pallets for other instruments. A metric camera experiment comprises a standard Zeiss aerial camera with 30 cm focal length and 23 × 23 cm format and this operates through a window in the manned module. From an altitude of 250 km, this can cover an area of 190 × 190 km (36 100 km^2) with a ground resolution of approximately 20 m. The camera is phase A of the Atlas camera development programme of the Deutsche Forschungs und Versuchsanstalt für Luft und Raumfahrt (DFVLR), the Federal Republic of Germany's research agency. Phase B is expected to involve:

(a) additional image motion compensation and installation of the camera on the exterior pallet;

(b) utilization of a similar camera, but with 60 cm focal length and image motion compensation;
(c) development of a new high resolution camera with image compensation for placement in the manned module.

Phase C will involve mounting the camera system in free flying satellites to operate independently of the Spacelab. In addition, DFVLR is building a Microwave Remote Sensing Experiment (MRSE) for the Spacelab.

In France the Centre National d'Etudes Spatiales (CNES) is developing SPOT, le Système Probatoire d'Observation de la Terre, and this spacecraft will be launched in 1984–85 by the ESA Ariane expendible launch vehicle into a sun-synchronous orbit at an altitude of 822 km. The mission control and data processing station is at Toulouse. The payload of two high resolution optical systems (HRV) with linear array detectors can operate in a 3-band multispectral mode with 20 m IFOV or in the panchromatic mode with 10 m IFOV. A rotating mirror will allow the acquisition of a scene from areas up to 400 km left or right of the spacecraft as well as in the vertical, hence facilitating side-to-side stereodata acquisition and more frequent scans of high priority areas. The first launch of Ariane in December 1979 was a great success, but the second launch in May 1980 failed through engine misfire. However the vehicle is now operational.

ESA is committed to launching an European remote sensing satellite and two approaches are proposed. One is a Coastal Ocean Monitoring Satellite System (COMSS) and the second is a Land Applications Satellite System (LASS). The objectives would have been to

(a) acquire data for oceanography, glaciology and climatology (COMSS);
(b) acquire data for crop inventorying and yield prediction, monitor forest productivity, land usage classification, inventorying water resources and explore mineral and energy resources.

However, all these functions have been assigned to a single spacecraft, the Earth Resources Satellite, ERS-1, which will utilize the SPOT spacecraft launched by Ariane in 1984–85 into a 650 km near-polar, sun-synchronous orbit. The sensors are:

(a) SAR, Synthetic Aperture Radar, with 30 m resolution and 100 km swath;

(b) OCM, Ocean Colour Monitor, with 10 spectral bands between 0·4 and 11·5 μm;
(c) IMR, Imaging Microwave Radiometer, operating in six frequencies;
(d) Scatterometer (two-frequencies) for wind direction and velocity;
(e) Radar altimeter for sea state determination.

The total data rate will be 117 megabits per second to the ESA Earthnet receiving stations.

In Japan, the Science and Technology Agency has effected a study programme for Marine Observation Satellites (MOS) and Land Observation Satellites (LOS). The initial spacecraft (MOS-1) would carry a Multispectral Electronic Self-Scanning Radiometer (MESSR) for measuring sea surface colour with a 50 m IFOV for a 100 km swath in 4 spectral bands between 0·51 and 1·10 μm. A Visible and Thermal Infrared Radiometer (VTIR) is designed to measure sea surface temperature over a 500 km swath with 1 band in the visible providing a 0·9 km IFOV and 3 bands in the infrared between 6·0 and 12·5 μm providing 2·6 km IFOV. The third instrument would be a two-frequencies Microwave Scanning Radiometer (MSR) to measure atmospheric water content. The spacecraft would be launched from Tanegashima about 1984–85 into a 909 km 99·1° inclination orbit. The second spacecraft, LOS-1, would comprise a payload with a two-linear array panchromatic camera with 25 m IFOV providing stereo coverage over a 50 km swath and also a Visible and Near Infrared Radiometer (VNIR) using linear arrays in 4 spectral bands from 0·45 to 1·10 μm with 25 m IFOV and 200 km swath for land classification. In addition a Visible and Infrared Radiometer (VIR) mechanical scanner with 50 m IFOV and 200 km swath for 5 bands in the visible and shortwave infrared and one band with 150 m IFOV in the thermal infrared for vegetation and geology plus an eight-channel infrared sounder with 25 IFOV and 750 swath to provide atmospheric correction. This spacecraft would be launched from Tanegashima about 1987 into a 700 km sun-synchronous orbit. The Japanese Ministry of International Trade and Industry has prepared a separate study for a Mineral and Energy Resources Exploration Satellite (MERES) the sensor payloads of which would include a linear array stereo camera with 30 m IFOV in 5 spectral bands between 0·51 and 1·10 μm, a mechanical scanner infrared radiometer with 2 bands of 50 m IFOV between 1·3 and 2·5 μm, 2 thermal bands of 130 m IFOV between 10·5 and 12·5 μm and an L-band synthetic aperture radar with 25 m resolution. NASA is evaluating this study in order to decide on the actual configurations to be built.

The Indian Space Research Organization (ISRO) has developed an Earth observation spacecraft, the Bhaskara, which was launched by the USSR in June 1979 into a 550 km 51° inclination orbit. This carried two television cameras providing 1 km resolution in 2 spectral bands for land observation and three microwave radiometers for ocean surveying. One television camera and two radiometers were still operational in November 1980 and a second Bhaskara was planned for launch in 1981. ISRO plans to develop a second generation Satellite for Earth Observation (SEO) and is negotiating with NASA and the USSR for launch facilities.

The Netherlands Agency for Aerospace Programmes (NIRV) is studying a remote sensing satellite to carry a Dutch multispectral linear array sensor in a near-equatorial orbit and the project would be effected jointly with Indonesia where it is intended to locate the ground data reception station.

The Canadian Center for Remote Sensing (CCRS) studied the applicability of NASA's Seasat data for monitoring ice conditions in polar seas. The conclusion was that the sensor should be a synthetic aperture radar capable of producing 25–30 m resolution for a swath width of approximately 100 km. For launch Canada would depend upon NASA and/or ESA.

Brazil stated that its Space Research Institute (INPE) is to develop a remote sensing satellite for possible launch in 1985 or 1986.

Communist China has stated its intention to develop a space programme including Earth observation satellites.

It is important to survey the impact of LANDSAT on water resources and land use evaluation programmes in order to determine precisely what its contribution has been to date. As stated earlier, it comprises the three original satellites, all virtually identical, LANDSATs 1, 2 and 3, launched respectively in June 1972, January 1975 and March 1978 on a sun-synchronous, near-polar orbit at ±920 km altitude. LANDSAT 4 was launched in July 1982 and its TM results are startling indeed, individual streets for instance being clearly discernible in Detroit. This is hardly surprising because the satellite represents a second generation replacement for the other three first generation ones, hence its ability to provide correspondingly better data for Earth resources surveys and monitoring programmes in terms of spatial, spectral and radiometric resolutions. The problem at the time of writing (1983) is that GSFC is not releasing much of the undoubtedly large mass of data accumulated. Also the LANDSAT D′ is not yet launched and may be regarded as being held in reserve.

Allusion should be made to the new Space Shuttle Challenger which made its initial flight from 5 to 9 April 1983. This is lighter than the Columbia and has fewer heat resistant tiles as well as engines built to operate at 9% greater thrust by Rocketdyne, the Rockwell International subsidiary. It carried and launched the TDSR (tracking and data relay) satellite which assumed an important role in transmission of TM information since a vital piece of equipment on LANDSAT 4 stopped functioning. At first there was trouble with the TDSR satellite orbit, but this appears to have been overcome. It may be stated that as of April 1983 a TM data tape covering 40 000 km^2 costs about US$3000. MS scanner data for the same area by comparison is much less expensive at about US$600. In June 1983 Challenger, with five astronauts aboard (one a woman, Dr Sally Ride) made a 6-day flight and had to land in California rather than the scheduled Florida because of fog and moisture in the latter which might have damaged the insulating tiles. The 146 h space trip is said to have achieved more than any previous shuttle mission in delivering two satellites, running experiments and operating a robot arm capable of retrieving satellites. This was shown to be feasible when a West German-built Shuttle PAllet Satellite (SPAS) was released and retrieved successfully. The next mission of the Shuttle was scheduled for August 1983 and in fact the schedule envisages approximately 15 flights by 1985.

Turning to already performed specific tasks using LANDSAT imagery, a significant dam inventory may be mentioned. After the Teton Dam, Idaho, and Töccoa Dam, Georgia, failures of 1976 and 1977, the then American President Jimmy Carter ordered an inventory of all sizeable dams and the job was assigned to the US Corps of Engineers. Time and funding did not permit a complete US aerial survey and States records were incomplete in a number of cases. Hence LANDSAT was utilized. NASA Houston wrote a computer programme such that, when surface water was encountered on the magnetic tape holding LANDSAT information, the computer printed out an 'X' at that location. The conventional approach listed about 50 000 dams in the USA, but the LANDSAT data identified about 10% more than this. Not to be over-enthusiastic, however, it must be added that because LANDSATS 1, 2 and 3 split up the Earth's surface into pixels of approximately 80 m^2 (approximately 1·1 acres) and treat each pixel as a point source of data, only impoundments larger than 10 acres can be identified by them. This led Harry Horton of the US Corps of Engineers to state that in Illinois, only half of the reservoirs known to exist were identified by LANDSAT at that time. However LANDSAT 4 and its successors will improve on this record considerably. This will go a long way towards answering the opinion of Godfrey[24] that LANDSAT as it then stood is not a panacea for mapping and resources inventorying problems. He indicated two major limitations, namely:

(a) Cloud concealment of the Earth's surface. This will be overcome.
(b) Inadequate resolution because the first generation was designed for an overview of the Earth rather than to provide fine details. This has been overcome by LANDSAT D.

No doubt other difficulties will be overcome and these include the US Geological Survey attempt to determine the locations of tree and shrub species in the Great Dismal Swamp of Virginia, an area of about 600 km^2 from the unsatisfactory results of which it was concluded that many wetland associations are too closely related spectrally to be separated accurately. Even with the first generation LANDSAT it was possible in Spokane, Washington, for the County Planning Department to use its data in preparing a land cover map quicker, cheaper and perhaps more accurately than would have been feasible in any other manner.

There is in the USA an EPA-mandated regional water quality management programme, the so-called '208'. This is named after the regulation under which non-point source pollution is to be controlled. The 208 programme is to some extent a land use planning measure aimed at reducing pollution such as silt runoff from bare soil construction areas or runoff from impervious city streets. In Washington DC, one of the goals is to improve the water quality of the Potomac River estuary which is tidal despite the distance from the Atlantic Ocean. The disadvantage of this is that polluted water flowing in from tributaries is not disposed of to the ocean, hence the pollution is more expensive to control. In the regional 208 programme, the Washington DC Council of Governments used LANDSAT in order to identify land cover types. The data for urbanized areas already available was better than LANDSAT's imagery, but for the rural areas where high resolution data were not necessary, LANDSAT proved to be valuable.

In fact this satellite system can be utilized to show rural land cover types from which can be predicted land impermeability which then permits the estimation of runoff rates. Consequently it has been possible to insert LANDSAT data into hydrological models in which pollution washoff is predictable. The model in question actually predicts peak flows for given storms. As regards hydrological modelling, the following observations should be made:

(a) Prediction accuracy using LANDSAT data appears to be good when compared with estimates of appropriate parameters such as runoff made by conventional means.
(b) As regards costs, in small basins it is cheaper to obtain land use data by such conventional means, but on a larger scale LANDSAT is superior. In fact for large scale basins, LANDSAT data appear to be so much more economical than their conventional equivalents that their utilization might reduce the cost of hydrological modelling quite considerably.

In view of the above discussion it can be seen that even first generation LANDSAT satellites have been useful in a number of projects and their imagery enables much information to be accumulated regarding regional geology. Actually this has been used to aid in siting some nuclear power plants by checking large areas for potential faults in lineaments. If such are found it is feasible to assess whether any earthquakes may arise in connection with them and so determine whether siting such a plant in the particular locality is advisable or not.

Earlier, the Stereosat was mentioned and if this is eventually launched it would be absolutely invaluable for geologists. Its sensors would look downwards, forwards and backwards. A stereo picture would be produced by matching two views of the same place on Earth taken some minutes apart, i.e. when the satellite would be at slightly different positions. Its resolution is envisaged as much greater than the first generation LANDSATs, in fact by a factor of 5 at least. Stereosat's stereo characteristics would assist in locating lineaments (often fault-associated) and topographical features as well as permitting 1:100 000 scale topographic maps of the Earth to be made. Such attributes would make the Stereosat of the highest importance in the developing countries which often lack topographic maps on this or sometimes any scale. Work of importance on this satellite was done at the Jet Propulsion Lab at the California Institute of Technology. There can be no doubt that the launching of this satellite with its proposed resolution of 15 m pixels would be highly desirable. This view is held by Paul Maughan of Comsat General, the private communications satellite company which has offered to finance, build and operate Stereosat by 1984 if the US Federal Government agrees to buy remote sensing data acquired by it.

It is interesting to note that satellite construction is becoming a competitive venture in the pursuit of which it appears that few holds are barred. Thus the weekly magazine *Newsweek* in its issue of 3 May 1982 reported that NASA officials were accusing France of employing 'questionable' tactics in order to induce operators of communications satellites to use the French-built Ariane rocket instead of the US Space Shuttle.[25] NASA stated that the French obtained customers by exaggerating the possible delays which may arise in the Shuttle schedule and also the possibility that customers paying for its services may be 'bumped' by US military payloads. This may have originated from the decision of Western Union, General Telephone and Electronics and Southern Pacific to switch five satellites to the French rocket. NASA added that the French are using trade leverage in order to pressurize Colombia and Brazil into launching their proposed satellites into orbit on Ariane.

10.6. USES OF DATA IN CONSTRUCTION

Although remote sensing is applied to construction today, there is no doubt that its future value will be immensely greater than at present. However some current instances are now cited.

(i) Site locating for a new town in the Jizan region of Saudi Arabia has been outlined by Landry.[26] Geotechnical mapping was effected and 'maps of factors' were produced and then synthesized into a 1:60 000 map over a region 100 km in diameter. The work was necessitated because of damage to the existing capital caused by deep salt dome stress. Remote sensing could have been very useful in this arid area.

(ii) In landform evaluation for construction purposes in South Africa, Howland[27] states that the relevant mapping was best effected from stereo-aerial photography at a suitable scale.

(iii) As regards the problems involved in the preparation of geotechnical maps at a scale of 1:25 000, Lopez Prado and Pena Pinto[28] have indicated the importance of aerial photography in prior studies.

Melnikov[29] has referred to the importance of aerial photoplans in the preliminary stages of effecting a national engineering geological survey in the USSR. The scales involved there were 1:100 000 to 1:500 000.

Aerial photography was also utilized in the engineering geology mapping of the west Carpathian landslide area as recorded by Malgot and Mahr.[30]

(iv) The utilization of terrestrial photogrammetry in a very rugged topographical area has been indicated by Venti and others.[31] This was in connection with a methodological proposal for an engineering geomorphological map relevant to forecasting rockfalls in the Alps.

(v) In mapping the after-effects of disastrous earthquakes, where it is feasible immediately after the seismic event, aerovisual land survey is desirable. Later an aerial photographic survey should be carried out prior to a ground investigation with detailed seismogeological mapping according to Solonenko.[32] These facilitate the estimation of seismic hazard for engineering constructions both from supposed isoseismal fields and possible types of residual and dynamic seismogenic deformations and related accompanying phenomena.

(vi) In Yugoslavia an appropriate methodology has been applied to the engineering geology mapping of folded mountain regions, the Outer Dinarides, and is applicable to other areas. The Outer Dinarides occupy an orogenic (mountain-building) region of the hinterland of the Yugoslavian coastline distinguished by complex and very variable engineering geological conditions arising from the dynamics of geological processes. Ivanovic[33] has indicated that the appropriate mapping scale is 1:25 000 and among the new methods proposed for the task has mentioned the phototheodolite and scanogram (satellite photo) analysis.

(vii) In India geotechnical mapping for river valley projects in the west utilizes aerial photography and also LANDSAT imageries. The delineation of lineaments by photogeological mapping is an important advantage of this technique, referred to by Srinavasan and Shenoi.[34]

Many other instances of the practical application of remote sensing to various problems of construction could be cited.

The basic objective of all the work is to produce useful engineering geology maps of value to the engineer; Radbruch-Hall *et al.*[35] have provided a valuable insight into the possibilities. They took conventionally derived data for the conterminous USA to create a base map at a scale of 1:7 500 000. Thereafter areas having geological and topographical conditions constituting constraints to surface construction were compiled using computer techniques. These facilitated satisfactory presentation and gave a method utilizable for rapid and economical black and white or colour comparison presentations or overlays of numerous maps of other data. The comparisons are made with digital imaging processing techniques originally developed for use with the multispectral images returned by both Earth-orbiting and interplanetary spacecraft. Much of the slow drafting and photography for the publication of maps can be eliminated or automated if these are digitized instead of drafted. A set of algorithms and software for the displaying of digital map data was developed using an Electrak (Trak 100) table digitizer, a PDP 11-45 computer with 28 K words core storage and an Optronics P-1500 Photowrite film output device.

The first stage in the making of the digital map necessitates outlining of the map units in digital form. Data encoded with the table digitizer are converted to raster formation (a regular array of numbers in rows and columns) for all the manipulations and displays. Software includes a programme specifically for the assignment of a specific attribute code to all image elements enclosed by lines, e.g. the area of a stated geological area, so that the resultant data base, e.g. a geological map, can be manipulated with a digital image-processing system. Of course one map may be added to another and map units can be encoded in terms of 256 'shades of grey'. A programme FILLIN completely fills each area with the value code assigned to it. There is also a second programme COLOR which assigns specific colours to density numbers and permits the creation of halftone colour separation plates by the Photowrite output device. A programme ZIP writes patterns within windows created by FILLIN. The set of engineering geology maps made by the application of the approach includes the following:

(i) A map showing, by shades of grey, the intensity of certain geological conditions which might present constraints to construction with a base comprising geological information related to the physical properties of soils and rocks.
(ii) A map showing which hazards exist at any point on the map.

(iii) A map showing where construction or land development may produce a worsening of existing hazards and/or secondary environmental effects. As noted, the shades of grey or patterns for all maps are computer-generated and the geological base compiled in conventional fashion. No doubt the up-coming improved LANDSAT D, etc., generation-derived data can be integrated to improve the resultant maps.

Figure 10.2 shows the LANDSAT 1 earth resources satellite (ERTS) launched in July 1972 (the LANDSATs 2 and 3 satellites being almost identical with this first generation type) with MSS. An allied technique which was developed in the 1950s for military purposes is the thermographic infra-red linescanning which has been used in aircraft and helicopters, in the latter case in South Africa as Warwick and others[36] have indicated. They mentioned the properties given in Table 10.2 for soils and rocks at 20°C using a Texas Instruments Model RS25 .

10.7. CONCLUDING REMARKS

The discussion above indicates the tremendous potential of the NASA and other instrumentation to engineering geology as well as in many other fields. Starting with Sputnik 1 (USSR, 1957) which orbited for three weeks, the sequence has continued with Explorer 1 (US, 1958) which discovered the van Allen belts, TIROS 1 (US, 1960) which transmitted images of terrestrial weather, Echo 1 (US, 1960), Telstar 1 (US, 1962), Syncom 2 (US, 1963), OAO-2 (US, 1968), Vela 6 (US, 1970), Intelsat IV (1971), ATS-6 (US, 1974), LAGEOS (LAser GEOdynamics Satellite; US, 1976), IUE (International Ultraviolet Explorer; US, UK, European Space Agency, 1978), OSCAR 8 (Orbiting Satellite Carrying Amateur Radio; US, 1978), LANDSAT 4 (US, 1982), TDRS (Tracking and Data Relay Satellite; US, 1983), Spacelab 1 (US, 1983), Space Telescope (US, for 1986), COBE (COsmic Background Explorer; US, for 1987), GPS (Global Positioning System or NavStar; US, for 1988), UARS (Upper Atmosphere Research Satellite; US, for 1989), OPEN (Origins of Plasmas in the Earth's Neighbourhood; US, for 1989) and no doubt will continue *ad infinitum*. Among the mighty and pertinent achievements to data have been those in the southern Egypt, northern Sudan area alluded to earlier as well as locating many unmapped lakes and geological features such as crustal fractures, discovering oil in the Sudan, tin in

TABLE 10.2

THERMAL PROPERTIES OF GEOLOGICAL MATERIALS AND WATER AT 20°C

Material	*Thermal conductivity* (*cal cm*$^{-1}$ *s*$^{-1}$°*C*$^{-1}$)	*Density* (*g cm*$^{-3}$)	*Thermal capacity,* (*cal g*$^{-1}$ °*C*$^{-1}$)	*Thermal diffusivity* (*cm*^{2} *s*$^{-1}$)	*Thermal inertia* *P*(*cal cm*$^{-2}$ *s*$^{-\frac{1}{2}}$°*C*$^{-1}$)	$\frac{1}{P}$[a]
1. Basalt	0·005 0	2·8	0·20	0·009	0·053	19
2. Moist clay soil	0·003 0	1·7	0·35	0·005	0·042	24
3. Dolomite	0·012 0	2·6	0·18	0·026	0·075	13
4. Gabbro	0·006 0	3·0	0·17	0·012	0·055	18
5. Granite	0·007 5 0·006 5	2·6	0·16	0·016	0·052	19
6. Gravel	0·003 0	2·0	0·18	0·008	0·033	30
7. Limestone	0·004 8	2·5	0·17	0·011	0·045	22
8. Marble	0·005 5	2·7	0·21	0·010	0·056	18
9. Obsidian	0·003 0	2·4	0·17	0·007	0·035	29
10. Peridotite	0·011 0	3·2	0·20	0·017	0·084	12
11. Loose pumice	0·000 6	1·0	0·16	0·004	0·009	111
12. Quartzite	0·012 0	2·7	0·17	0·026	0·074	14
13. Rhyolite	0·005 5	2·5	0·16	0·014	0·047	21
14. Sandy gravel	0·006 0	2·1	0·20	0·014	0·050	20
15. Sandy soil	0·001 4	1·8	0·24	0·003	0·024	42
16. Sandstone	0·012 0 0·006 2	2·5	0·19	0·013	0·054	19
17. Serpentine	0·006 3 0·007 2	2·4	0·23	0·013	0·063	16
18. Shale	0·004 2 0·003 0	2·3	0·17	0·008	0·034	29
19. Slate	0·005 0	2·8	0·17	0·011	0·049	20
20. Syenite	0·007 7 0·004 4	2·2	0·23	0·009	0·047	21
21. Welded tuff	0·002 8	1·8	0·20	0·008	0·032	31
22. Water	0·001 3	1·0	1·01	0·001	0·037	27

[a] Often utilized as the thermal inertia value. Thermal inertia has an important effect on the imagery of thermal infrared line scanning as rocks with high thermal inertia, quartz or dolomite for example, are relatively cool in the day and warm at night. Those soils and rocks with low thermal inertia such as gravels or shales are warm in the day and cool at night, i.e. the variation in temperature of materials with high thermal inertia during the diurnal cycle is far less than those with low thermal inertia. Thermal inertia is directly related to the density of a material, rising with this parameter.

The above data are derived from ref. 5.

Brazil and uranium in Australia, finding a new islet off the Atlantic coast of Canada (christened Landsat island) as well as an uncharted reef in the Indian Ocean and in addition mapping routes for railways, pipelines and electric power rights-of-way (these latter relating directly to engineering geology and projects involving its applications).

REFERENCES

1. Dearman, W. R. and Fookes, P. G., 1974. Engineering geology mapping for civil engineering practice in the Kingdom. *Q. J. Eng. Geol.*, **7,** 223–56.
2. UNESCO–IAEG, 1976. *Engineering Geology maps. A Guide for Their Preparation*, UNESCO Press, Paris.
3. Matula, M., 1979. Regional engineering geological evaluation for planning purposes. *Int. Assn Eng. Geol., Bull.*, **19,** 18–24.
4. Rodriguez Ortiz, J. M. and Prieto, C., 1979. A proposal for quantitative terrain evaluation for highway construction. *Q. J. Eng. Geol.*, **12,** 139–46.
5. Reeves, R. G. (Ed.), 1975. *Manual of Remote Sensing*, Vols I, II. The Am. Soc. of Photography, Falls Church, Va, USA.
6. Hempenius, S. A., 1969. *Wall Chart of Image Formation Techniques for Remote Sensing from a Moving Platform*, International Institute for Aerial Survey and Earth Sciences, ITC, Enschede, The Netherlands.
7. Hempenius, S. A., 1976. Critical review of the status of remote sensing. *Bildmessung und Luftbildwesen*, **44**(1), 29–42.
8. Blachut, T. J., 1971. *Mapping and Photo-interpretation System Based on StereoOrthophotos*, Nat. Res. Counc., Canada.
9. Visser, J., Bouw, T., Grabmaier, K., Graham, R., Howard, E., Kunji, B., Lorenz, R. and Van Zuylen, L., 1980. Orthophotos: production and application. *ITC Jour.*, **4** (Special Issue: *Proceedings International Institute for Aerial Survey and Earth Sciences, Post-Congress Seminar, Hamburg*, following *XIVth Congress of the International Society for Photogrammetry*), pp. 638–59.
10. Eden, I. A., 1964. The art of taking air photographs. *Photogrammetric Record*, April.
11. Finsterwalder, R., 1979. Zur Genauigkeit der Kartierung mittels Stereo-orthophotos. *Bildmessung und Luftbildwesen*, 47.
12. Rengers, N. and Sosters, R., 1980. Regional engineering geological mapping from aerial photographs. *IAEG Bull.*, **21,** 103–11.
13. Doyle, F. J., 1979. A large format camera for the space shuttle. *Photogrammetric Engineering*, 1.
14. Reuss, D. H., 1974. Numerisch-photogrammetrische Kluftmessung. *Rock Mechs*, Supp. 3, 5–15.
15. Verstappen, H. Th., 1977. *Remote Sensing in Geomorphology*. Elsevier Scientific Publishing Company, Amsterdam.
16. Cluff, L. S. and Brogan, G. E., 1974. Investigation and evaluation of fault activity in the USA. *Proc. 2nd International Congress, Int. Soc. Eng. Geol.*, Vol. I, Theme II, Sao Paolo, Brazil.
17. Rengers, N., 1979. Remote sensing for engineering geology: possibilities and limitations. *ITC J.*, **1,** 46–67.
18. Donker, N. H. W. and Mulder, N. J., 1977. Analysis of MSS digital imagery with the aid of principal component transfer. *ITC J.*, **3,** 434–66.
19. Donker, N. H. W. and Meijerink, A. M. J., 1977. Digital processing of LANDSAT imagery to produce a maximum impression of terrain ruggedness. *ITC J.*, **4,** 683–704.

20. Beaumont, T. E., 1979. Remote sensing for the location and mapping of engineering construction materials in developing countries. *Q. J. Eng. Geol.*, **12,** 147–58.
21. Mekel, J. F. M., 1972. The geological interpretation of radar images. In: *ITC Textbook of Photo-interpretation*, ITC, Enschede, The Netherlands, Chapter VIII.
22. Leberl, F., 1978. Current status and perspectives of active microwave imaging for geoscience application. *ITC J.*, **1,** 167–90.
23. Doyle, F. J., 1981. Satellite systems for cartography. *ITC J.*, **4,** 169–80.
24. Godfrey, K. A., 1979. What future for remote sensing from space? *Civil Engrg, ASCE*, 62–5.
25. Anon., 1982. Satellite sales war. *Newsweek*, 3 May, 9.
26. Landry, J., 1979. Recherche de sites favourables a l'implantation d'une ville nouvelle dans la region de Jizan (Arabie Saoudite). *IAEG Bull.*, **19,** 57–61.
27. Howland, A. F., 1979. Landform evaluation as a method of road construction investigation in South Africa. *IAEG Bull.*, **19,** 25–30.
28. Lopez Prado, J. and Pena Pinto, J. L., 1979. Problems involved in the preparation of geotechnical maps at a scale of 1:25 000. *IAEG Bull.*, **19,** 84–7.
29. Melnikov, E. S., 1979. The main principles of procedure for the national engineering geological survey in the USSR. *IAEG Bull.*, **19,** 93–5.
30. Malgot, J. and Mahr, T., 1979. Engineering geological mapping of the west Carpathian landslide areas. *IAEG Bull.*, **19,** 116–21.
31. Venti, V., Silvano, S. and Spagna, V., 1979. Methodological proposal for an engineering geomorphological map. Forecasting rockfalls in the Alps. *IAEG Bull.*, **19,** 134–8.
32. Solonenko, V. P., 1979. Mapping the after-effects of disastrous earthquakes and estimation of hazard for engineering construction. *IAEG Bull.*, **19,** 138–42.
33. Ivanovic, S., 1979. On the methodology of engineering geological mapping in folded mountain regions as applied to the Outer Dinarides in Yugoslavia. *IAEG Bull.*, **19,** 142–4.
34. Srinavasan, P. B. and Shenoi, R. S., 1979. Concepts of geotechnical mapping for river valley projects in western India. *IAEG Bull.*, **19,** 226–33.
35. Radbruch-Hall, D. H., Edwards, K. and Batson, R. M., 1979. Experimental engineering geological maps of the conterminous United States prepared using computer techniques. *IAEG Bull.*, **19,** 358–63.
36. Warwick, D., Hartopp, P. G. and Viljoen, R. P., 1979. Application of the thermal infra-red line scanning technique to engineering geological mapping in South Africa. *Q. J. Eng. Geol.*, **112,** 159–79.

Author Index

Numbers in italic type indicate those pages on which references are given in full.

Subject Index

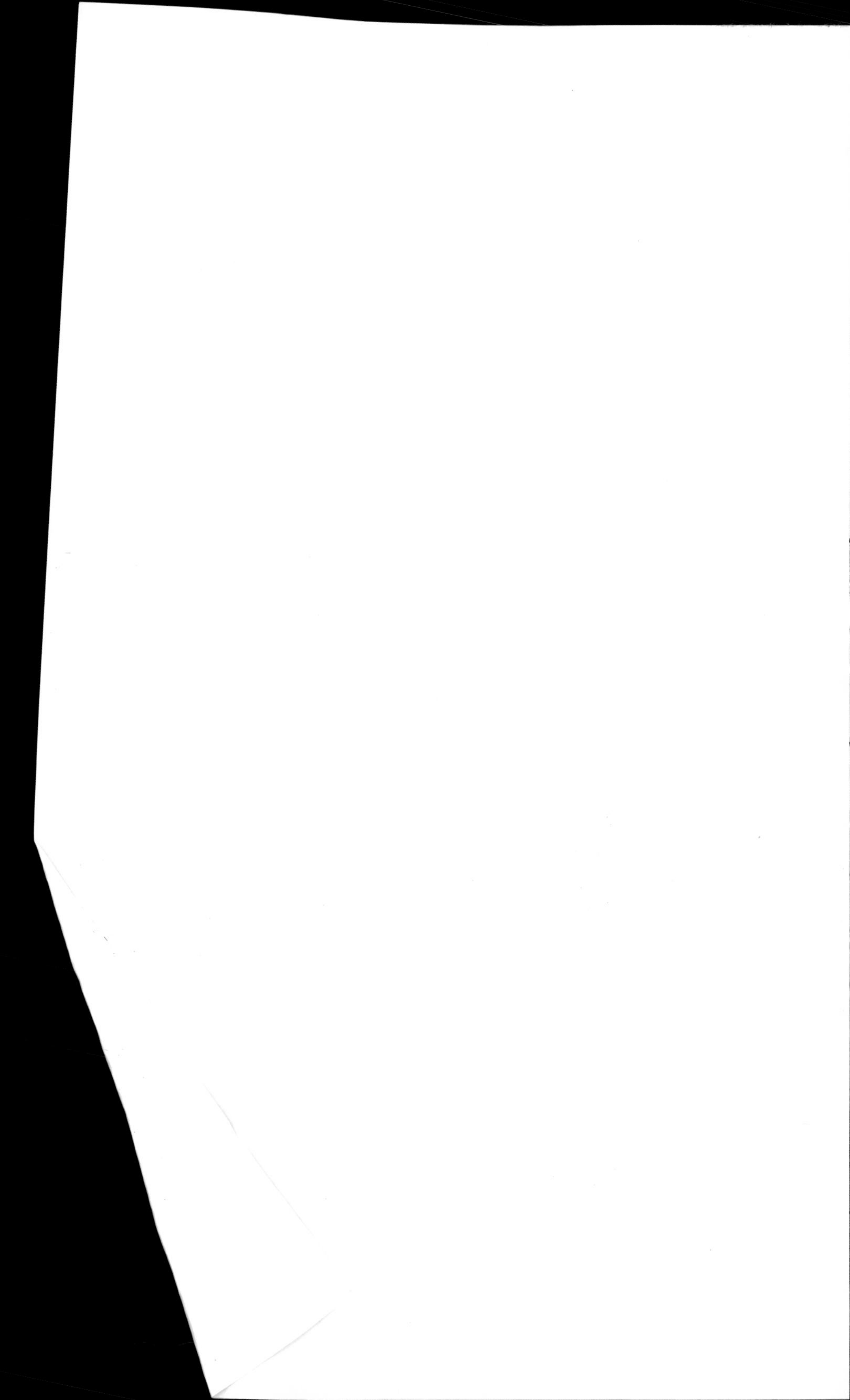

The National Poetry Series was established in 1978 to ensure the publication of five collections of poetry annually through five participating publishers. Publication is funded by the Lannan Foundation, the Amazon Literary Partnership, the Poetry Foundation, Barnes and Noble, the Gettinger Family Foundation, Bruce Gibney, HarperCollins Publishers, Stephen King, Newman's Own Foundation, News Corp, Anna and Olafur Olafsson, the O. R. Foundation, Laura and Robert Sillerman, Elise and Steven Trulaske, Amy R. Tan and Louis De Mattei, the PG Family Foundation, the Betsy Community Fund, and the Board of NPS. For a complete listing of generous contributors to the National Poetry Series, please visit www.nationalpoetryseries.org.

2016 COMPETITION WINNERS

I Know Your Kind
by William Brewer of Brooklyn, New York
Chosen by Ada Limon for Milkweed

For Want of Water
by Sasha Pimentel of El Paso, Texas
Chosen by Gregory Pardlo for Beacon Press

Civil Twilight
by Jeffrey Schultz of Los Angeles, California
Chosen by David St. John for Ecco

MADNESS
by Sam Sax of Austin, Texas
Chosen by Terrance Hayes for Penguin Books

Thaw
Chelsea Dingman of Tampa, Florida
Chosen by Allison Joseph for the University of Georgia Press

THAW

POEMS BY CHELSEA DINGMAN

The University of Georgia Press *Athens*

Published by the University of Georgia Press
Athens, Georgia 30602
www.ugapress.org

Designed by Erin Kirk New
Set in New Caledonia
Printed and bound by Thomson-Shore, Inc.
The paper in this book meets the guidelines for permanence and durability of the Committee on Production Guidelines for Book Longevity of the Council on Library Resources.

Most University of Georgia Press titles are available from popular e-book vendors.

Printed in the United States of America
21 20 19 18 17 P 5 4 3 2 1

Library of Congress Cataloging-in-Publication Data

Names: Dingman, Chelsea, author.
Title: Thaw : poems / by Chelsea Dingman.
Description: Athens : The University of Georgia Press, [2017] | Series: The National Poetry Series
Identifiers: LCCN 2016056639 | ISBN 9780820351315 (pbk. : alk. paper) | ISBN 9780820351308 (e-book)
Subjects: LCSH: Families—Poetry.
Classification: LCC PR9199.4.D565 A6 2017 | DDC 811/.6—dc23
LC record available at https://lccn.loc.gov/2016056639

for my father

This is how it continues:

The cold, the snow, the slight trembling in your hands.

—Daniel Simko, "Homage to George Trakl"

Contents

Acknowledgments

Grateful acknowledgments to the following editors and the presses where these poems first appeared:

The Adroit Journal: "Letters from a War"
burntdistrict: "The Suicide"
Dialogist: "For My Son: A Forest of Stars"
The Fourth River: "After the Tornado: Summer 1989"
Carolina Quarterly: "Amid Reports of a Blizzard & Black Ice on the Coquihalla Highway" (formerly, "Clan of Fatherless Children") and "Hands, I've Had"
Grist Journal: "Hunting" and "The Gulf"
The MacGuffin: "Hiraeth"
Milk Journal: "Sunset" and "When My Mother and I Speak about the Weather"
Mom Egg Review: "Dog Days"
Raleigh Review: "The Windsurfer" and "Any Other Sun"
Red Sky: Poetry on the Global Epidemic of Violence against Women: "Sirens"
RHINO Poetry Journal: "Little Hell"
Slipstream: "Ancestry" and "Hungry Season"
So to Speak: "Current" and "Elegy for My Child"
South Dakota Review, "Daughter, Released"
Sou'wester: "Sirens," "Woman, Disarmed," and "Burning"
Stone Highway Review: "Billy"
Vermillion Literary Journal: "From a Morgue in Minnesota"
Yemassee: "Felled Pine"

I want to thank Hunter and Sawyer for being so patient, even though these words took me away from them while I was in the same room. Laine and Shelby, thank you for always making the trek across worlds to see me. I may be gone, but I'm never far. To my mother: though we don't always agree, thank you for what you've given me, for the years you struggled. You taught me the resilience of women.

Also, thank you to the amazing writers and mentors to whom I am indebted: John A. Nieves, not only for helping me to conceive of myself as a poet but for helping me turn this project into a book. Also, for always believing. I owe you everything. Jay Hopler: without your teaching and guidance, this book would never have been written. You made me the poet that I am. "Thank-you" is too small a word. Rita Ciresi, Ira Sukrungruang, Heather Sellers, John Henry Fleming, Karen Brown, and Jarod Rosello: thank you for everything. I was lucky to be part of something great. To my cohort at USF: thank you for giving me your eyes and your words.

I want to thank Allison Joseph for believing in this work, for her praise, and for making this wild dream a reality. I am also incredibly grateful to the National Poetry Series, its donors, and Beth Dial, who was a great support in this process. To the University of Georgia Press: thank you for making books, for making this book, and to each person I worked with, for this wonderful experience. I couldn't have asked for better.

To my husband, Chris: this was all your idea, and I'm making the most of it. Thank you for giving me this chance, for the endless laundry and grocery shopping that you've done while I've been away. You made this possible.

And to my father: thank you for the years we had, too few, but ingrained in me all the same. For a few minutes, each time these words are read, you are alive in the world.

1.

PROOF OF DISAPPEARING

Hunting (circa 1985)

What we grieve is
not how death can be
dispelled in a photo, or a dream
on our hip we carry
like a child. But a man's eyes,
blackened by the butt of a rifle.
Stars fading in the crosshairs
of the sun. A phantom
trigger, his finger
hooked through its heart.

At a glance, the blood
could belong to a deer, breath
escaping in the chill fall
air, just smoke.

Like the camera, our eyes fail
to see what falls outside
the frame—twisted limbs
like a bird's wings
broken on the ground. How a bullet
can enter so quietly as to leave
a skull almost intact. How,
afterwards,
a body glitters
like the cherry
still burning
in someone else's fingers.

From a Morgue in Minnesota

Ataxia Telanglectasia is a rare, neurodegenerative, autosomal recessive disease causing severe disability and appears in early childhood.—A-T Children's Project

We bury a child each year, sometimes
two, bones like cell bars, bland food

fit only for a baby's slick gums. Their mothers
wait years for a chest to flutter

closed, herringbone ribs
barely lining the canvas. False

light, a shaft, hangs children
in moving hands like miners

digging six feet down. Each child
withers—an anorexic who refuses

water, heart protesting
long after the body is lost

hope in our mouths. Crowning,
they wrap their hands in a fist. Then,

at age three, they can't. Bodies turn
inward, a flower's petals folding

regardless of the rain, tender
ground, the unwavering desire

that forces our eyes away.

Felled Pine

Behind black curtains, dogwood
blossoms scattered from seam to seam, I
asked after you, a secret

squirreled away like a moment
I can no longer see. Children
crying in the wet street, you pretended

to control the rain, a ruse
to ease our fear. I wonder
where you'll wake this morning, sickness

sticking to your skin. We didn't
drag you to the nearest bed, stay until the fever
broke. It's what I'd do if you were my child,

but you are the sibling
four years behind me, tiny scars
we carved into our arms like names

in wood, little hearts. I always thought
you'd age, rings on an old pine
in a churchyard in Revelstoke, soft

and rotting from its cuts, bark crumbling
beneath my fingers. But you lost yourself
in the sky, ruined by years

of rain and snow, dirty
hands on the saw before you fell.

Sirens

Say someone will come, but no
one comes. The echo

filters through buildings, palm fronds
feathering tile and stone,
                    yet I remember

only mountains, snow circling our house
like crows' wings. How I long for the metal

sting numbing my hands. I had
a mother then. I held the wind
in my throat like a song. A coal-

black sky, blue-lit by morning
that arrives too late, somewhere
north. I lay down on the bed to sirens

  like a loon's calls. Like a warning. The torn pocket
  of skin dangling where the blade exited

my mother's back. Each staple
making a new hole. Tonight,
no one comes

          with sounds that carry, with water
that will tumble from the sky. That starved town
in the distance is the gash

where she was torn from
breast plate to shoulder

blade. If I could unzip cold
skin, maybe I'd know
how to stop reaching

for snow, dark blue
mountains haloed by stars.

Little Hell

I can't breathe this
morning—my dead father, smoking
at the kitchen table, marred

by the scenery, trees
on the flatbed of his truck
going somewhere. He's

a terrible dream I had. Empty
rooms, dust in a corner. He
disappeared near Hope, ice

beneath his tires, and I
went south, to vultures
feeding on entrails

in the yard, dead
armadillos. I escaped
the snow, not its secrets. They follow,

whispering of mold
in the grass, corpses
of trees, naked and shivering.

Testimony of Hinges

This is how we stay: tongues
on fire, two people dancing
to our own screams. This is how

we dance: you, on the phone,
begging to come back to a house
I flee on foot. To the dagger

in my mother's smile. This is how
you love: fingers trawling the skin
under my shirt as I pretend to sleep. If not me,

who else do you skin? My father
sleeps in a body I can't touch. You are someone new
who abandons hair on our pillows, the day

turning us against ourselves. If I leave
now, no one will know
I broke my wrists to give you my hands,

sawed clean through the bone. I dreamt
new hands, pink-tipped fingers
to drag over the knobs of your spine. All I have left,

dear stepfather, is my mouth: a blade
I draw across the petals of your flesh, bruised
blood rising in ragged blooms.

Athanasia

This body is in tatters, worn
by winter's long argument
with the snow. Perhaps it's time

          to speak of rusty hinges:
your eyes closed while you sleep,
hands steepled, a pale mouth's

flattened horizon. Violet sheets
swell the space
you once lay next to me

in the dark. It comes to this: a shudder
in your chest, my breath
seeking a warm place where

there is only a body
of snow. The words I hold back,
you can't have now. But on this skin's scroll,

          we wrote with our hands, words
that can't be ruined by cold,
empty streets, a second hand. Perhaps we are gods

now, black ink spelling us out—
new blood threading the bodies

of the damned.

Elegy for Empty Rooms

There will be no more sons. Bodies
twisted in the shade
of a canopied crib. I know

the shape of their blood, the long
wait for their faces
to materialize. What exists now

behind a closed door? In me, a hollow is
ice-cold. I exist to exist, a street
covered in snow. I dreamt

more faces at dinner, the table
stretched by wooden wings. Instead, I endure
parlor games: at a party, a chain

dangling over my navel, pressed flat
as a penny. How the chain spins, then
stills. How stillness tells me

I'll only know spring's passing
as a field somewhere
the skies are all red.

Dog Days

I can't stop the dawn
to catch my breath. You expand
while my eyes rest, as if you're taking

on water. I'm unsure of your best
interests, so vast, your childhood
is a foreign country, this language

I struggle to master. An accident
is in the street up ahead
unless each mile is driven

with careful attention
to the signs, an eye
on the highway behind. I drive

to follow you, it seems, rather
than little steps I remember
at my heels. My burden

is a landmark, a turn.
Your life becomes the road
away. The quiet sores I carry.

Amid Reports of a Blizzard & Black Ice on the Coquihalla Highway

The forecaster says it will snow
and I tell my son *yes* when he asks
for waffles, peanut butter. I can't mention
your truck in a gully
somewhere. How we used to sling songs
into the night, windows
rolled down. I didn't know you were close
to death. That my cupped hands waited
only for what falls. I want
to tell you everything. How my greatest fear is leaving
my child behind. How, afterward, I ran from the world
like snow from the sky. They say snow is harmless,
yet I know what it is to be lost
when touched. There is so much I want to say to you,
but I choose to stay quiet like the stars
amidst the sky's falling. I choose to live.
The weatherman keeps talking and I wait
to hear him say there's been a crash. For you
to sprout wings from your back
on the side of a road. I wait
in this room with my son. The world
outside, obliterated. Only we are left, less
whole. The walls are winged
beasts that fold themselves over us.
You are the snow.

After the Tornado, Summer 1989

Like a branch felled
from the oak in our front yard, hunger harkens
my spine, stretched

  toward the sky. For a mother, I'd kneel
all day. I'd go back: the trailer
park, empty hands, palsied

bodies waiting for something
else. I wait when there is no one
to tell me how to lay down again,

to sleep. I scratch the mud-
soaked earth with a stick until it bleeds
water. Until it covers every inch of skin

I can muster. Some women
carry children over mountains
on their backs. I do the walking.

I learn how to fall. Behind
closed eyes, I can't see
the sky's struggle, dust

made of the dead. Mothers, many times,
bear more than one heart-
beat. But, my mother's breath is

misplaced somewhere. The broken
leg on the last chair
left at the table, only I can know

I live like this windswept plain:
because the sky is done
                    starving.

After the Accident

You told me you'd be
here. The restaurant fills

and refills, yet dozens of faces
are not yours. Heat from the kitchen

reaches glass doors
where I stand waiting, a fog

so thick it reminds me of the forest
when we used to sneak out

at night. We couldn't see anything: the moon,
pine needles like thousands of tiny fingers.

As I wait, I wonder if you were playing
when you fell. The week before

last, we sat on the roof amidst the stars
as if we belonged to the sky,

bright and unbidden. You wanted to feel
the night become otherwise. Please know, little

brother: I watched over you like a reflection, skin
sealed by old bruises. Still—they say

your palms were burnt, a star
caught in your throat. I came here, tonight,

to keep a promise. To tell you
I can no longer see

the pines, a stitch
of moon through their fingers.

To the God I Prayed to as a Child:

are you working against us
children's eyes and hands
unfurling like petalled wings
little bones bare
flanked in blue flannel
plastic bands printed with their births
they curl on their sides
they wither—and you stay away
from wet lips spitting your name
ribs petalling closed around a breath
the sun that rises anyway
is this silence a sign? if it were crimson
I'd know to stop speaking
does their stalled breath make an enemy
of you? I'm tired of putting my faith in
the skin's pinked surface a ceiling
of sky that never fails to surface
the clouds' dirty gutter after a murderous rain
I won't lie I've had enough waiting
while you push against us
without showing your hands [sigh] I've had
enough of your petty tricks   I'll save myself
for someone else tonight

Erebus

Sometimes you slip inside
my skin without an invitation.
Maybe it's my own need

to forget the sounds of night
pulling away, your fingers
shining like precious stones

on a distant sky. Some lives are like this:
four heads at the dinner table,
a fable before bed. Why now,

do I long for someone new? Is it because
you're like the dark tide, manipulating
sand? My chest stings shut, collecting our bones

too great a task for the earth, silver
night. How do I trust myself to go underground
with you, a region railed by wind? I want different seasons,

yet I favour fall's dark days, lost leaves,
the trees' bare bodies untouched
by any other sun. Leave behind the chaos

you come from, a fierce wind
hugging your ribs. We made someone
once, our bodies sewn together

only to part. The mouth
of a god bellowing
in the bed between us.

Onshore Years Later

Water wells below all surfaces and, yet,
everything on the surface
looks whole. My father's lost

skin, an olive veil I pull
over my body when I want to
remember. His voice

stains my tongue. I want to know how
I sound, lips pressed against an ear
at dusk, drunk

with sun-fall. A shadow
drags its knuckles along the Gulf. This sea
is not the sea I was born to. Does water

return or escape? From a pier, swimmers
spill themselves into the sea
between breaks. I hold what remains

of my father, grainy between my fingers,
and imagine his blue eyes, burning. In my fists,
I begin to unfasten ashes

from my skin, salt
from the sea, the grave
way we drown.

On Nights When I Am a Mother

Lord, make these black waters blaze,
bolted down, a bright beam
to which I'm bound. Make me an ocean

to empty into, limbs and wombs and ash-
black plumes speaking of night
somewhere else.

I can't own this world, clenched
in your closed fists. Can you

hear me on my knees? It that you: rain
rising in the gutters, the moon's scaly skinned head,
the cough caught in my son's throat?

I encircle his body, lungs racked
with fluid, wondering which of us you'll save
from drowning.

The hush of your hands buries me
in sleep's calm dark. Will you come for him
as you came for my father, before

I'm ready? Wrapped like petals around
his slim shadow, I hold tightly, as if you won't be able to

pry him away with a hammer and a crowbar.
With a gasp of your blazing breath.

Prayer of the Wolf

Forgive these fangs. How I kneel
for nothing anymore. Retreating rock ruins
my knees, elbows bearing the scarlet bloom
of roses running ramshackle over
a wooded hillside. I have only scars. Above,
the sky is so clear. Should I have to beg
to remain that way? Your name, the growl
in my throat, I confuse you for a friend in the dark
and gentle snow. Snow savages a man
somewhere in the distance, a last twitch
as his cheek hitches itself to unfailing
eyes. The sky still so clear. How quickly
we become something else. Is it wrong
not to kneel when I know the snow
will fall anyway? Behind the walls
of every wood, a child. This girl is no more
a little girl as winter kicks a hole in the sky. I'd howl
your name, but it sounds like a saw
to the skinned pines. Please tell me
where you go when cows low
in the fields, bellies dragging over tall grass,
black soil. Night falls and you take the moon
out of its tin cage, only to parade it
before us. I beg you—the hours
want nothing except to reclaim themselves. I
have hunted you like the white
moon, yet I'm lost. I hear bleating,
but it leads me further away
from the sky's clenched fist. I long for
wool, rubbing against my limbs, bloody as the slain
body in the snow. What can I hope to become
when a lone lamb can't be salvaged
and hunger is all I have?

Sunset

Do you forget the roar
of tiny lungs unsettling
your sleep? How I was once, a sparrow
lost in the yard? Or was it safer
for you to let him
open new seams, your scars still
so raw? I used to wake
and run all of the faucets,
as if we lived inside the falls. The house, lit
like a constellation. Was I ever yours
after that? There is such violence
in the sunset. You wanted me
to beg, but I held my breath as I wanted
to be held. I should have said I wanted sky
to claim the stars. That I understood
to be a good girl I had to lie
low like the aging hardwood floor. But, when I return,
it's to stand in the soft light
on the sun-porch. To admit the sunsets
are drawn by my hand.

Woman, Disarmed

Above flat grasses, red
skies like a torn dress. The streets pass
without returning. Unknown,
now, who will call you
mother & sister &
daughter before
you're the expanse of milk-
weed & memory under our tongues? Tell God
you prefer green. How it brings
out your eyes. Is your body
the prayer, skin
peeled back to reveal the pearls of
your knuckles? Like children,
I want to line your promises
along a wall, take
measure. Where you go,
fields may be fallow. Yellowed
and shorn. But the bow
you left to me.
Abandoned building, I look through
shattered windows
to learn what life
you gave up. I remember
night, taunting me
into your room. The light I crawled toward
like a war. Is this mercy?
Reckless world, held captive
inside a snow
globe. Our beauty,
in the shadows:

the drawers we kept
shut, moving shapes
not yet named. I don't want to know
how your voice fails, the shadow
that will hold you
apart.

Hungry Season

I asked for a demonstration
of faith, not a ring or ceremony

built on strange words. A ritual
that only we would know—song

written on a sleeve, body
glowing against the late afternoon, silky

with sweat, shutting out noise
rising from the street. I wanted

a promise to carry like a stone, smooth
between my thumb and forefinger, days

when you fell away, and I couldn't feel
the heat on my skin. We are

the secret I hunger: old initials
on a sidewalk after the sun

sealed our prints together, proof
your hands once ached mine.

Daughter, Released

You empty from my body like a song
on the radio that ends
right after I find the station. Then,
a different song. I dream you
in a pink frothy dress, curls
wicked around your ears. I wake
and the house is silent,
save for the cries
behind every door. Is there no end
to the dark's wanting? There was once
a child, before snow melted
inside my bones. Before
I could say I wanted to be
a mother. How few seconds it takes
to be otherwise. Outside, snow
climbs metal siding. The sun, white
like a dandelion as it's dying. I try to whisper
Go into the wind, as if you will
listen. They tell me
you weren't ready, but I fear
they mean me. That I have to find
what comes from darkness
before anything good can stay.

Billy

For my father

You wear a little girl's shoes, laced
over the ankle, in the picture dated
1963, waiting for your father to come

home from the fields, already drunk,
three-year-old daughter under a makeshift
cross of sticks behind a tree swing. Lips

and teeth nasty brown, tobacco tucked
in his gum, you wanted to join him,
not knowing when he made you quit

school, you'd pay for the son before
you, stillborn, carrying the weight
of extra place settings, your mother's

arthritis tying her to a chair, watching
disease ebb at him, ninety pounds. He
gave his last nights to a dog, instead,

licking his cold bones to sleep.

Waiting for Winter's End

Why did you bring snow
to cover the pavement's grey
beaded coat? I drive past
lilies' petals petrified by cold
licking their smooth skin. You aren't

beyond the glass sky, a blue mountain's snow-
filled mouth, the slithering creek on its belly.

Can I get some direction? I call your name,
out of the car, snow crushing my crown,
wind threading water. Is that your fingers
dripping blood? Aren't we all
your children? Lost on this road:
low hanging sky, inked water
split by silver-shelled stones, the moon's
scarred face. O, how can I trust in
the things my eyes can't see?

Even in the dark, sun-spoiled glaciers empty
into a stream.

If the red sun refuses to rise soon,
I'll know you can't salvage any of us,
winter's dark hands at our throats.

Borderlands

You hop in your pickup, drive south
to Spokane, Seattle, places
you'll need a gun at camp. Not to hunt
deer or elk, but to pretend at being young

boys again. You don't stop
at the town you grew up in. You only know
how to want: a house somewhere
new, somewhere you can stretch

and storm. Aren't we also the road's
lure: three kids around the kitchen
table, swing set in the yard? You want
to leave long before you load the flatbed,

September's teeth in my mother's
knees. Which one of you understands
sacrifice? Her nights spent at a motel,
cleaning filthy sheets. You, driving

the TransCanada highway, black coffee
in hand, windows unrolled. But my mother
doesn't yet know she will be a widow,
raising us alone. By the phone, she waits

to hear about a house you've found
us, the day she'll get to own something
for the first time. You could leave her
before your truck swerves. She's still

young, daydreaming about a new dress
to wear out with her friends. Instead,
you don't have to grow older, see the mess
you left on the highway outside town.

Hit and Run

Ahead, taillights disappear
  like pieces of a red dress shirt. His body

lies to the left of twisted yellow lines,
  a god who has fallen
    too far from sky. You kneel, small

coin of your ear covering a mouth, smoky
  breath leaking into the night. Your son,
    framed by the window at your back,
      tells you not to go, not to

bend and break
  like waves around a beach. Not to bear the weight
    of another man's prayers.

    His black pupils.
  Your slowed breath.
    A single set of headlights
      shining. The sky and the ground: two black seas.

Traffic streams somewhere beyond the scene. Brakes squeal
      like the body that flew, not a bird with wings
  but a man fallen before you. Is it a comfort
      that his eyes can't look away? Flashing

lights and sirens crush the blacktop, a last hymn
      that means nothing to you, a body
        more bird than god. The wind,
  like breath, unburying its heart.

The Suicide

There are birds, pedaling
through the open-throated
morning, the sky
still flushed. A horn, sirens,
in the distance. But not rain.
Just mist, burning off
like a woman disrobing: first
one shoulder, then
the whole. The birds,
black-feathered, circle
slowly, as if they hang
suspended by invisible strings.
On the ground, something
damaged—tied together, beaten,
pieces of collar and coffin
bones. Does the animal's
name not matter anymore?
They land, attendant. But only
after it seems there is no
movement. Should this dark
creature roll over, suddenly,
hoarse with pain, the birds
would flee for the trees
like children caught in a
strange house. Like sirens,
chasing something lost.
Not like bodies, mid-fall,
wings tucked close at their
sides. Following sound,
not a beast that can still rise.

Immortality

Wind hollows the wheat
chaff, howl of a stray
hungering morning. This terrible north
collects pieces you don't
recognize. Your mother, grey
two-story house, singing
through a distant night, *lay down*
your sweet head. Outside
the chapel, under a streetlamp,
you draw a picture of God
in the snow, where He isn't
merely a man, lost
in this human hour, body
weeping in the thaw.

In Ten Years

My mother forgets she's a mother,
voices trickling down a drain
in a lockdown ward. She waits
for people she no longer knows, sandpaper
skin scaly and weathered. Her bags
packed at her feet each morning, I hold her
hands, put her clothes away.
She calls me by her sister's name.
Her body, another country
tarnished by dry winter air.
South of her borders, I suffer
the loss of my children's tiny bones,
high voices. At my mother's greying
house, we're kind strangers, no history
hiding in the corners of the room. She's lost
as a forest of pines are lost: one tree cut
from it's roots, then another, naked
patches of ground dotting the mountainside.
In these spaces, I find her: young,
before my father didn't make it
home. Before motherhood
buried her under record snowfalls.
My eight-year-old feet skip
to school through the woods, skirt
willowing in the glacial wind. Her eyes
follow, turn
away.

In the Absence of Sky

The trees shiver. When my neighbor calls me
to the woods, summer idles
on the street. The sun, bare-lidded. I am

a new game under his gaze.
It isn't until he folds both knees
in fallen leaves that I notice

the sky, missing. Only the oaks' arms fly
overhead, a canopy of green. Wind swells
through us like thunder. I can't focus on

anything but the absence of blue,
the forest's edge. Longing to run,
I stand still, arms thrust high

above. His voice is soft and
trembling, but I can't refuse
the storm creeping in like dark

wallpaper. The crows circle,
borrowing heat from the earth
below before day's

end. Perhaps, as children, we are only
spared by degrees—the water's
depth, a storm's brute

force, the hands willing
to unspool branches,
to let light in.

While He Is Missing

There is a door
in the dream

for him to enter
the way water enters

the body,
but doesn't stay.

He smells of cigarettes
and coffee, of vodka and the forest's damp

musk coiled in his hair. Smoke
is the only thing to escape

his mouth. You wait to hear
if the end sounds like rain

running through the trees, praying
not to hit hard ground—

to find shelter, instead, in a body
of water that lays bare

across land. Look—
he sways in the inches

stretched between shadows. His name,
dancing

on your lips like broken shutters. You leave him
like this each time: a father. Stranded

in a place where nothing can forgive
the coming rain.

Another Genesis

A tribeswoman, now: my mother is
the summer sun, resting her body
on the flat-backed sea. She slips

slowly below the water line, as my otherness
grows. Long ago, we belonged. Without warmth,
I shiver as she submits—yellow bones

braiding the sea, a blurred sky's wings
wavering in the wind. Will anyone see the sky
without the sun? What will water become

without a body stretched blue
to reflect? No one prepares for this
wound. A daughter can't know how to stop

reaching. The water ripples
at midday: a pair of cerulean skies. I can't see
the way I came to be here. A tribe of stray bodies surfs

the shore, waves dragging themselves
through sand. Sun-bare, I wait for night
to unnail her body from the sea.

The Last Place

A few hard drops strike
soil. Petrichor rises and I can't
remember how far away
you lived. Years become galaxies. You are not
with me, as people suggest. I haven't felt you
anywhere but where I saw you last,
so many skies from here. Who holds a child's memories,
now that I am no longer a child? Some man
made of clouds and wind, perhaps. Or maybe
it has been you all along, rolling them
like cigarettes in your fingers: days
at the lake, in the cab of your truck,
singing on the highway as trees flee
our speeding bodies. As stars
hitch themselves like flags
to our tailgate in a blue night.

Ancestry

You squeeze your eyes shut
on the Rockies, as if you could ever forget. Water

threads itself through land, blood
beneath the skin, always in danger

of spilling. You wear this place.
The heavy snow. Sorrowing sky. Wildfire

burning your nostrils. It is woven
in your psyche like a tattoo. A reminder

of wanting. The past,
another child you carry on your hip,

needing nurturing. Years
thin, blood left

on bending highway, reeds,
pine trees hunted and skinned, white

of their bones raking the sky—
undone. Dense pine

forests outlast withering frames, fragile
temperaments. They will not die

with you. You carry them south, roots
still buried in the Columbia River valley, yet growing

in your marrow all the same, moss
over knuckles like decaying stumps, thick

scent of earth bringing you home.
At midnight, afraid, you talk to wind, missing

bodies, places that fathered you.

Letters from a War

After another man's name
found your mouth, after
the bodies laid down

until they couldn't rise, after
I began to see men
as streets and mountains

and moving skies—I starved
just enough to stay
hungry. Not to kneel

before deep voices, reaching
for any word that didn't
force its way into my mouth. Why

is my body still empty
with another inside? I used to think
I'd go thirsty to see you

break like a highway's bones
under the winter snow. Maybe
mornings you forget braiding

thick bundles of hair over
old bruises. But I've forgiven
how your whispers sound

like regret. How a mother leaves
when the night is long. My belly brims
with someone, slight and soundless,

who I can't refuse. I know now
    how briefly we are beautiful. How the first
        death, for women, is our own.

2.

PROOF IN DISAPPEARING

Epilogue to Drowning

You rub your thumbs
between my shoulder blades, a lie
in each stroke. How did we get here? The first time

you feathered my skin
with your tongue, I decided to drown.

I didn't want our limbs to ever break
the surface of water, indecent
sky. But, I'm not sure anything can be done

now. My body can't outrun your fingers,
falling like flames. At my back, new

wings—open doors of my ribs
spread wide to expose kidneys,
a heart. When your hands travel

these wounds, will your cuticles catch
like blades? I never thought

it would come to this: two people undone
by their hands, open
mouths. Our downfall will not be

emptiness, darling boy, but the water
we dare to fill our lungs.

Winter in the Rockies

Is this heaven? Hidden
highways. An avalanche. Nothing
around for miles to hear tires
leave the road. The mountains
bow, back-lit by white skies. I walk
& wonder if I walk for any reason
except to walk. My father,
drenched in drifting snow, was left
here. Yet I can't say I'm closer
to the truth about loss than I was
as a child when the world I saw
was a world that doesn't ache
to be anything else. It's funny
how easy it is to forget
the sound of water in winter. I lay down
on the banks alongside the frozen
lake. Its long body, still. But
I'm listening now, as water
like a sleeping child wrestles
with the blankets pulled over
its face, waiting to see
which one of us will wake.

Athena

When I last saw him, he could've been
anyone's father. The good daughter, dragging him
from snow banks in a dream, his black hair

curled in my fingers. His hair hadn't had time
to grey. I turned him over. Put my mouth to his,
tasted diesel fuel, a blazing tail-

pipe between my lips. My feathers,
fallen at our feet. Awake, I stand
on a four-lane highway, canyon

walls like crows' wings
curling over a creek. No sign of snow
that held him in its white womb

until he was someone else. His mouth
is not mine but one my son wears
to kiss me goodnight, to tell me

the dark is only a temporary lapse
between bodies: earth and sun,
a god and the child he armours.

Prayer for an Unnamed Child

Tonight, I'm empty, a Ford
pickup stalled at the lights,
the shoulder of town, thin-skinned
streets bare of any sidewalks. A plea

scorches the night's black-stitched ear,
licked dry by daylight. Perhaps

it is only tires that shriek,
the pavement's gasp,
glass on the road.

I stand between the stop sign
and an outdoor pool, veiled in blue
tarps, highway at my back.
Can you see my bones
stretch to meet
the torn sky? I want

to be a man,
forget my womb
ever held tiny fingers
in its palm, my spine
swayed like snow,
pinning the pines'
thinned arms down.

For My Son: A Forest of Stars

After Ocean Vuong

Before
we give up our country to give you
a country of your own. You can't have

my eyes, but we'll speak
with the same tongue.
Someday,

you'll hear sand in my voice
when searching for what is lost

onshore. I want you to have the truth,
but truth is as seductive as a storm

settling over the earth. Many nights,
my mother cried in another room.
I heard lashing. I saw fire—

how scalded skin, to some, is
love. I hope you don't need to pray

for cinderblock walls. For the sky's colours to steal
our words. You can't yet know

that men can be bent
like branches. That women must kneel
too often, their eyes on the red shore

behind. There is a moment when I touch you
and you are real, not a shadow

of shed skin as we turn to dust
on the hardwood. If you go somewhere
without me, my face candling

into melted wax, you must understand—I fell
in love while we were falling, the devil

wind tearing at my mouth and nose.
All I could do was smile.

When you step away, look
at the ground I've given you. Dig

until your shovel hits
the trees' twisted roots, spring

water gurgling below. This is the proof:
our hands on these shafts,

my words held in your mouth
              like stars that forget

                        they have to fall.

girl, unfinished

You wish you didn't have a name. You wish
he didn't bring flowers,

a casserole—his tires rutting the gravel
driveway, the morning

police came to say your father died. Silence
& grief. Branches bent
under the snow. Silence
like a wound

in your brother's mouth. The scent
of cedar: a memory

chest, cushioned pews, trees
behind the church. Silence
& snow. But he was there, with his wife,
before your mother was his

wife. Then, his cigarettes, flicked
on the lawn. Summer,

caught in your throat. He
took your clothes, changed the locks, left you
in the yard. Silence

all around—hungry sky,
a broken tricycle rusting

the sidewalk, front wheel
twisted. Then, the wings
your skin grew over

as you stood wishing
for a second bloom.

Wedding Dress

In the back of my mother's closet,
it hangs: a cloud, the vapour
from our lungs the night
we met. Snow
covered this town like a shroud. I imagine
its silk in the half-moon
light, dirt scarring delicate seams,
the dusty floor. The way we smiled,
standing in the July heat. I slid thin fabric
past my hips afterwards, abandoned
to a hanger, the space
above an old .22. Perhaps
a dress is only silk and lace. A sweetheart
neckline, the varying whiteness
of its bones. Shut away
there, the corset gapes, tulle
browned with age, as it waits
for warmth—to know the moment
when winter turns to spring
and it will be able to twirl
in the floor-length mirror, light
and new and impossible to harm.

Revenant

I want the river to live
in shades of chalk
and moon, bones

on its bed. I see it clearly
in my mind, yet it hides itself
there, alongside the street

with a green house where my parents
lived. Great pines resting their heads
against the sky. The colours

at dawn, sweet chill in the summer
grass before early snows. When I return,
the water is no longer

hidden. Its teeth bear down
on the banks. But I notice only
how low it sits in the mouth

of the gorge, an unforgiving pull
at my ankles—the way it rushes forth
to become something else.

The Gulf

Rain refuses to stop falling
as if it can dissolve all things—
a road, its name. Inside my belly,
another belly, burning. Small
hands change shape in the dark
night. Can I give you tomorrow
when I've lost hold of today? A man lays next to me,
sweltering, and I'm stupid
with the need to be small. To reduce my body
to a war. For you to crawl headfirst into
the fanged mouth
of this world, abandoned
on the gravel roadside in a storm-surge. I forget
not to sacrifice my bones like gods
make a woman forget praise
without suffering. Still, the sky
can't be outrun. A hurricane sweeps
sand and sea, bodies torn
away from the nearest shore. Shouldn't one rest
upon the other? Darling child, show me
why I swell, your name
cut from fire. That you are
not crawling, but running headlong
into your own skin. When you get here,
show me the sky—
the earth's skin
as changed as our own
by sea levels, a sharpening storm.

In My Father's Voice

It wasn't pain that surfaced
when you knew we were listening. How
you were made to quit
school in eighth grade, heavy
machinery in your hands
on the side of a half-built mountain
dam. The stutter when my mother

asked you to sign anything. It wasn't the pen,
but your fingers, fumbling, the shape
of letters you couldn't find. There

was no accent from the old country—
Ukrainian you spoke on the farm,
your father and two siblings buried
by a creek, the red barn's bones
cracked under the sun. You never mentioned

your plans, or your struggle—
                                        how the sky hung
so low, you were crippled. Only
that I could find water
in every wood. That the pines
would grow tall enough to
pierce the sky. But, at night, I still hear

                    muffled cries through the walls—
how a man on his knees sounds
like lightning. How the door echoes
long after it closes. I want

more than anything, now, to tell you
I understand, mouth bent to my son's
small ear—what it took to quiet
    cries born of hunger, of the cold
            night on your skin.

The Afternoon I Was Abandoned, Sparrows

flew into the yard. It was spring, but cold. The man
changing the locks while I stood in the street
was a man my mother knew. Like powder, wings

raised above fresh snow. Like arms
swarming the walls of a house, they flew
into the horizon's long frame. The man

turned his back, closed the door
when he finished. But the sparrows sat
on the snow-laden lawn, refusing

shelter. Refusing to flee white fire. Their wings,
beating the air, spread like small
crosses racing over the ground.

Winter in Sodertalje

Not even a street light. Forgotten,

a baby lays on the floor,
bald and screaming. I dig out the car,
haul wood. At one point,
I turn, looking for proof in disappearing
prints. With the sky
in shards, who will notice if I am lost?

In the bed, my arms spoon
a pillow, old imprint of your face
clinging to blue cotton. I don't ask you
to name this silence
when you appear, bedraggled and beaten,
winter woven inside moon-pale skin. Instead, I build
a fire in each hearth, large enough
to burn through the black

night. In sleep, you reach for me,
a spruce to a stark sky, aching
limbs sheathed with snow.

As the Daughter Recites Psalm 91

He says he'll kill her

if she runs, but she doesn't
run. She sits still, deep

inside the planet
of her bathtub, water
roaring in place

of her mouth. Maybe
it's for the sound. Maybe it's a wish
to pour herself into another
body, release the parts

grown soft. She bares her skin
to cold, white tile,
soaking wounds in Epsom
salts, purpling
like foxgloves. Here,

she can close her eyes
and forget she's a mother.
How his mouth in her ear sounds
like penitence. That she still longs
his fingers on her skin
as soon as the water

cools. I stand outside, my hands
pressed to the door
in surrender, as I once slept,

unfinished, in the cradle
of her bones. Yet,
in my mouth, a prayer:

this time, let her not forget.

Hiraeth

Pine-rimmed water, winding
highway, heaving skies. You slipped
your skin like an old housedress,
humid summer nights we were alone.
Left it, pooling the cream rug
in your bedroom. West,
the sky rested on the shoulders of mountains,
and I lost my breath as we climbed
over rock, dirty stumps,
soft with decay. Deep inside a twisted wood,
water we could only hear
breathed cool air on our damp skin.
I was lost then too.
Why have you shown me
ragged ridges, a river,
if they weren't mine to keep?
Tonight, somewhere else,
the sky bends over my face,
pink lace, low clouds.
I want to be a lodgepole
pine, planted in the ruins
of a logged peak,
head straining
to touch the tilted sky.

Autumn Wars

I wait to see how we'll leave
    paradise. Queen palms

sway over porticos and porches. Our small sons
sleep surrounded by treed slopes they
            may not remember. We are

long miles from childhood, the cold north we fled. Why
do we always want better? Is rising
not enough? I blanket you

against this new season, long-tendriled limbs
curled into commas, the wind crowing. Perhaps

we should be less like the pines we've left
to petrify, roots stifled in frozen ground. Less like

                        whispers of dogwood
blossoms that never peek through snow
banks, that never claw

through my mother's thinned voice
calling me home. You are the only place
I come back to. The sun,
                perishing in a storm

can't be seen amidst the rain, the same way
it can't survive the snow's white

mask. Once, we armed ourselves and drove
over tundra in a twining womb
                                        of white fields and sky

to get out. I saw then that some things can never be
                        made beautiful.

Current

I ask if we're still
friends, our second child between us
in the bed. Spent,

someone I can't dream has left
this body, a cathedral squandered
where mountains fall flat at our feet. I look back
on ruins, the sky galaxying south, always

pulling away from the earth
as if it knows better. Suffering is faithful
work. We kneel, our mouths

closer to the ground than the sky. Maybe
our child is the window
and we are merely walls. Our parents have been
lost to us like rainfall is lost
to the dry earth in drought, red sun

bleeding in brushstrokes. This house is
a wild forest now. There is no city. Reaching for pine
trees in your steel-dark eyes, the taste of

rain in your mouth, I begin
the dutiful act of drowning
in everything not yet lost

like water.

Burning

Because you used to be
made of winter: snow

and cold and sleet. Because their hands
let go and you tired of freezing. Because

you're building yourself
from fire—secrets

are only safe when someone is
underground. You were a woman running

until you forgot you were
a woman. The fire, nailed inside you

like a sword. Why would a child make you less
than the horses, saddled

next to the barn? Their reins
dangle and, yet—. I once saw fire

spread through a swamp
during a storm, after lightning

struck. It's not restraint
that is feared, but being left

to burn. How a fire can
starve, surrounded by air.

Elegy for My Child

It didn't take long for you
to go missing. I roll over
and tap the window. Each fat whisker
of rain stains the glass black,
a skinny-streamed feather. Once,
your body was so small
we couldn't wrap you tight enough
in a blanket. How you howled to return
to a womb. Sometimes, when pregnant, I wondered
how long I could hold you inside
before my skin's heft would falter.

What do I do with this rain
drilling the roof like nails into planks?

I'd gladly give up this night,
close my eyes to forget your cries
are an ocean. This ache isn't
the ache I imagine when I think about losing you
in the grocery store, running madly
between aisles, shallow panes, registers. Outside,

the house cranes its neck
into the sky, wind hurling itself
against the frame. It shudders,
body stripped bare, standing
empty to face the street.

It's in the fall—

The way the fork
clatters to the empty plate. An open hand
to ruined skin. How snow
can't help but reach
for the ground, too heavy to pretend
to fly. How will we leave each other

next? Unfasten the clouds—
lend them to the earth. Let us lie
on our backs, the winter

night rising around us. Could we fly
if nothing before had fallen? Imagine

every petal, water droplet, everything
that was once part of the sky,
surging like the sun
from the horizon. Yet these hands,

lined with ghosts: rivers, children,
your bare skin—they still fall
like snow. Next to you,

I wait for the wreck, two cars
on a one lane road. For the wind
to pick up. In your hands,
to be perfectly ruined.

Nocturne

I can't see the whole sky
at once. Black-ribboned water
churning across the lake. Pines behind

mountain peaks. When the sun lowers
its jaw to the earth, I finally know
how to disappear. Shadow-flames

flicker, the way I remember your voice. The moon's
husk is already bright, but distant. It won't come
into focus, the way stubble blurred your chin

in my fingers. I replay how I failed
to make you love me. The hours it takes
to know darkness. As night arrives,

instead of folding myself in its chest
like a child, I sit on the porch and listen
to wind in the leaves, squirrels foraging,

the water baying at large rocks
on riverbanks. I remember
when we were tethered to light

like flies. But I can't see
the whole sky at once. How bright
the bodies it holds.

The Windsurfer

My husband is made of cloth, the sea
staining his skin with salt
as it exhales. Pale, he's used
to falling from the sky, wings
tangled in the wind. Each tear
is sewn shut by deft hands
and yet, his skin is threadbare,
dragged over rocks onshore
when he makes his descent. Like an arrow,
I draw a needle across the wounds,
waiting for the skin to give, the night
too perilous to excise, to run from. As long
as he's made to fly without being able to see
the ground, I learn how to glide, to tread
water. But, we're never freer than
when we're falling. Death, take me
first, the wind in every hollow. I'm not afraid
of the sound a body makes
when it dips below sea level. I fear
steering the stretched sky alone, wind
tearing through space
he's left like a song
to hang the hungry sea.

The Conversation

Bones shatter against a hull's open
mouth. I crane to see beyond
fists flinging me down, brute
breaths trapped behind bared teeth

like the clasped tines of a zipper.
You cut me out of a lace dress:
slit of skin, as the knife fangs

from tail to throat, a still-waving rainbow
trout fresh from white waters.

Would you take my head, too,
if you could?

Belly emptied black
into billowing water, you lay me down,
carcasses collecting like rinds
along the banks' soft lips.

Remove my womb, the poppy
red eggs, my stacked spine's silver

boning. I can't be a man any more
than you can wrest the slim waist
of a cloud. Maybe,

before the blade sinks, you let
go, a memory of pretty bones blistering
the ribbon of your pink throat.

I'll swim upstream through a sky of water,
gather myself in the river's palm.

Hands, I've Had

To disappear, I close the front door
on my mother, as a man's hands thrash
worn yellow walls. The winter

sun, heavy-lidded, hangs
halfway over our house until the night sky appears,
star-shot, holes where its eyes should be.

How I long for blindness, the sky
holding itself above.

&

Maybe, I didn't see you
until a gasp of tires, black
ice-slick highway, the truck's tail-

spin. Your arm shot
across my chest like a brick
barrier. Before the meridian met steel.

Sometimes we're still sliding,
a prairie sky crouched overhead.

Sometimes, there is only falling:
fingers clasping steel like snow.

&

Tonight, we're still
breathing. Your hand cleaves
  to the skin under my shirt, a cigarette

cupped in the other. The moon
bares its teeth, swallows
  the dark. We sit on the deck,

kids sleeping in the bowels behind
us. I imagine falling, clothes
  strewn on your parent's floor. I'm afraid, bare-

skinned. Our bodies break
the wind, another war
  whittling us down. Your hands,

like pink-tipped lilies, are the only things
pinning me to the earth,
  proof in disappearing prints.

Winter in Florida

Tracks in the snow sing
somewhere else, the cicadas
singing beyond our back porch

tonight. You're heavy-lidded,
legs kicked out in front of a lawn
chair. The light, always changing—

do you see? I sit so still
but my eyes are open. Maybe
north, we would at least leave

tracks. Is this a better world I hear
singing, as we wait, withering
in the southern heat? December,

trapped in our throats. Either way,
there's no going back—can't you
see? Look quickly: the sun

dives behind a swamp. Behind
the house. Behind, voices
colour the wind. But it's you I want

to sing us away now, swallow
to a spring morning. We've been lost in
a string of seasons, in every sky.

Holding the Sky

It isn't night that draws us
apart. Nor burials: our small bodies

up to our necks
in snow. We left those children there,

fair-haired and fat-fingered,
dreaming of spring, daisy fields

behind our old green house. Even now,
you run as if you can't help it. Tonight,

I find you, skin pinned and prodded, pupils
nailed down to a prick. I drape myself in heavy clothes

as if it will keep you warm. As smoke wafts
from your mouth into the night

sky. I try to run my fingers through it,
but it changes as it leaves

your body, and I am left
with only this, brother:

dusk rippling in front of us,
my empty hands.

When My Mother and I Speak about the Weather

She holds a knife
to the apple's skin, green
and smooth. The veins

in her hands raise. I keep trying to
forget that she wants to open
my body like a constellation. To disassemble

the stars. She gestures with the tip
toward my neck, a slit of mirror. It catches
light from the windows. I want

to remind her that the light outside
can be anything she wants. But, she can't reach
the stars. And so. There is nothing else

to do, except open the sky's chest
and let the stars fall
like seeds, like so much debris.

Live Oaks

languish amid parcels of land,
pink and tan houses. I should know

paradise, but all I see, sometimes,
are the forests I knew before, covered in snow

as we waited for the thaw, for pines
to spread their arms under

  the white sun. Like horses
waiting in the stables, I ran

in the spring—longing a body's weight
against me, longing to run

free. In Florida, the ground sinks wet
under my feet. The sun, heavy

on the oaks' backs. I don't know
if I dreamed other places now. If

the trees I once knew
survived as I survived—-

  hoping not to be cut down
  young, needing nothing more than light

rain, the night
      to part ways with the sky.

In Hindsight

You slept with your fists
balled next to your ears,

as if you had to fight
    even then. I wanted you

like rainfall that wouldn't subside. Your father
belonged less to me than to you

afterwards. Perhaps that's true of us
    both. But he became a face

on a flyer, posted on wooden poles
strung with wire.   Some days, I imagine you

on the banks of a river, wet with white-
water-numbed limbs. You deserve

to hear the gurgle in its throat, before
it dries up in the sun

scalding your shoulders. Some days,
I come into focus: my head

in your neck, my hands
            in your hands. When I fade

with the cicadas, sing for me. Tell me
    all of this is an accident—you are not

my body's ruin, but the ocean
I empty myself into.

Any Other Sun

Long seconds your leaving sews
in my skin: a daughter's gasp, trapped
in the space between bodies. The map unfurls

and I keep running. What is it I hope to find? A war
breaks out in this borrowed home
as I recline in the Gulf's clean palm, adrift

under the sun. Tell me I'm not
a disappointment. The struggle
to get to this country. You couldn't even read.

I wish I could carry you now, as your mother did,
show you this world. This world,
a churning river. In the morning, a child can be

almost grown. How would it have been
inside your skin, your language
lapping the shores of these shallow bones? I saw snow

crowning silver sills. I saw you disappear
like the sun. Here, my hands belong
to someone. Thin, like my mother's

blood. I'm ready to be powerless against
the night. To leave behind skin
pinned beneath my palms like sheets
on a line, swaying in a southern wind.

Notes

"Little Hell." The title references a song by City and Colour.

"On Nights when I Am a Mother." This poem is loosely after Pablo Neruda's "Funeral in the East" and "Walking Around" and Tomas Tranströmer's "Prelude."

"Hit and Run." This poem is after Kevin Prufer's "Fallen from a Chariot."

"Athena." This poem references a line from Ocean Vuong's "Telemachus."

"For My Son: A Forest of Stars." References a poem by Ocean Vuong called "To My Father / To My Future Son."

"Hiraeth." (Welsh) ". . . a homesickness for a home you cannot return to, or that never was."—Oxford and Merriam-Webster

"Current." The last line references a poem by Meg Day called "San Francisco / October 17, 1989."

"Nocturne." This poem is after Chloe Honum's "Come Back" and loosely follows the form there, mirroring some language.